Encyclopedia of Alternative and Renewable Energy: Advances in Solar Cell Technology

Volume 30

Encyclopedia of Alternative and Renewable Energy: Advances in Solar Cell Technology

Volume 30

Edited by **Terence Maran and David McCartney**

New York

Published by Callisto Reference,
106 Park Avenue, Suite 200,
New York, NY 10016, USA
www.callistoreference.com

Encyclopedia of Alternative and Renewable Energy:
Advances in Solar Cell Technology
Volume 30
Edited by Terence Maran and David McCartney

International Standard Book Number: 978-1-63239-204-6 (Hardback)

Printed in the United States of America.

Contents

Permissions

List of Contributors

Preface

This book aims to highlight the current researches and provides a platform to further the scope of innovations in this area. This book is a product of the combined efforts of many researchers and scientists, after going through thorough studies and analysis from different parts of the world. The objective of this book is to provide the readers with the latest information of the field.

This book examines topics related to photovoltaic generations, its unique scientific ideas and advances made in the solar cell technology. Extensive attention has been given to the significant effects in solar cells, both in common and on particular cases such as those of quantum dots, super-lattices, etc. Usage of advanced materials, like the oxide of copper as an active substance for solar cells, InP in solar cells with MIS arrangement, AlSb for use as an absorber layer, among others, have been included. Furthermore, other significant topics such as the analysis of both the status and prospects of organic photovoltaics such as photovoltaic textiles, photovoltaic fibers, etc. have also been discussed.

I would like to express my sincere thanks to the authors for their dedicated efforts in the completion of this book. I acknowledge the efforts of the publisher for providing constant support. Lastly, I would like to thank my family for their support in all academic endeavors.

Editor

1

Solar to Chemical Conversion Using Metal Nanoparticle Modified Low-Cost Silicon Photoelectrode

Shinji Yae
University of Hyogo
Japan

1. Introduction

The storage of solar energy is one of the key technologies in making solar energy a major energy resource for humans (Turner et al., 2004). Solar hydrogen, which is produced by splitting water using solar energy, can be considered the perfect renewable energy. Hydrogen is an energy medium that can be easily stored and transported. It can generate not only thermal energy but also electricity using fuel cells and mechanical energy using hydrogen engines. Hydrogen returns to water via such energy generation. Therefore, solar hydrogen can form the basis of a clean, renewable energy cycle.

1.1 Solar hydrogen production with photoelectrochemical solar cells

Solar hydrogen production, that is, water splitting by photoelectrochemical solar cells equipped with a titanium dioxide (TiO_2) photoelectrode has been attracting much attention since Fujishima and Honda's report in 1972 (Fujishima & Honda, 1972). Photoelectrochemical solar cells have important and unique features (Licht, 2002, Nakato, 2000). They are fabricated simply by immersing a semiconductor electrode and a counterelectrode into an electrolyte solution (Fig. 1). They can convert solar energy not only to electricity (Fig. 1a) but also directly to storable chemical energy (Figs. 1b and 2) such as water splitting into hydrogen and oxygen (Arakawa et al., 2007, Fujishima & Honda, 1972, Grätzel, 1999, Khaselev & Turner, 1998, Lin et al., 1998, Miller et al., 2005, Park & Bard, 2005, Sakai et al., 1988), the decomposition of hydrogen iodide into hydrogen and iodine (Nakato et al., 1998, Nakato, 2000, Takabayashi et al., 2004, 2006), and the reduction of carbon dioxide to hydrocarbons or carbon monoxide (Hinogami et al., 1997, 1998). For the photovoltaic photoelectrochemical solar cell (Fig. 1a), two electrodes are immersed in a redox electrolyte solution. Opposite reactions such as the oxidation of the reductant to an oxidant on an n-type semiconductor electrode and the reduction of the oxidant to the reductant on the counterelectrode occur through solar illumination. Thus, the composition of redox electrolyte solution does not change, and only electricity is obtained as usual solid-state solar cells. For the photo to chemical conversion type of photoelectrochemical solar cells (Fig. 1b), different reactions occur on the electrodes, for example, the oxidation of iodide ions to iodine (triiodide ions) and hydrogen evolution. This decomposition of hydrogen iodide is an 'up-hill' reaction. Thus, solar energy is directly converted to chemical energy using this type of photoelectrochemical solar cells. The junction of an electrolyte solution and a semiconductor can generate a high-energy barrier, thus reaching a high photovoltage level even with a low-cost, low-quality semiconductor. However, water splitting using titanium dioxide encounters serious difficulty in achieving hydrogen evolution. There are three solutions to this difficulty: using another semiconductor with an energy band gap that is wider than titanium dioxide; using a multi-photon system equipped with multi-photoelectrodes in series or a tandem-type photoelectrode; and using a hydrogen-producing semiconductor, such as silicon (Si), and an oxidation reaction other than oxygen evolution, such as oxidation of iodide ions into iodine.

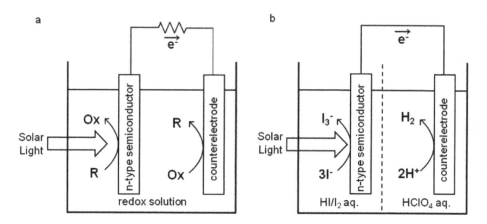

Fig. 1. Schematic illustrations of PEC solar cells. a) photovoltaic type; b) solar-to-chemical conversion type.

The Gibbs energy change for decomposition of hydrogen iodide (HI) into hydrogen (H_2) and iodine (I_2) (triiodide ion (I_3^-)) in an aqueous solution (see equations 1, 2 and 3) is smaller than that for water splitting. Thus, silicon photoelectrodes, which have a narrower energy band gap than titanium dioxide, can decompose hydrogen iodide with no external bias (Figs. 1b, 2 and equations (1)-(3)). The efficiency of a solar-to-chemical conversion via the photoelectrochemical decomposition of hydrogen iodide using single-crystalline silicon electrodes reached 7.4% (Nakato, 2000, Nakato et al., 1998, Takabayashi et al., 2004, 2006). Fuel cells using hydrogen gas and iodine solution can generate electricity via the reverse reaction of hydrogen iodide decomposition (equation (3)). Therefore, this system can form a solar energy cycle in a similar manner as the cycle consisting of water splitting and hydrogen-oxygen fuel cells.

$$\text{Anode} \qquad 3I^- + 2h^+ \rightarrow I_3^- \qquad\qquad (1)$$

$$\text{Cathode} \qquad 2H^+ + 2e^- \rightarrow H_2 \qquad\qquad (2)$$

$$\text{Total reaction} \qquad 2HI \rightarrow H_2 + I_2\,(I_3^-) \qquad\qquad (3)$$

Fig. 2. Photoelectrochemical solar cell produces hydrogen gas via photodecomposition of hydrogen iodide with no external bias.

1.2 Metal nanoparticle modification of semiconductor electrode

Unfortunately, bare semiconductors can easily corrode or be passivated in aqueous solutions, and do not have enough catalytic activity for electrochemical reactions. Modifying the semiconductor surface with metal nanoparticles eliminates these problems without lowering the high energy barrier feature (Allongue et al., 1992, Hinogami et al., 1997, 1998, Jia et al., 1996, Nakato, 2000, Nakato et al., 1988, 1998, Nakato & Tsubomura 1992, Takabayashi et al., 2004, 2006, Yae et al., 1994a). The operation principle of this type of solar cells is explained as follows (Nakato et al., 1988, Nakato & Tsubomura 1992). Figure 3 shows a schematic illustration of cross section of a platinum (Pt)-nanoparticle modified n-type silicon (n-Si) photoelectrode. Photogenerated holes in n-Si transfer to the redox solution through the Pt particles, thus leading to a steady photocurrent. With no Pt particle, the photocurrent decays rapidly. The surface band energies of n-Si are modulated by the deposition of Pt particles. However, the effective barrier height is nearly the same as that for bare n-Si in case where the size of the Pt particles (or more correctly, the size of the areas of direct Pt-Si contacts) is small enough, much smaller than the width of the space charge layer. Thus, a very high barrier height, nearly equal to the energy band-gap, can be obtained if one chooses a electrochemical reaction with an enough high potential. Also, a major part of the n-Si surface is covered with a thin Si-oxide layer and passivated, and hence the electron-hole recombination rate at the n-Si surface is maintained quite low. For these reasons, very high photovoltage can be generated.

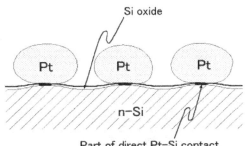

Fig. 3. Schematic illustration of cross section of a Pt-nanoparticle-modified n-Si photoelectrode.

In this study, Pt-nanoparticle modified multicrystalline Si wafers and microcrystalline Si (μc-Si:H) thin films (Yae et al., 2007a, b) are used as photoelectrodes in the photodecomposition of hydrogen iodide for low-cost and efficient production of solar hydrogen. We also attempt solar water splitting using a multi-photon system equipped with the microcrystalline Si thin film and metal-oxide photoelectrodes in series (Yae et al., 2007b).

2. Platinum nanoparticle modified porous multicrystalline silicon

Multicrystalline silicon is the most common material of the present conventional solar cells. We applied multicrystalline Si to solar hydrogen production by the photodecomposition of hydrogen iodide. To improve the conversion efficiency and stability of photoelectrochemical solar cells, multicrystalline n-type Si wafers were catalyzed with platinum nanoparticles and antireflected with a porous Si layer. The Pt nanoparticles were deposited on the multicrystalline Si substrates by electroless displacement deposition immersing substrates in a metal salt solution including hydrofluoric acid (Yae et al., 2006a, 2007c, 2008). The porous Si layers were formed by metal-particle-assisted hydrofluoric acid etching immersing metal-particle modified Si wafers in a hydrofluoric acid solution (Yae et al., 2003, 2006a, 2009).

2.1 Platinum nanoparticle formation by electroless displacement deposition

The metal nanoparticles play a tremendously important role in photoelectrochemical solar cells for photovoltaic and photochemical conversion of solar energy (Nakato et al., 1988, Nakato & Tsubomura 1992). The kind, particle density and size of nanoparticles influence the stability and conversion efficiency of solar cells. Thus, controlling the size and distribution density (particle density) of metal nanoparticles on semiconductor, multicrystalline Si in this case, is important for obtaining efficient photoelectrochemical solar cells. As well-controlled and low-cost methods of depositing metal nanoparticles on Si, we have been utilizing several methods such as simply dropping a solution of colloidal metal particles (Yae et. al., 1994a), preparing a Langmuir-Blodgett layer of nanoparticles (Jia et al., 1996, Yae et. al., 1994b, 1996), electrodeposition (Yae et. al., 2001, 2008), and electroless displacement deposition (Yae et al., 2007c, 2008). We choose the electroless displacement deposition to prepare Pt nanoparticles on multicrystalline Si wafers (Yae et al., 2006a). The electroless deposition of metals is a simple process involving only the immersion of substrates into metal-salt solutions. The electroless deposition is classified into autocatalytic and displacement depositions (Paunovic & Schlesinger, 2006). The electroless displacement deposition of metals onto Si uses a simple solution containing metal-salt and hydrofluoric acid. The deposition reaction is local galvanic reaction expressed by equations (4) and (5) consisting of the cathodic deposition of metal on Si accompanying positive hole injection into the valence band of Si and the anodic dissolution of Si. Various kinds of metals can be deposited on Si using this method (Chemla et al., 2003, Gorostiza et al., 1996, 2003, Nagaraha et al., 1993, Yae et al., 2007c, 2008).

Deposition of metal on Si:

$$M^{n+} \rightarrow M + nh^+ \qquad (4)$$

Anodic oxidation and dissolution of Si in an hydrofluoric acid solution:

$$Si + 4HF_2^- + 2h^+ \rightarrow SiF_6^{2-} + 2HF + H_2 \qquad (5)$$

or

$$Si + 2H_2O + 4h^+ \rightarrow SiO_2 + 4H^+ \tag{5}$$

$$SiO_2 + 6HF \rightarrow SiF_6^{2-} + 2H^+ + 2H_2O \tag{5}$$

Non-polished multicrystalline n-type Si wafers (cast, ca. 2 Ω cm, 0.3 mm thick) were washed with acetone, and etched with 1 mol dm^{-3} sodium hydroxide or potassium hydroxide solution at 353 K for saw damage layer removal. Before deposition of Pt particles, the wafers were treated by one of two pretreatment methods (hereafter, we call these pretreatments method A and method B). Method A: the wafers were washed with acetone, immersed in a CP-4A (a mixture of hydrofluoric acid, nitric acid, acetic acid, and water) solution for three min, and then immersed in a 7.3 mol dm^{-3} hydrofluoric acid solution for two min. Method B: the wafers were immersed in 14 mol dm^{-3} nitric acid at 353 K for 30 min after method A treatment. The multicrystalline Si wafer was immersed in a 1.0x10^{-3} mol dm^{-3} hexachloroplatinic (IV) acid solution containing 0.15 mol dm^{-3} hydrofluoric acid at 313 K for 30-120 s.

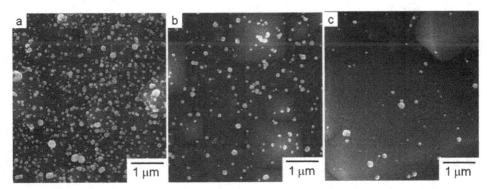

Fig. 4. Typical scanning electron microscopic (SEM) images of multicrystalline n-Si wafers pretreated by method A (image a) or B (b and c) and immersed in the Pt displacement deposition solution for 120 (a and b) or 30 s (c).

Figure 4 shows typical scanning electron microscopic (SEM) images of Pt deposited multicrystalline Si wafers. Spherical Pt-nanoparticles were sparsely scattered on the multicrystalline Si surfaces. Thin Si oxide layer formed by immersing the multicrystalline Si wafers in nitric acid solution (method B) decreased the particle density from 4x10^9 to 0.9x10^9 cm^{-2} (Figs. 4a and b). Shortening the immersion time from 120 to 30 s decreased the average particle size from 87 to 67 nm (Figs. 4b and c). The size and particle density of electrolessly deposited Pt nanoparticles on multicrystalline Si can be controlled by changing the deposition conditions. This is consistent with our previously reported results on single-crystalline silicon aside from high particle density (Yae et al., 2007c).

2.2 Porous silicon formation by metal-particle-assisted hydrofluoric acid etching

The antireflection of the semiconductor surface is an effective method for improving the energy conversion efficiency of solar cells (Sze, 1981, Nelson, 2003). The surface texturization by anisotropic etching is a common antireflection method for single crystalline Si solar cells. However, multicrystalline Si cannot be uniformly texturized by the anisotropic etching caused by its variety of orientations of crystallites. The metal-particle-assisted hydrofluoric acid etching can form both macroporous and microporous layers on multicrystalline Si wafers, which are modified with fine metal particles, by simply immersing the wafers in an hydrofluoric acid solution without a bias and a particular oxidizing agent (Yae et al. 2006a, 2009). In previous papers, we reported that porous layer formation by this etching for 24 h decreased the reflectance of Si and increased the solar cell characteristics, which are not only photocurrent density but also photovoltage (Yae et al. 2003, 2005, 2006a, 2009).

2.2.1 Etching mechanism

The metal-particle-assisted hydrofluoric acid etching of Si proceeds by a local galvanic cell mechanism requiring photoillumination onto Si or dissolved oxygen in the solution (Yae et al. 2005, 2007d, 2009, 2010). Figure 5 shows a schematic diagram of n-Si and electrochemical reaction (equations (5), (6) and (7)) potential in a hydrofluoric acid solution. The local cell reaction consists of anodic dissolution of Si (equation (5)) and cathodic reduction of oxygen (equation (6)) and/or protons (equation (7)) on catalytic Pt particles. Under the photoillumination, photogenerated holes in the Si valence band anodically dissolve Si on the whole photoirradiated surface of Si. Under the dark condition, the etching proceeds by holes injected into the Si valence band with only cathodic reduction of oxygen on Pt particles, and thus the etching is localized around the Pt particles. The localized anodic dissolution produces macropores, which have Pt particles on the bottom, on the Si surface as shown in Fig. 6. We previously revealed two points about metal-particle-assisted hydrofluoric acid etching of Si: 1) the etching rate increased with photoillumination intensity on Si wafers and dissolved oxygen concentration in hydrofluoric acid solution; and 2) the time dependence of photoillumination intensity on the Si sample in the laboratory, which is ca. 0.2 mW cm^{-2} illumination for 6 h, dark condition for 12 h and then ca. 0.2 mW cm^{-2} illumination for 6 h, is suitable to produce the macro- and microporous combined structure effective for improving

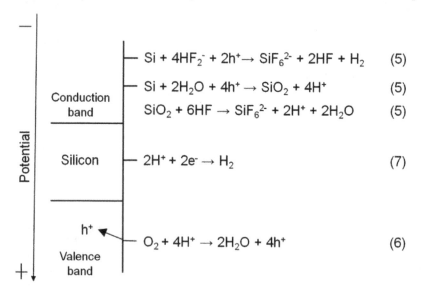

$$Si + 4HF_2^- + 2h^+ \rightarrow SiF_6^{2-} + 2HF + H_2 \quad\quad (5)$$

$$Si + 2H_2O + 4h^+ \rightarrow SiO_2 + 4H^+ \quad\quad (5)$$

$$SiO_2 + 6HF \rightarrow SiF_6^{2-} + 2H^+ + 2H_2O \quad\quad (5)$$

$$2H^+ + 2e^- \rightarrow H_2 \quad\quad (7)$$

$$O_2 + 4H^+ \rightarrow 2H_2O + 4h^+ \quad\quad (6)$$

Fig. 5. Schematic diagram of silicon and electrochemical reaction potential in a hydrofluoric acid solution.

the solar cell characteristics (Yae et al. 2005, 2006b, 2009). In this section, we applied this method to the Pt-nanoparticle-modified multicrystalline n-Si to improve the solar cell characteristics, and attempted to shorten the etching time by controlling etching conditions such as the photoillumination intensity and the dissolved oxygen concentration.

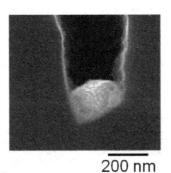

200 nm

Fig. 6. Typical cross-sectional scanning electron micrograph of silicon macropore having a Pt particle on the bottom.

2.2.2 Porous structure control

The Pt-nanoparticle modified multicrytalline n-Si wafers were immersed in a 7.3 mol dm^{-3} hydrofluoric acid aqueous solution at 298 K. In some cases, oxygen gas bubbling was applied to the solution, and/or the n-Si wafers were irradiated with a tungsten-halogen lamp during immersion in the solution in a dark room. The reflectance of Si wafers was measured using a spectrophotometer in the diffuse reflection mode with an integrating sphere attachment.

Preparation conditions	Pretreatment	Pt deposition time (s)	Prorous layer formation (matal-particle-assisted hydrofluoric acid ethcing) conditions	Total etching time (h)
a	A	120	without light control for 24 h	24
b	B	120	without light control for 24 h	24
c	B	120	under 40 mW cm^{-2} with no bubbling for 3 h	3
d	B	120	40 mW cm^{-2} with no bubbling for 2 h and then in the dark with oxygen bubbling for 4 h	6
e	B	120	adding under 40 mW cm^{-2} with oxygen bubbling for 0.5 h to condition d	6.5
f	B	60	40 mW cm^{-2} with no bubbling for 2 h and then in the dark with oxygen bubbling for 4 h	6
g	B	60	adding under 40 mW cm^{-2} with oxygen bubbling for 0.5 h to condition f	6.5

Table 1. Preparation conditions of Pt nanoparticle modified porous multicrystalline n-Si

The deposition conditions of Pt-nanoparticles and metal-particle-assisted hydrofluoric acid etching conditions are listed in Table 1. Figure 7 shows typical scanning electron microscopic images of multicrystalline n-Si wafers that were pretreated by method A (image a) or B (image b) and metal-particle-assisted hydrofluoric acid etching without light control for 24 h (conditions a and b in Table 1). Macropores, whose diameter is 0.3–1 µm, were formed on whole surfaces of multicrystalline n-Si wafers. The density of pores, i.e. porosity, of n-Si wafer pretreated by method B is lower than that for method A. This is consistent with the Pt particle density on multicrystalline Si surface before etching (Fig. 4a and b). Both samples showed an orange photoluminescence under UV irradiation, thus microporous layers were formed on both samples.

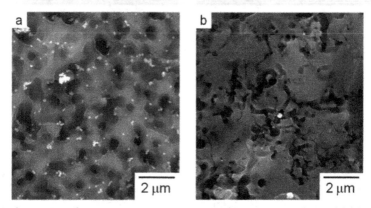

Fig. 7. Typical scanning electron microscopic images of Pt nanoparticle modified porous multicrystalline n-Si. Preparation conditions: images a and b are for conditions a and b in Table 1, respectively.

Figure 8 shows typical scanning electron microscopic images of multicrystalline n-Si that were pretreated by method B and metal-particle-assisted hydrofluoric acid etching under control of the photoillumination and the dissolved oxygen concentration (conditions c to g in Table 1). A microporous layer giving photoluminescence and no macropores was formed by etching under photoillumination without any gas bubbling estimated dissolved oxygen concentration of solution is ca. 5 ppm (Fig. 8a, condition c). The etching under the dark condition with oxygen gas bubbling (the solution was saturated with oxygen) after the etching under photoillumination produced macro- and microporous combined structure on the multicrystalline n-Si wafer (Fig. 8b, condition d). The morphology of the Si surface is similar to that formed by the etching without light control and gas bubbling for 24 h (Fig. 7b, condition b). Addition of the photoillumination with oxygen bubbling to the preceding conditions enlarged the macropore size and microporous layer thickness (Fig. 8c, condition e). Shortening the immersion time of multicrystalline n-Si wafers in the Pt displacement deposition solution, i.e. reduction of particle size and particle density of Pt on the wafers, reduced the number of macropores on the etched n-Si wafers (Figs. 8d and e, conditions f and g, respectively). The structure change in the porous layer of multicrystalline n-Si by changing the photoillumination intensity and dissolved oxygen concentration is consistent with our previously reported results on single crystalline n-Si (Yae et al., 2005, 2006b, 2009).

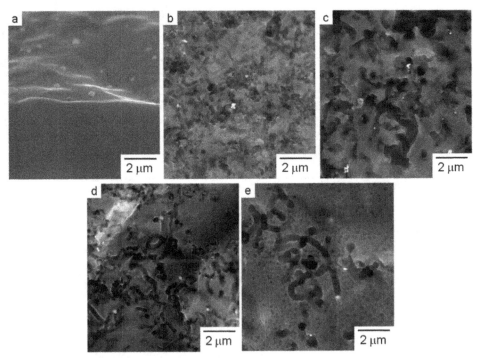

Fig. 8. Typical scanning electron microscopic images of Pt nanoparticle modified porous multicrystalline n-Si. Preparation conditions: images a, b, c, d, and e are for conditions c, d, e, f, and g in Table 1, respectively.

2.2.3 Antireflection effect

The macroporous layer formation changed the surface color of multicrystalline n-Si wafers to dark gray. Figure 9 shows the reflectance spectra of multicrystalline n-Si wafers. The porous layer prepared by the etching without light control and gas bubbling for 24 h reduced the reflectance from over 30% to under 6.2% (curves a and b) (Yae et al., 2006a, 2009). The porous layers prepared by the etching under the conditions d and g of Table 1 gave reflectance between 8 and 17% (curves c and d). This value is higher than that of the wafer prepared under the non-controlled conditions, but much lower than the non-etched wafer.

2.3 Photovoltaic photoelectrochemical solar cells

To evaluate electrical characteristics of photoelectrodes, we prepared photovoltaic photoelectrochemical solar cells (Fig. 1a) equipped with the Pt-nanoparticle modified porous multicrystalline n-Si photoelectrode. The multicrystalline n-Si electrode and Pt-plate counterelectrode were immersed in a redox electrolyte solution. Just before measuring the solar cell characteristics, the multicrystalline n-Si electrode was immersed in a 7.3 mol dm^{-3} hydrofluoric acid solution for two min under the elimination of dissolved oxygen by bubbling pure argon gas into the solution. This treatment is important to obtain high photovoltage caused by halogen atom termination of Si surface as mentioned below. A mixed solution of 7.6 mol dm^{-3} hydroiodic acid (HI) and 0.05 mol dm^{-3} iodine (I$_2$) was used as a redox electrolyte solution of the photovoltaic photoelectrochemical solar cell. Photocurrent density versus potential (j-U) curves were obtained with a cyclic voltammetry tool. The potential of the n-Si wafer was measured with respect to the Pt counterelectrode. The multicrystalline n-Si was irradiated with a solar simulator (AM1.5G, 100 mW cm^{-2}) through the quartz window and a redox electrolyte solution ca. 3 mm thick.

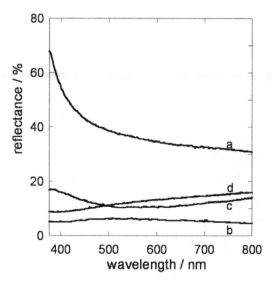

Fig. 9. Reflectance spectra of multicrystalline n-Si wafers: curve a after immersion in sodium hydroxide solution for saw damage layer removal; b, c, and d prepared under the conditions a, d, and g in Table 1, respectively.

2.3.1 Effect of particle density and size of platinum nanoparticles

Figure 10 show typical photocurrent density versus potential (j-U) curves of Pt-nanoparticle modified multicrystalline n-Si photoelectrodes having no porous layer pretreated under the same conditions as the specimens of Fig. 4. The decrease in particle density and size of Pt-nanoparticles increased the open-circuit photovoltage (V_{OC}) and short-circuit photocurrent density (j_{SC}) of photovoltaic photoelectrochemical solar cells from curve a to curve c of Fig. 10. Thus, the conversion efficiency (η^S) of the solar cells increased from 3.8% to 5.0%.

The reason for the increase in photocurrent density of the photoelectrochemical solar cells is the decrease of surface coverage of Pt-nanoparticles on Si. The surface coverage is 20% and 5% for Fig. 4a and b, respectively. This decrease is expected to increase the intensity of solar light reaching the Si surface by 19%. This is almost consistent with the increase in the short-circuit photocurrent density by 17%. The average open-circuit photovoltage of 12 samples is 0.42 V. This is lower than that for Pt-nanoparticle-electrolessly-deposited single crystalline n-Si electrodes (0.50 V in the average of 76 samples). This is explained by the following two reasons. 1) Lower quality of multicrystalline Si than single crystalline: The characteristics of multicrystalline Si solar cells are commonly lower than those of single crystalline. Thus, not only photovoltage but also the short-circuit photocurrent density and fill factor (F.F.) of photoelectrochemical solar cells are 12.1 mA cm^{-2} and 0.57 lower than those of single crystalline (18.3 mA cm^{-2} and 0.60 on average, respectively). 2) Insufficient density of termination of Si surface bonds with iodine atoms: The termination of Si surface bonds with iodine atoms shifts the flat band potential of Si toward negative, and thus increases the photovoltage of photoelectrochemical solar cells using hydroiodic acid and iodine redox electrolyte (Fujitani et al., 1997, Ishida et al., 1999, Yae et al., 2006a, Zhou et al., 2001). An electrolyte solution of 8.6 mol dm^{-3} hydrobromic acid (HBr) and 0.05 mol dm^{-3} bromine (Br$_2$) has sufficient negative redox potential to generate high open-circuit photovoltage without the termination. Using the hydrobromic acid and bromine electrolyte solution increases the photovoltage by 0.06 V for multicrystalline and 0.03 V for single-crystalline n-Si electrodes from those using hydroiodic acid and iodine electrolyte solution. This result indicates that the density of the termination of multicrystalline n-Si surface bonds with iodine atoms is insufficient for generating high photovoltage.

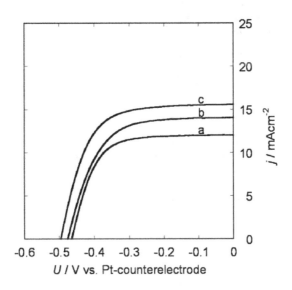

Fig. 10. Photocurrent density versus potential (j-U) curves of photovoltaic photoelectrochemical solar cells equipped with Pt-nanoparticle modified multicrystalline n-Si photoelectrode having no porous layer pretreated under the same conditions as the specimens of Fig. 4. Pretreatment: method A (image a), B (b and c); Pt deposition time: 120 (a and b), 30 s (c).

2.3.2 Effect of porous layer

Table 2 and Figure 11 indicate the average characteristics and typical photocurrent density versus potential (j-U) curves of photovoltaic photoelectrochemical solar cells equipped with a Pt-nanoparticle modified porous multicrystalline n-Si electrode prepared under the conditions listed in Table 1. The characteristics of photoelectrodes prepared under the conditions a and b as those for the wafers indicated in Fig. 7 show that the combination of the controlling particle density and size of Pt particles, and the formation of porous layer using metal-particle-assisted etching obtained a large increase in the conversion efficiency (η^s) from 3.8% for curve a in Fig. 10 and 2.9% in average of 12 samples to 5.1% in the average (Table 2). The formation of continuous microporous layer (Figs. 8a and 11a, and condition c in Table 1) increased photovoltage (V_{OC}), and decreased fill factor (F.F.) of the solar cells. The formation of macro- and microporous combined structure (Figs. 8b and c, and conditions d and e in Table 1, respectively) increased photocurrent density (j_{SC}) and fill factor (F.F.), and thus increased the conversion efficiency (η^β) of solar cells (Fig. 11b, and conditions d and e in Table 2). The decrease of particle density and size of Pt particles (Figs. 8d and e, and conditions f and g in Table 1, respectively) increased photocurrent density (j_{SC}) and conversion efficiency (η^β) (Fig. 11c, and conditions f and g in Table 2). The conversion efficiency of solar cells reached 7.3% of curve c in Fig. 11 and 6.1% in the average of 4 samples (Table 2), and the etching time was shortened to 6.5 h from 24 h by controlling the photoillumination intensity and the dissolved oxygen concentration during etching (condition g in Table 1 and 2).

Preparation conditions see Table 1	No. of tested samples	Open-circuit photovoltage V_{OC} (V)	Short-circuit photocurrent density j_{SC} (mA cm^{-2})	Fill factor F.F.	Efficiency η^β (%)
a	21	0.47	13.8	0.60	3.9
b	7	0.50	16.6	0.62	5.1
d	17	0.46	17.6	0.60	4.9
e	3	0.50	17.4	0.63	5.5
f	3	0.49	18.0	0.66	5.8
g	4	0.50	19.5	0.63	6.1

Table 2. Characteristics of photovoltaic photoelectrochemical solar cells equipped with Pt-nanoparticle modified porous multicrystalline n-Si electrode prepared under the conditions in Table 1. Average values are indicated.

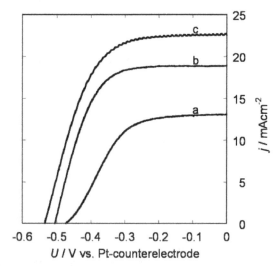

Fig. 11. Photocurrent density versus potential (j-U) curves of photovoltaic photoelectrochemical solar cells equipped with a Pt-nanoparticle modified porous multicrystalline n-Si electrode. Preparation conditions: curves a, b, and c, are for conditions c, d, and g listed in Table 1, respectively.

The increase in photocurrent density of photoelectrochemical solar cells equipped with Pt-nanoparticle modified multicrystalline n-Si electrode by the Pt-particle-assisted hydrofluoric acid etching is ca. 15% lower than the 30-40% estimated with reduction of the reflectance from 33% to 5-14% at the light wavelength of 700 nm. This difference can be explained by the difference in the refractive index between air (1.000), water (1.332 at 633 nm) and Si (3.796 at 1.8 eV (689 nm)) (Lide, 2004). The reflectance of Si is calculated at 34% in the air and 23% in the water. Using 23% as the initial value of reflectance estimates the increase in photocurrent density by the etching at 12-23%. This value is consistent with the experimental result of ca. 15%.

The photovoltage of photoelectrochemical solar cells equipped with Pt-nanoparticle modified multicrystalline n-Si electrode was improved by formation of the porous layer by Pt-particle-assisted hydrofluoric acid etching (Table 2). The photovoltage increase by the etching in dark conditions for 24 h was 0.01 V (V_{OC}: 0.43 V) in the average of eight samples, much lower than the 0.05 V (V_{OC}: 0.47 V) by the etching in a laboratory without light control (condition a in Table 1 and 2). These results show that the microporous layer effectively increases the photovoltage of such photoelectrochemical solar cells. This increase is explained by the following two possible mechanisms. 1) Screening Pt-nanoparticles' modulation of Si surface band energies by the microporous layer: The photovoltage of an n-Si electrode modified with metal particles depends on the distribution density of metal particles and the size of the direct metal-Si contacts. While metal particles are necessary as electrical conducting channels and catalysts of electrochemical reactions, the particles modulate the Si surface band energies. Thus, larger direct metal-Si contacts than a suitable size and/or a higher distribution density of metal particles than a suitable value reduce the effective energy barrier height, and then reduce the photovoltage of solar cells. The presence of a moderately thick microporous layer between the metal particles and bulk n-Si screens the modulation and thus raises the energy barrier height of the n-Si electrode, as discussed in the previous paper (Kawakami et al., 1997). 2) Increase in density of termination of Si surface bonds with iodine atoms: As we discussed in the previous section, the low open-circuit photovoltage (0.42 V) of the flat (nonporous) multicrystalline n-Si electrodes can be caused by the insufficient density of the termination of Si surface bonds with iodine atoms. Using the hydrobromic acid and bromine electrolyte solution increased the average open-circuit photovoltage of porous n-Si electrodes prepared under the condition a in Table 1 by 0.03 V for multicrystalline and 0.02 V for single-crystalline n-Si from those of using hydroiodic acid and iodide electrolyte solution. This result indicates that the density of the termination of the multicrystalline n-Si surface bonds with iodine atoms is increased to sufficient value for generating high V_{OC} by forming the microporous layer.

2.4 Solar to chemical conversion (solar hydrogen production)

In the preceding section, we prepared the efficient photovoltaic photoelectrochemical solar cells using the Pt-nanoparticle modified porous multicrystalline n-Si electrode. In this section, these electrodes were used for solar to chemical conversion via the photoelectrochemical decomposition of hydrogen iodide (HI) to iodine (I_2 or I_3^-) and hydrogen gas (H_2), that is, solar hydrogen. A two-compartment cell was used (Fig. 1b). The multicrystalline n-Si electrode was used as a photoanode in the mixed solution of hydroiodic acid and iodine of the anode compartment. A platinum plate was used as a counterelectrode in the perchloric acid ($HClO_4$) solution of the cathode compartment. Both compartments were separated with a porous glass plate. Figure 12 shows the typical photocurrent density versus potential (j-U) curve for the multicrystalline n-Si electrode prepared under the condition g in Table 1 and 2. The potential (U) of the electrode was measured versus the Pt-plate counterelectrode in the perchloric acid solution of the cathode compartment (Fig. 1b). The short-circuit photocurrent density of 21.7 mA cm^{-2} was obtained. The solution color at the Si surface darkened, and gas evolution occurred at the Pt cathode surface. These results clearly show that the photoelectrochemical solar cell equipped with the Pt-nanoparticle modified porous multicrystalline n-Si electrode can decompose hydrogen iodide into hydrogen and iodine with no external bias, as shown in the equations (1), (2) and (3) in the section 1.1.

The dashed curve in Fig. 12 shows the current density versus the potential (j-U) curve of Pt electrode, which was in the anode compartment, instead of the Si electrode of the above cell for hydrogen iodide decomposition (Fig. 1b). The onset potential of the anodic current was 0.25 V versus the Pt-counterelectrode in the cathode compartment. This value indicates that the Gibbs energy change for the hydrogen iodide decomposition in the present solutions is 0.25 eV. The energy gain of solar to chemical conversion using the photoelectrochemical solar cell is calculated at 5.4 mW cm^{-2} by the product of the Gibbs energy change per the elementary charge and the short-circuit photocurrent density of 21.7 mA cm^{-2} under simulated solar illumination (AM1.5G, 100 mW cm^{-2}). Thus, we calculate the efficiency of solar to chemical conversion (solar hydrogen production) via the photoelectrochemical decomposition of hydrogen iodide at 5.4%. The average in solar-to-chemical-conversion efficiency of five samples was 4.7%.

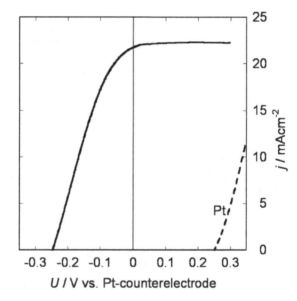

Fig. 12. Photocurrent density versus potential (j-U) curve (solid line) for solar-to-chemical conversion type of photoelectrochemical solar cell equipped with Pt-nanoparticle modified porous multicrystalline n-Si electrode prepared under condition g in Table 1. The two-compartment cell for photodecomposition of hydrogen iodide (Fig. 1b) was used. Dashed line: Pt electrode measured in the anode compartment of the two-compartment cell instead of the Si photoelectrode.

In Section 2, it was described that platinum-nanoparticle modified porous multicrystalline silicon electrodes prepared by electroless displacement deposition and metal-particle-assisted hydrofluoric acid etching can generate hydrogen (solar hydrogen) and iodine through the photoelectrochemical decomposition of hydrogen iodide in aqueous solution with no external bias at the solar-to-chemical conversion efficiency of 5.4%. The control of particle density and size of Pt particles by changing the initial surface condition of Si and deposition condition of Pt, and the control of porous layer structure by changing the etching conditions improve the conversion efficiency.

3. Platinum nanoparticle modified microcrystalline silicon thin films

Hydrogenated microcrystalline silicon (μc-Si:H) thin films are promising new materials for low-cost solar cells. The microcrystalline Si thin film approach has several advantages, including minimal use of semiconductor resources, large-area fabrication using low-cost chemical vapor deposition (CVD) methods, and no photodegradation of the solar cell's characteristics (Matsumura, 2001, Meier et al., 1994, Yamamoto et al., 1994). We applied microcrystalline Si thin films to solar hydrogen production by the photodecomposition of hydrogen iodide (Yae et al., 2007a, 2007b) and solar water splitting(Yae et al., 2007b). Figure 13 schematically shows a cross-section of the microcrystalline silicon thin-film photoelectrode. Photoelectrochemical solar cells require neither a p-type semiconductor layer nor a transparent conducting layer, which is necessary to fabricate solid-state solar cells.

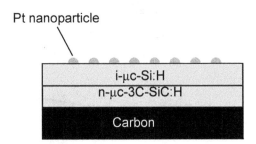

Fig. 13. Schematic cross-section of Pt-nanoparticle modified microcrystalline Si thin-film photoelectrode.

3.1 Preparation of photoelectrodes and photovoltaic photoelectrochemical solar cells

Hydrogenated microcrystalline silicon thin films were deposited onto polished glassy carbon (Tokai Carbon) substrates by the hot-wire catalytic chemical vapor deposition (cat-CVD) method (Matsumura et al. 2003). A 40-nm-thick n-type hydrogenated microcrystalline cubic silicon carbide (n-μc-3C-SiC:H) layer was deposited on the substrates using hydrogen-diluted monomethylsilane and phosphine gas at temperatures of 1700°C for the rhenium filament. An intrinsic hydrogenated microcrystalline silicon (i-μc-Si:H) layer, with thickness of 2-3 μm, was deposited on the n-type layer using monosilane gas at 1700°C for the tantalum filament. The microcrystalline silicon thin film electrodes were prepared by connecting a copper wire to the backside of the substrate with silver paste and covering it with insulating epoxy resin.

We deposited the Pt nanoparticles on the microcrystalline silicon surface using electroless displacement deposition as for the multicrystalline Si photoelectrodes (section 2.1). Figure 14 shows an scanning electron microscopic (SEM) image of the microcrystalline silicon film's surface after immersion in the Pt deposition solution for 120 s. Platinum nanoparticles of 3-200 nm in size and 1.5×10^{10} cm^{-2} in particle density were scattered on the film. The size and distribution density of Pt particles varied with the deposition conditions, such as oxide layer formation on the films before deposition and the immersion time of films in the deposition solution. The distribution density is much higher than that for a single-crystalline n-Si wafer, but the changing behaviors of the size and distribution density are similar to those of the single crystalline (Yae et al., 2007c, 2008).

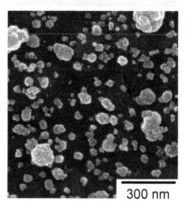

300 nm

Fig. 14. Scanning electron microscopic image of Pt-nanoparticle modified microcrystalline Si thin film surface.

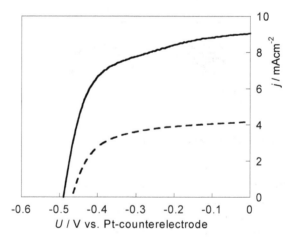

Fig. 15. Photocurrent density versus potential (j-U) curves for photovoltaic photoelectrochemical solar cell equipped with the Pt-nanoparticle modified microcrystalline Si photoelectrode measured in the 7.6 mol dm^{-3} (M) hydroiodic acid (HI)/0.05 M iodine (I_2) (dashed line) and 3.0 M HI/0.002 M I_2 (solid line) redox solutions.

For the photovoltaic photoelectrochemical solar cell (Fig. 1a), the Pt-nanoparticle-modified microcrystalline silicon thin film electrode and Pt-plate counterelectrode were immersed in a hydroiodic acid and iodine redox electrolyte solution as for the multicrystalline Si photoelectrodes (section 2.3). Figure 15 shows the photocurrent density versus potential (j-U) curves for the photovoltaic solar cell. The microcrystalline silicon film was stably adherent to the glassy carbon substrate after completing the photoelectrochemical measurements in these highly acidic solutions. The open-circuit photovoltage was 0.47-0.49 V. This is higher than the 0.3 V value obtained for the microcrystalline silicon thin film electrode covered with a continuous 1.5-nm-thick Pt layer, which was deposited using the electron-beam evaporation method. These results clearly indicate that the Pt-nanoparticle-modified microcrystalline silicon thin film electrodes work by using the same mechanism as the Pt-nanoparticle-modified single-crystalline n-Si electrodes, which work as ideal semiconductor photoelectrodes for generating high photovoltage and stable photocurrent described in previous sections 1.2 and 2.3.1. The reduction of redox electrolyte concentration increased the short-circuit photocurrent density to 9.1 from 4.2 mA cm^{-2} (Fig. 15, solid line). This increase is caused by a decrease in the visible light absorption of the triiodide (I_3^-) ion in the redox solution. The increased photocurrent raised open-circuit photovoltage to 0.49 V, and thus the photovoltaic conversion efficiency reached 2.7%.

3.2 Solar to chemical conversion (solar hydrogen production) via hydrogen iodide decomposition

The Pt-nanoparticle modified microcrystalline Si thin film electrode were used for solar to chemical conversion via the photoelectrochemical decomposition of hydrogen iodide to iodine and hydrogen gas as the multicrystalline Si photoelectrodes (section 2.4). For the photoelectrochemical decomposition of hydrogen iodide, a two-compartment cell was used (Fig. 1b and 2).

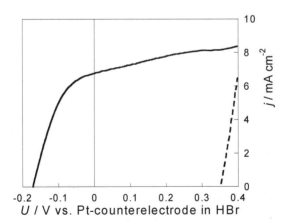

Fig. 16. Photocurrent density versus potential (j-U) curve (Solid line) for the Pt-nanoparticle modified microcrystalline Si thin film electrode measured in the hydroiodic acid and iodine mixture solution of the anode compartment of the two-compartment cell for solar to chemical conversion (solar hydrogen production, Fig. 1b). Dashed line: Pt electrode measured in the anode compartment of the two-compartment cell instead of the Si photoelectrode. Electrolyte solutions: anode compartment: 3.0 M HI/0.002 M I_2; cathode compartment: 3.0 M HBr.

The solid line in Fig. 16 shows the photocurrent density versus potential (j-U) curve for the Pt-nanoparticle-modified microcrystalline Si thin film electrode measured in the hydroiodic acid and iodine mixture solution of the anode compartment of the two-compartment cell. The potential of the electrode was measured versus the Pt counterelectrode in the hydrobromic acid solution of the cathode compartment. In the short-circuit condition under the simulated solar illumination, we obtained a shirt-circuit photocurent density of 6.8 mA cm^{-2}, the solution color on the photoelectrode surface darkened, and gas evolution occurred at the Pt cathode surface. These results clearly show that the photoelectrochemical solar cell equipped with the Pt-nanoparticle-modified microcrystalline Si thin film electrode can decompose hydrogen iodide into hydrogen gas and iodine with no external bias with 2.3% of solar-to-chemical conversion efficiency.

3.3 Hydrogen production via solar water splitting using multi-photon system

A multi-photon system equipped with the microcrystalline Si thin film and titanium dioxide (TiO$_2$) photoelectrodes in series (Fig. 17) was prepared based on a work in literature using a dye-sensitization-photovoltaic cell and a tungsten trioxide (WO$_3$) photoanode (Grätzel, 1999). A titanium dioxide photoanode and a Pt cathode (counterelectrode) were immersed in a perchloric acid (HClO$_4$) aqueous solution in a quartz cell. A photovoltaic photoelectrochemical solar cell equipped with the Pt-nanoparticle-modified microcrystalline Si electrode (section 3.1) was connected to the titanium dioxide photoanode and Pt cathode in series. Simulated solar light irradiated to the titanium dioxide photoelectrode. The titanium dioxide, which has a 3-eV energy band gap, absorbs the short-wavelength part (UV) of the solar light. The long-wavelength part of the solar light transmitted by the titanium dioxide and quartz cell reaches the Pt-nanoparticle-modified microcrystalline Si thin-film of the photovoltaic photoelectrochemical solar cell. The photovoltaic cell applies bias between the titanium dioxide photoanode and the Pt cathode in a perchloric acid aqueous solution for splitting water to hydrogen and oxygen.

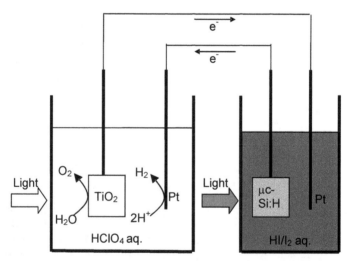

Fig. 17. Schematic illustration of multi-photon system equipped with titanium dioxide and microcrystalline Si photoelectrodes for solar water splitting.

The titanium dioxide photoanode was prepared as follows. Transparent conductive tin oxide (SnO_2)-coated glass plates were used as substrates. Titanium dioxide powder (P-25, average crystallite size: 21 nm) was ground with nitric acid, acetyl acetone, surfactant (Triton X-100), and water in a mortar. The obtained paste was coated on the substrate and dried. The titanium dioxide-nanoparticle film was heated in air at 500°C for three hours. The titanium dioxide electrode was prepared by connecting a copper wire to the bare part of the conductive tin oxide film with silver paste and covering it with insulating epoxy resin.

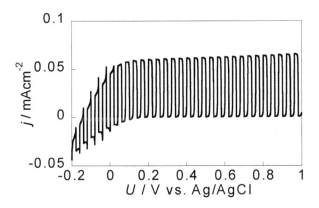

Fig. 18. Photocurrent density versus potential (j-U) curve for the titanium dioxide photoelectrode in a perchloric acid aqueous solution under chopped simulated solar illumination.

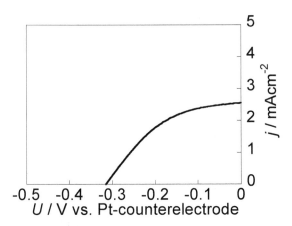

Fig. 19. Photocurrent density versus potential (j-U) curve for the photovoltaic photoelectrochemical solar cell equipped with a Pt-nanoparticle-modified microcrystalline Si electrode in the redox solution under simulated solar light illumination through the titanium dioxide photoelectrochemical cell.

Figure 18 shows the photocurrent density versus potential (j-U) curve for the titanium dioxide photoelectrode in a perchloric acid aqueous solution under simulated solar illumination. The dissolved oxygen in the solution was eliminated by using argon gas flow into the solution before the measurement. The anodic photocurrent starts to generate at -0.14 V vs. Ag/AgCl. This onset potential is more positive than -0.24 V vs. Ag/AgCl for hydrogen evolution, and thus this electrode cannot split water into hydrogen and oxygen without external bias. Figure 19 shows the photocurrent density versus potential (j-U) curve for the photovoltaic photoelectrochemical solar cell equipped with a Pt-nanoparticle-modified microcrystalline Si electrode in the redox solution under simulated solar light illumination through the titanium dioxide photoelectrochemical cell. The shirt-circuit photocurrent density was decreased from 5.3 mA cm^{-2} for the cell under direct solar light illumination to 2.6 mA cm^{-2} by light attenuation with the titanium dioxide cell. The multi-photon system (Fig. 17) using the same titanium dioxide and Pt-nanoparticle-modified microcrystalline Si electrodes as those in Figs. 18 and 19 indicated the photocurrent density versus potential (j-U) curve of Fig. 20. This system generated anodic photocurrent at a potential that was more negative than -0.24 V vs. Ag/AgCl for hydrogen evolution. Figure 21 shows that steady photocurrent was obtained for the multi-photon system in the short-circuit condition (Fig. 17). Tiny gas bubble formed on the Pt cathode during measurement under the short-circuit condition. These results show that this multi-photon system can split water into hydrogen and oxygen with no external bias with solar light. Since two photoelectrodes of titanium dioxide and Pt-nanoparticle-modified microcrystalline Si were connected in series, photovoltage was the sum of the two electrodes' values and photocurrent was the lower of the two electrodes' values. Therefore, the photocurrent density for water splitting was determined by that of the titanium dioxide electrode and very low. The photocurrent density, and thus hydrogen production by solar water splitting, is expected to increase by using a semiconductor with a narrower band gap, such as tungsten trioxide, instead of titanium dioxide. The theoretical simulation obtained 8 mA cm^{-2} of shirt-circuit photocurrent density, that is, 10% of solar-to-chemical conversion efficiency for solar water splitting for the tungsten trioxide and Si multi-photon system.

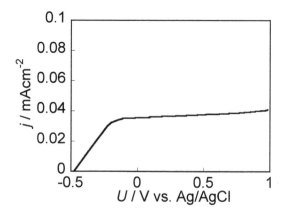

Fig. 20. Photocurrent density versus potential (j-U) curve for the multi-photon system (Fig. 17) using the same titanium dioxide thin film and Pt-nanoparticle modified microcrystalline Si photoelectrodes and electrolyte solutions as those in Figs. 18 and 19 under simulated solar light illumination.

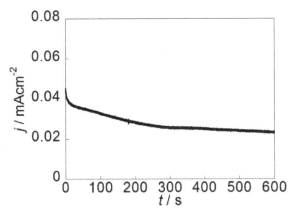

Fig. 21. Short-circuit photocurrent density (j) as a function of time (t) for the multi-photon system of Fig. 20 under simulated solar illumination.

4. Conclusion

Multicrystalline silicon wafers and microcrystalline silicon thin films, which are common and prospective low-cost semiconductor materials for solar cells, respectively, were successfully applied to produce solar hydrogen via photodecomposition of hydrogen iodide and solar water splitting. These photoelectrochemical solar cells have the following advantages: 1) simple fabrication of a cell by immersing the electrode in an electrolyte solution; 2) there is no need for a p-type semiconductor or a transparent conducting layer; and 3) direct solar-to-chemical conversion (fuel production). Modification of silicon surface with platinum nanopartilces by electroless displacement deposition and porous layer formation by metal-particle-assisted hydrofluoric acid etching improve solar cell characteristics. The solar-to-chemical conversion efficiency reached 5% for the photodecomposition of hydrogen iodide, and hydrogen gas evolution was obtained by the solar water splitting with no input of external electricity.

5. Acknowledgment

The author is grateful to Prof. H. Matsuda, Dr. N. Fukumuro (University of Hyogo), Dr. S. Ogawa, Prof. N. Yoshida, Prof. S. Nonomura (Gifu University), Mr. S. Sakamoto (Nippon Oikos Co., Ltd.), and Prof. Y. Nakato (Osaka University) for co-work and valuable discussions. The author would like to thank the students who collaborated: H. Miyasako, T. Kobayashi, K. Suzuki, and A. Onaka. The author is grateful to Prof. Y. Uraoka of Nara Institure of Science and Technology for the simulation of the solar water splitting using the multi-photon system. The present work was partly supported by the following programs: Grants-in-Aid for Scientific Research (C) from the JSPS (17560638, 20560676, and 23560875), Grants-in-Aid for education and research from Hyogo Prefecture through the University of Hyogo, Core Research for Evolutional Science and Technology (CREST) from the Japan Science and Technology Agency (JST), and Research for Promoting Technological Seeds from JST. The author wishes to thank Nippon Sheet Glass Co., Ltd. for donating transparent conductive tin oxide coated glass plates. Figures 15 and 16 were reprinted from ref. Yae et al., 2007a, copyright Elsevier (2007).

6. References

Allongue, P., Blonkowski, S., & Souteyrand, E. (1992). *Elecrochim. Acta,* Vol. 37, 781.

Arakawa, H., Shiraishi, C., Tatemoto, M., Kishida, H., Usui, D., Suma, A., Takamisawa, A., & Yamaguchi, T. (2007). *Proc. SPIE,* Vol. 6650, *Solar Hydrogen and Nanotechnology II,* Guo, J. (Ed.), San Diego, 665003.

Chemla, M., Homma, T., Bertagna, V., Erre, R., Kubo, N., & Osaka, T. (2003). *J. Electroanal. Chem.,* Vol. 559, 111.

Fujishima, A. & Honda, K. (1972). *Nature,* Vol. 238, 37.

Fujitani, M., Hinogami, R., Jia, J. G., Ishida, M., Morisawa, K., Yae, S., & Nakato, Y. (1997). *Chem. Lett.,* 1041.

Grätzel, M. (1999). *Cattech,* Vol. 3, 4.

Gorostiza, P., Servat, J., Morante, J. R., & Sanz, F. (1996). *Thin Solid Films,* Vol. 275, 12.

Gorostiza, P., Allongue, P., Díaz, R.; Morante, J. R., & Sanz, F. (2003). *J. Phys. Chem. B,* Vol. 107, 6454.

Hinogami, R., Nakamura, Y., Yae, S., & Nakato, Y. (1997). *Appl. Surf. Sci.,* Vol. 121/122, 301.

Hinogami, R., Nakamura, Y., Yae, S., & Nakato, Y. (1998). *J. Phys. Chem. B,* Vol. 102, 974.

Ishida, M., Morisawa, K., Hinogami, R., Jia, J. G., Yae, S., & Nakato, Y. (1999). *Z. Phys. Chem.,* Vol. 212, 99.

Jia, J.-G., Fujitani, M., Yae, S., & Nakato, Y. (1996). *Electrochim. Acta,* Vol. 42, 431.

Kawakami, K., Fujii, T., Yae S., & Nakato, Y. (1997). *J. Phys. Chem. B,* Vol. 101, 4508.

Khaselev, O. & Turner, J. A. (1998). *Science,* Vol. 280, 542.

Licht, S. (Vol. Ed.). (2002). *Semiconductor Electrodes and Photoelectrochemistry,* Bard. A. J. & Stratmann, M. (Series Eds.), *Encyclopedia of Electrochemistry,* Vol. 6, Wiley-VCH, Weinheim.

Lide, D. R. (Ed.). (2004). *CRC Handbook of Chemistry and Physics,* CRC Press, Boca Raton, 85th Ed., pp. 10-232, 10-234 and 12-150.

Lin, G. H., Kapur, M., Kainthla, R. C., & Bockris, J. O'M. (1989). *Appl. Phys. Lett.,* Vol. 55, 386.

Matsumura, H. (2001). *Thin Solid Films,* Vol. 395, 1.

Matsumura, H., Umemoto, H., Izumi, A., & Masuda, A. (2003). *Thin Solid Films,* Vol. 430, 7.

Meier. J., Flückiger, R., Keppner, H., & Shah, A. (1994). *Appl. Phys. Lett.,* Vol. 65, 860.

Miller, E. L., Marsen, B., Paluselli, D., & Rocheleau, R. (2005). *Electrochem. Solid-State Lett.,* Vol. 8, A247.

Nagahara, L. A., Ohmori, T., Hashimoto, K., & Fujishima, A. (1993). *J. Vac. Sci. Technol. A,* Vol. 11, 763.

Nakato, Y., Ueda, K., Yano, H., & Tsubomura, H. (1988). *J. Phys. Chem.,* Vol. 92, 2316.

Nakato, Y. & Tsubomura, H. (1992). *Elecrochim. Acta,* Vol. 37, 897.

Nakato, Y., Jia, J. G., Ishida, M., Morisawa, K., Fujitani, M., Hinogami, R., & Yae, S. (1998). *Electrochem. Solid-State Lett.,* Vol. 1, 71.

Nakato, Y. (2000). Photoelectrochemical Cells, In: *Wiley Encyclopedia of Electrical and Electronics Engineering Online,* Webster, J. (Ed.), John Wiley & Sons, Available from: http://www.interscience.wiley.com

Nelson, J. (2003). *The Physics of Solar Cells,* Imperial College Press, London, pp. 276-279.

Paunovic, M., Schlesinger, M. (2006). *Fundamentals of Electrochemical Deposition 2nd. Ed.,* John Wiley & Sons, New York.

Park, J. H. & Bard, A. J. (2005). *Electrochem. Solid-State Lett.*, Vol. 8, G371.

Sakai, Y., Sugahara, S., Matsumura, M., Nakato, Y., & Tsubomura, H. (1988). *Can. J. Chem.*, Vol. 66, 1853.

Sze, S. M. (1981). *Physics of Semiconductor Devices*, John Wiley & Sons, New York, 2nd Ed., pp. 811-816.

Takabayashi, S., Nakamura, R., & Nakato, Y. (2004). *J. Photochem. Photobiol. A*, Vol. 166, 107.

Takabayashi, S., Imanishi, A., & Nakato, Y. (2006). *Comptes Rendus Chimie*, Vol. 9, 275.

Turner, J. A., Williams, M. C., & Rajeshwar, K. (2004). *Interface*, Vol. 13, No. 3, 24.

Yae, S., Tsuda, R., Kai, T., Kikuchi, K., Uetsuji, M., Fujii, T., Fujitani, M., & Nakato, Y. (1994). *J. Electrochem. Soc.*, Vol. 141, 3090.

Yae, S., Nakanishi, I., Nakato, Y., Toshima, N., & Mori, H. (1994). *J. Electrochem. Soc.*, Vol. 141, 3077.

Yae, S., Fujitani, M., Nakanishi, I., Uetsuji, M., Tsuda, R., & Nakato, Y. (1996). *Sol. Energy Mater. Sol. Cells*, Vol. 43, 311.

Yae, S., Kitagaki, M., Hagihara, T., Miyoshi, Y., Matsuda, H., Parkinson, B. A., & Nakato, Y. (2001). *Electrochim. Acta*, Vol. 47, 345.

Yae, S., Kawamoto, Y., Tanaka, H., Fukumuro, N., & Matsuda, H. (2003). *Electrochem. Comm.*, Vol. 5, 632.

Yae, S., Tanaka, H., Kobayashi, T., Fukumuro, N., & Matsuda, H. (2005). *Phys. Stat. Sol. (c)*, Vol. 2, 3476.

Yae, S., Kobayashi, T., Kawagishi, T., Fukumuro, N., & Matsuda, H. (2006). *Solar Energy*, Vol. 80, 701.

Yae, S., Kobayashi, T., Kawagishi, T., Fukumuro, N., & Matsuda, H. (2006). *The Electrochemical Society Proceedings Series*, Vol. PV2004-19, *Pits and Pores III: Formation, Properties and Significance for Advanced Materials*, Schmuki, P., Lockwood, D. J., Ogata, Y. H., Seo, M., & Isaacs, H. S. (Eds.). 141.

Yae, S., Kobayashi, T., Abe, M., Nasu, N., Fukumuro, N., Ogawa, S., Yoshida, N., Nonomura, S., Nakato, Y., & Matsuda, H. (2007). *Sol. Energy Mater. Sol. Cells*, Vol. 91, 224.

Yae, S., Onaka, A., Abe, M., Fukumuro, N., Ogawa, S., Yoshida, N., Nonomura, S., Nakato, Y., & Matsuda, H. (2007). *Proc. SPIE*, Vol. 6650, *Solar Hydrogen and Nanotechnology II*, Guo, J. (Ed.), San Diego, 66500E.

Yae, S., Nasu, N., Matsumoto, K., Hagihara, T., Fukumuro, N., & Matsuda, H. (2007). *Electrochim. Acta*, Vol. 53, 35.

Yae, S., Abe, M., Kawagishi, T., Suzuki, K., Fukumuro, N., & Matsuda, H. (2007). *Trans. Mater. Res. Soc. Jpn.*, Vol. 32, 445.

Yae, S., Fukumuro, N., & Matsuda, H. (2008). Electrochemical Deposition of Metal Nanoparticles on Silicon, In: *Progress in Nanoparticles Research*, Frisiras, C. T. (Ed.), pp. 117-135, Nova Science Publishers, Inc., New York.

Yae, S., Fukumuro, N., & Matsuda, H. (2009). Porous Silicon Formation by Metal Particle Enhanced HF etching, In: *Electroanalytical Chemistry Research Trends*, Hayashi, K. (Ed.), pp. 107-126, Nova Science Publishers, New York.

Yae, S., Tashiro, M., Abe, M., Fukumuro, N., & Matsuda, H. (2010). *J. Electrochem. Soc.*, Vol. 157, D90.

Yamamoto, K., Nakashima, A., Suzuki, T., Yoshimi, M., Nishio, H., & Izumina, M. (1994). *Jpn. J. Appl. Phys.* Vol. 33, L1751.

Zhou, X., Ishida, M., Imanishi, A., & Nakato, Y. (2001). *J. Phys. Chem. B*, Vol. 105, 156.

Ultrafast Electron and Hole Dynamics in CdSe Quantum Dot Sensitized Solar Cells

Qing Shen[1] and Taro Toyoda[2]
[1]PRESTO, Japan Science and Technology Agency (JST)
[2]The University of Electro-Communications
Japan

1. Introduction

A potential candidate for next-generation solar cells is dye-sensitized solar cells (DSSCs). Much attention has been directed toward DSSCs employing nanostructured TiO_2 electrodes and organic-ruthenium dye molecules as the light-harvesting media. The high porosity of nanostructured TiO_2 film enables a large concentration of the sensitizing dye molecules to be adsorbed. The attached dye molecules absorb light and inject electrons into the TiO_2 conduction band upon excitation. The electrons are then collected at a back conducting electrode, generating a photocurrent. DSSCs exhibit high photovoltaic conversion efficiencies of about 11% and good long-term stability. In addition, they are relatively simple to assemble and are low-cost (O'Regan & Grätzel, 1991; Grätzel, 2003; Chiba et al., 2006). However, in order to replace conventional Si-based solar cells in practical applications, further effort is needed to improve the efficiency of DSSCs. A great amount of work has been done on controlling the morphology of the TiO_2 electrodes by employing ordered arrays of nanotubes, nanowires, nanorods and inverse opal structures (Adachi et al., 2003; Paulose et al., 2006; Law et al., 2005; Song et al., 2005; Nishimura et al., 2003) in order to improve the electron transport and collection throughout the device. Another important factor in improving the performance of DSSCs is the design of the photosensitizer. The ideal dye photosensitizer for DSSCs should be highly absorbing across the entire solar light spectrum, bind strongly to the TiO_2 surface and inject photoexcited electrons into the TiO_2 conduction band efficiently. Many different dye compounds have been designed and synthesized to fulfill the above requirements. It is likely that the ideal photosensitizer for DSSCs will only be realized by co-adsorption of a few different dyes, for absorption of visible light, near infrared (NIR) light, and/or infrared (IR) light (Polo et al., 2004; Park et al., 2011). However, attempts to sensitize electrodes with multiple dyes have achieved only limited success to date.

Narrow-band-gap semiconductor quantum dots (QDs), such as CdS, CdSe, PbS, and InAs, have also been the subject of considerable interest as promising candidates for replacing the sensitizer dyes in DSSCs (Vogel et al., 1990, 1994; Toyoda et al., 1999, 2003; Peter et al., 2002; Plass et al., 2002; Shen et al., 2004a, 2004b, 2006a, 2006b, 2008a, 2008b, 2010a, 2010b; Yu et al., 2006; Robel et al., 2006; Niitsoo et al., 2006; Diguna, et al., 2007a , 2007b; Kamat, 2008, 2010; Gimenez et. al., 2009; Mora-Sero et al., 2009, 2010). These devices are called QD-sensitized solar cells (QDSCs) (Nozik, 2002, 2008; Kamat, 2008). The use of semiconductor QDs as sensitizers has some unique advantages over the use of dye molecules in solar cell applications (Nozik, 2002, 2008). First, the energy gaps of the QDs can be tuned by controlling their size, and therefore the absorption spectra of the QDs can be tuned to match the spectral distribution of sunlight. Secondly, semiconductor QDs have large extinction coefficients due to the quantum confinement effect. Thirdly, these QDs have large intrinsic dipole moments, which may lead to rapid charge separation. Finally, semiconductor QDs have potential to generate multiple electron-hole pairs with one single photon absorption (Nozik, 2002; Schaller, 2004), which can improve the maximum theoretical thermodynamic efficiency for photovoltaic devices with a single sensitizer up to ~44% (Hanna et al., 2006). However, at present, the conversion efficiency of QDSCs is still less than 5% (Mora-Sero´, et al., 2010; Zhang, et al., 2011). So, fundamental studies on the mechanism and preparation of QDSCs are still necessary and very important.

In a semiconductor quantum dot-sensitized solar cell (QDSC), as the first step of photosensitization, a photoexcited electron in the QD should rapidly transfer to the conduction band of TiO_2 electrode and a photoexcited hole should transfer to the electrolyte (Scheme 1). Thus charge separation of the photoexcited electrons and holes in the semiconductor QDs and the electron injection process are key factors for the improvement of the photocurrents in the QDSCs. In this sense, the study on the photoexcited carrier dynamics in the QDs is very important for improving the conversion efficiency of the solar cell. To date, the information on the carrier dynamics of semiconductor QDs adsorbed on TiO_2 electrode is limited, although a few studies have been carried out for CdS, CdSe and InP QDs using a transient absorption (TA) technique (Robel et al., 2006, 2007; Tvrdy et al., 2011; Blackburn et al., 2003, 2005). Most of them focused on the electron transfer process and the measurements mostly were carried out in either dispersed colloidal systems or dry electrodes. In recent years, the authors' group has been applying an improved transient grating technique (Katayama et al., 2003) to study the photoexcited carrier dynamics of semiconductor nanomaterials, such as TiO_2 nanoparticles with different crystal structures and CdSe QDs absorbed onto TiO_2 and SnO_2 nanostructured electrodes (Shen et al., 2005, 2006, 2007, 2008, 2010). The improved TG technique is a simple and highly sensitive time-resolved optical technique and has proved to be powerful for measuring various kinds of dynamics, such as population dynamics and excited carrier diffusion. Comparing to the TA technique, the improved TG technique has a higher sensitivity and measurements under lower light intensity are possible (Katayama et al., 2003; Shen et al., 2007, 2010a). This fact is very important for studying the carrier dynamics of QDs used in QDSCs under the conditions of lower light intensity similar to sun light illumination. The improved TG technique is also applicable to samples with rough surfaces, like the samples used in this study.

This chapter will focus on the ultrafast photoexcited electron and hole dynamics in CdSe QD adsorbed TiO_2 electrodes employed in QDSCs characterized by using the improved TG technique. CdSe QDs were adsorbed on TiO_2 nanostructured electrodes with different adsorption methods. The following issues will be discussed:

1. Pump light intensity dependence of the ultrafast electron and hole dynamics in the CdSe QDs adsorbed onto TiO_2 nanostructured electrodes;
2. Separation of the ultrafast electron and hole dynamics in the CdSe QDs adsorbed onto TiO_2 nanostructured electrodes;
3. Electron injection from CdSe QDs to TiO_2 nanostructured electrode;
4. Changes of carrier dynamics in CdSe QDs adsorbed onto TiO_2 electrodes versus adsorption conditions;
5. Effect of surface modification on the ultrafast carrier dynamics and photovoltaic properties of CdSe QD sensitized TiO_2 electrodes.

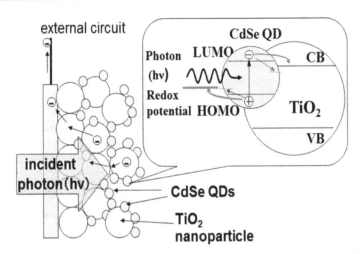

Scheme 1. Electron- hole pairs are generated in semiconductor QDs after light absorption. Then photoexctied electrons in the semiconductor QDs are injected to the conduction band of TiO$_2$ and/or trapped by surface or interface states. The photoexcited holes are scavenged by reducing species in the electrolyte and/or trapped by surface or interface states. The nanostructured TiO$_2$ is employed as an electron conductor and electrons transport in TiO$_2$ to a transparent conductive oxide (TCO) substrate, while the electrolyte is used as a hole transporter and hoels are transported to a counter electrode.

2. Electron and hole dynamics in CdSe QDs adsorbed onto TiO$_2$ electrodes

2.1 Preparation methods of CdSe QD adsorbed TiO$_2$ nanostructured electrodes

The method for preparing the TiO$_2$ electrodes has been reported in a previous paper (Shen et al., 2003). A TiO$_2$ paste was prepared by mixing 15 nm TiO$_2$ nanocrystalline particles (Super Titania, Showa Denko; anatase structure) and polyethylene glycol (PEG) (molecular weight (MW): 500,000) in pure water. The resultant paste was then deposited onto transparent conducting substrates (F-doped SnO$_2$ (FTO), sheet resistance = 10 Ω/sq). The TiO$_2$ films were then sintered in air at 450 °C for 30 min to obtain good necking. The highly porous nanostructure of the TiO$_2$ films (the pore sizes are of the order of a few tens of nanometers) was confirmed through scanning electron microscopy (SEM) images.

CdSe QDs can be adsorbed onto the TiO$_2$ nanostructured electrodes by using the following methods:

1. Chemical bath deposition (CBD) method (Hodes et al., 1994; Shen et al., 2008)
   Firstly, for the Se source, an 80 mM sodium selenosulphate (Na$_2$SeSO$_3$) solution was prepared by dissolving elemental Se powder in a 200 mM Na$_2$SO$_3$ solution. Secondly, 80 mM CdSO$_4$ and 120 mM of the trisodium salt of nitrilotriacetic acid (N(CH$_2$COONa)$_3$) were mixed with the 80 mM Na$_2$SeSO$_3$ solution in a volume ratio of 1:1:1. The TiO$_2$ films were placed in a glass container filled with the final solution at 10 °C in the dark for various times to promote CdSe adsorption.

2. Successive ionic layer adsorption-reaction (SILAR) method (Guijarro et al., 2010a)
   In situ growth of CdSe QDs using the SILAR method was carried out by successive immersion of TiO$_2$ electrodes in ionic precursor solutions of cadmium and selenium.

Aqueous solutions were employed in all cases. A 0.5 M $(CH_3COO)_2Cd$ (98.0%, Sigma-Aldrich) solution was used as the cadmium source, while a sodium selenosulfate (Na_2SeSO_3) solution was used as the selenium precursor. The sodium selenosulfate solution was prepared by heating under reflux for 1h a mixture of 1.6 g of Se powder, 40 mL of 1 M Na_2SO_3 (98.0%, Alfa Aesar) and 10 mL of 1 M NaOH. The resulting solution was filtered and mixed with 40 mL of 1M CH_3COONa (99.0+%, Fluka) solution, and finally was stored in the dark. The pH of the sodium selenosulfate solution was optimized to improve the QD deposition rate by using 0.25 M H_2SO_4 and/or 0.1 M NaOH stock solutions.

3. Direct adsorption (DA) of previously synthesized QDs (Guijarro et al., 2010b)
    Colloidal dispersions of CdSe QDs capped with trioctylphosphine (TOP) were prepared by a solvothermal route which permits size control. DA of CdSe QDs was achieved by immersion of TiO_2 electrodes in a CH_2Cl_2 (99.6%, Sigma Aldrich) CdSe QD dispersion, using soaking times ranging from 1 h to 1 week.

4. Linker assisted adsorption (LA) of previously synthesized QDs (Guijarro et al., 2010b)
    LA was performed employing p-mercaptobenzoic acid (MBA; 90%, Aldrich), cysteine (97%, Aldrich), and mercaptopropionic acid (MPA; 99%, Aldrich) as molecular wires. First, the linker was anchored to the TiO_2 surface by immersion in saturated toluene solutions of cysteine (5 mM) or MBA (10 mM) for 24 h. Secondly, these electrodes were washed with pure toluene for ½ h to remove the excess of the linker. Finally, the modified electrodes were transferred to a toluene CdSe QD dispersion for 3 days, to ensure QD saturation. The procedure for modification of TiO_2 with MPA has previously been reported (Guijarro et al., 2009).

After the CdSe QD adsorption, the samples were coated with ZnS by alternately dipping them three times in 0.1 M $Zn(CH_2COO)$ and 0.1 M Na_2S aqueous solutions for 1 minute for each dip (Yang et al., 2003; Diguna, et al., 2007a; Shen et al., 2008).

2.2 An improved transient grating (TG) technique

The transient grating (TG) method is a well-established laser spectroscopic technique of four wave mixing (Eichler et al., 1986; Harata et al., 1999). In the TG method, two time-coincident short laser pulses (pump beams) with the same wavelength and intensity intersect with an angle in a sample to generate an optical interference pattern at the intersection. Interaction between the light field and the material results in a spatially periodic modulation of the complex refractive index, which works like a transient diffraction grating for a third laser pulse (probe beam) incident to the photoexcited region. Then, by measuring the time dependence of the diffraction light of the probe beam, dynamics of the transient grating produced in the sample can be monitored. The TG technique is a powerful tool for detecting population dynamics (Rajesh et al., 2002), thermal diffusion (Glorieux et al., 2002), diffusion of photoexcited species (Terazima et al., 2000), energy transfer from photoexcited species to liquids (Miyata et al, 2002), structural or orientational relaxation (Glorieux et al., 2002), the sound velocity of liquids (Ohmor et al., 2001), and so on. Although this technique provides valuable information, it presents some technical difficulties for general researchers (Harata et al., 1999). First, the three beams must overlap on a small spot, typically within a spot diameter of less than 100 μm on a sample, and each beam must be temporally controlled. This is very difficult for pulsed laser beams, especially for those with a pulse width of ~100 femtoseconds. Secondly, since the diffraction of the probe beam, namely the signal, is quite weak, it is difficult to find the diffraction beam during the measurements. Thirdly, for a solid sample, the surface must be optically smooth. It is almost impossible to measure a sample with a rough surface by using the conventional TG technique.

In 2003, Katayama and co-workers (Katayama et al., 2003; Yamaguchi et al., 2003) proposed an improved TG technique (it was also called a lens-free heterodyne TG (LF-HD-TG) or a near field heterodyne TG (NF-HD-TG) technique in some papers), which overcomes the difficulties that exist in the conventional TG technique. The improved TG technique features (1) simple and compact optical equipment and easy optical alignment and (2) high stability of phase due to the short optical path length of the probe and reference beams. This method is thought to be versatile with applicability to many kinds of sample states, namely opaque solids, scattering solids with rough surfaces, transparent solids, and liquids, because it is applicable to transmission and reflection-type measurements. The principle of the improved TG technique has been explained in detail in the previous papers (Katayama et al., 2003; Yamaguchi et al., 2003) and is only described briefly here. Unlike the conventional TG technique, only one pump beam and one probe beam without focusing are needed in the improved TG technique. The pump beam is incident on the transmission grating. Then, the spatial intensity profile of the pump beam is known to have an interference pattern in the vicinity of the other side of the transmission grating, and the interference pattern has a grating spacing that is similar to that of the transmission grating. When a sample is brought near the transmission-grating surface, it can be excited by the optical interference pattern. The refractive index of the sample changes according to the intensity profile of the pump light and the induced refractive index profile functions as a different type of transiently generated grating. When the probe beam is incident in a manner similar to that of the pump beam, it is diffracted both by the transmission grating (called a reference light) and the transiently generated grating (called a signal light). In principle, the two diffractions progress along the same direction; therefore, these two diffractions interfere, which is detected by a detector positioned at a visible diffraction spot of the reference beam.

In the improved TG technique used for studying the ultrafast carrier dynaimcs of semiconductor QDs, the laser source was a titanium/sapphire laser (CPA-2010, Clark-MXR Inc.) with a wavelength of 775 nm, a repetition rate of 1 kHz, and a pulse width of 150 fs. The light was separated into two parts. One of them was used as a probe pulse. The other was used to pump an optical parametric amplifier (OPA) (a TOAPS from Quantronix) to generate light pulses with a wavelength tunable from 290 nm to 3 µm used as pump light in the TG measurement. The probe pulse wavelength was 775 nm.

2.3 Power dependence of carrier dynamics in CdSe QDs adsorbed onto TiO$_2$ electrodes

As described in depth in previous papers (Katayama et al., 2003; Shen et al., 2005, 2007), the TG signal is directly proportional to the change in the refractive index occurring in the sample ($\Delta n(t)$) upon photoexcitation. In the timescale of these experiments less than 1 ns, assuming Drude's model, the refractive index change, i.e., the TG signal intensity for CdSe QD adsorbed samples, will be a linear function of the concentration of photogenerated carriers (electrons and holes in CdSe QDs, i.e., $N_{e,CdSe}$ and $N_{h,CdSe}$, and injected electrons in TiO$_2$, i. e., $N_{e,TiO2}$) , as follows (Guijarro et al., 2010a, 2020b):

$$\Delta n(t) = \frac{1}{2n_{0,CdSe}} \left(\frac{-N_{e,CdSe}(t)e^2}{m_{e,CdSe}\omega_p^2\varepsilon_0} \right) + \frac{1}{2n_{0,CdSe}} \left(\frac{-N_{h,CdSe}(t)e^2}{m_{h,CdSe}\omega_p^2\varepsilon_0} \right) + \frac{1}{2n_{0,TiO_2}} \left(\frac{-N_{e,TiO_2}(t)e^2}{m_{e,TiO_2}\omega_p^2\varepsilon_0} \right) \quad (1)$$

where ω_p is the radial probe frequency, e is the elementary charge, ε_0 is free-space permittivity, and $n_{0,CdSe}$ (2.7) and $n_{0,TiO2}$ (2.5) are the refractive indices of CdSe and TiO$_2$, respectively. The important feature of the TG signal is that both the photoexcited electron and hole carrier densities contribute to the signal. In principle, the exact contribution on $\Delta n(t)$ by each carrier depends inversely on its carrier effective mass. According to the Drude theory (Kashiski et al., 1989; Kim et al., 2009), it can be considered that only free photoexcited electrons and holes are responsible for the population grating signals. For CdSe, the effective masses of electrons and holes are $0.13m_0$ and $0.44m_0$ (m_0 is the electron rest mass), respectively (Bawendi et al., 1989), so both the photoexcited electron and hole carrier densities in the CdSe QDs contribute to the signal. It is known that the effective mass of electrons for TiO$_2$ is about 30 m_0, which is about two orders larger than that for CdSe. Therefore, the TG signal due to the injected electrons in TiO$_2$ (no holes injected into TiO$_2$) can be ignored (Shen et al., 2005, 2006a, 2007, 2008a, 2010a).

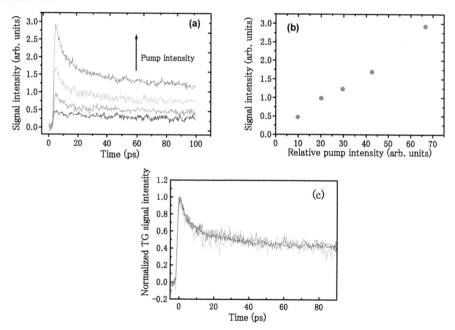

Fig. 1. Dependence of the TG kinetics of CdSe QDs adsorbed onto nanostructured TiO$_2$ films (CBD method and the CdSe adsorption time was 24 h) on the pump light intensity (a); Dependence of the TG peak intensity on the pump light intensity (b); Normalized TG kinetics measured with different pump light intensities (c) (Shen et al., 2010a).

For a semiconductor material, usually, there are three kinds of photoexcited carrier relaxation dynamics. The first one is a one-body recombination, which is trapping and/or transfer of photoexcited electrons and holes. In this case, the lifetimes of the photoexcited electrons and holes are independent of the pump light intensity. The second one is a two-body recombination, which is a radiative recombination via electron and hole pairs. The third one is a three-body recombination, which is an Auger recombination via two electrons and one hole, or via one electron and two holes. In the latter two cases, the lifetimes of photoexcited electrons and holes are dependent on the pump light intensity. In order to determine what kinds of photoexcited carrier dynamics are reflected in the TG kinetics, we first confirmed many-body recombination processes such as the Auger recombination process existed or not under our experimental conditions. For this purpose, we measured the dependence of the TG kinetics on the pump intensity (from 2.5 to 16.5 µJ/pulse) for CdSe QD adosrobed TiO_2 electrodes (Figure 1) (Shen et al., 2010a). As shown in Fig. 1, we found that the dependence of the maximum signal intensity on the pump intensity was linear, and that the waveforms of the responses overlapped each other very well when they were normalized at the peak intensity. These results mean that the decay processes measured in the TG kinetics were independent of the pump intensity under our experimental conditions, and many-body recombination processes could be neglected. Therefore, it is reasonable to assume that the decay processes of photoexcited electrons and holes in the CdSe QDs are due to one-body recombination processes such as trapping and transfer under our experimental conditions.

2.4 Separation of the ultrafast electron and hole dynamics in the CdSe QDs adsorbed onto TiO₂ nanostructured electrodes

Figure 2 shows a typical kinetic trace of the TG signal of CdSe QDs adsorbed onto a TiO_2 nanostructured film (prepared by CBD method for 24 h adsorption) measured in air. The vertical axis was plotted on a logarithmic scale. Three decay processes (indicated as A, B, and C in Fig. 2) can be clearly observed. We found that the TG kinetics shown in Fig. 2 could be fitted very well with a double exponential decay plus an offset, as shown in eq. (2):

$$y = A_1 e^{-t/\tau_1} + A_2 e^{-t/\tau_2} + y_0 \tag{2}$$

where A_1, A_2 and y_0 are constants, and τ_1 and τ_2 are the time constants of the two decay processes (A and B in Fig. 2). Here, the constant term y_0 corresponds to the slowest decay process (C in Fig. 2), in which the decay time (in the order of ns) is much larger than the time scale of 100 ps measured in this study. The time constants of the fast (τ_1) and slow (τ_2) decay processes of photoexcited carriers in air are 6.3 ps and 82 ps, respectively (Table 1). As mentioned above, τ_1 and τ_2 are independent of the pump intensity under our experimental conditions, so the three decay processes are mostly due to one-body recombination such as carrier trapping and carrier transfer.

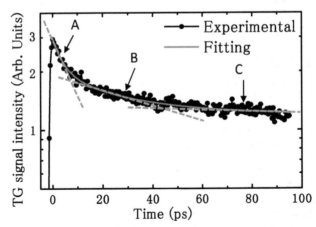

Fig. 2. TG kinetics of CdSe QDs adsorbed onto a nanostructured TiO_2 film measured in air. The vertical axis is plotted on a logarithmic scale. Three decay processes A, B and C can be clearly observed (Shen et al., 2010a).

In order to separate the photoexcited electron dynamics and hole dynamics that make up the TG kinetics, the TG kinetics of the same sample was measured both in air and in a Na_2S aqueous solution (hole acceptor) (1 M) (Shen et al., 2010a). As shown in Fig. 3, a large difference can be clearly observed between the TG responses measured in air and in the Na_2S solution. By normalizing the two TG responses at the signal intensity of 90 ps, we found that they overlapped with each other very well for time periods of longer than 15 ps, but the fast decay process apparently disappeared when the time period was less than 10 ps in the TG kinetics measured in the Na_2S solution (hole acceptor). This great difference can be explained as follows. In air, both hole and electron dynamics in the CdSe QDs could be measured in the TG kinetics. In the Na_2S solution, however, photoexcited holes in the CdSe QDs will transfer quickly to the electrolyte and only electron dynamics should be measured in the TG kinetics. Therefore, the "apparent disappearance" of the fast decay process in the Na_2S solution implies that the hole transfer to S^{2-} ions, which are supposed to be strongly adsorbed onto the CdSe QD surface, can be too fast in these circumstances as indicated by Hodes (Hodes, 2008) and therefore could not be observed under the temporal resolution (about 300 fs) of our TG technique. This observation is particularly important, because the result directly demonstrated that the transfer of holes to sulfur hole acceptors that are strongly adsorbed on the QD surface could approach a few hundreds of fs. An earlier study on the dynamics of photogenerated electron-hole pair separation in surface-space-charge fields at GaAs(100) crystal/oxide interfaces using a reflective electro-optic sampling method

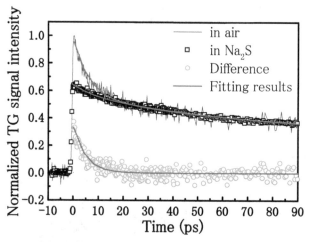

Fig. 3. TG responses of the CdSe QDs adsorbed onto nanostructured TiO_2 films measured in air (-) and in Na_2S solution (□) as well as their "difference response" (○) (Shen et al., 2010a).

showed that the hole carrier transit time was faster than 500 fs (Min et al., 1990). We believe an ultrafast hole transfer time from the QDs to hole acceptors that are strongly adsorbed on the QD surface is a more feasible and reasonable explanation, since photoexcited holes can more easily reach the surface of QDs with diameters of a few nm. The TG response measured in the Na_2S solution, which is considered to only relate to electron dynamics as mentioned above, can be fitted well with eq. (2). As shown in Table 1, besides the slower decay process, a faster decay process with a decay time of 9 ps was also detected in the TG response measured in the Na_2S solution. Such a faster decay process with a characteristic time of a few picoseconds in the TG response measured in the Na_2S solution was considered to correspond to electron transfer from the QDs in direct contact with the TiO_2 (first layer of deposited QDs) (Guijarro et al., 2010a, 2010b). It is worth noting that the relative intensity A_1 (0.07) measured in the Na_2S solution is much smaller compared to the A_1 (0.39) measured in air and it could be ignored here. The slower relaxation process in the TG response was not influenced by the presence of the Na_2S solution, as shown in Fig. 3. The decay time τ_2 (85 ps) and the relative intensity A_2 (0.31) for the slower decay process in the TG response measured in the Na_2S solution are almost the same as those measured in air (Table 1). The slower electron relaxation mostly corresponds to electron transfer from the CdSe QDs to TiO_2 and trapping at the QD surface states, in which the decay time depends to a great extent on the size of the QDs and the adsorption method that is used (Guijarro et al., 2010a, 2010b; Shen et al., 2006, 2007; Diguna et al., 2007b). The slowest decay process (with a time scale of ns) may mostly result from the non-radiative recombination of photoexcited electrons with defects that exist at the CdSe QD surfaces and at the CdSe/CdSe interfaces.

The difference between the two TG responses measured in air and in Na_2S solution (normalized for the longer time), which was termed the "difference response", is believed to correspond to the photoexcited hole dynamics in the CdSe QDs measured in air. As shown in Fig. 3, the difference response decays very fast and disappears around 10 ps and can be fitted very well with a one-exponential decay function with a decay time of 5 ps (Table 1). Thus, we did well in separating the dynamics of photoexcited electrons and holes in the CdSe QDs and found that the hole dynamics were much faster than those of electrons. Some papers have also reported that the hole relaxation time is much faster than the electron relaxation time in CdS and CdSe QDs (Underwood et al., 2001; Braun et al., 2002). In air, the fast hole decay process with a time scale of about 5 ps can be considered as the trapping of holes by the CdSe QD surface states. This result is in good agreement with the experimental results obtained by a femtosecond fluorescence "up-conversion" technique (Underwood et al., 2001).

Thus, by comparing the TG responses measured in air and in a Na_2S solution (hole acceptor), we succeeded in separating the dynamic characteristics of photoexcited electrons and holes in the CdSe QDs. We found that charge separation in the CdSe QDs occurred over a very fast time scale from a few hundreds of fs in the Na_2S solution via hole transfer to S^{2-} ions to a few ps in air via hole trapping.

TG kinetics	A_1	τ_1 (ps)	A_2	τ_2 (ps)	y_0
In air	0.39± 0.01	6.3±0.4	0.29 ±0.01	82 ±7	0.27± 0.01
In Na_2S	0.07± 0.01	9 ±1	0.31± 0.01	85 ±1	0.25± 0.01
Difference	0.33± 0.01	5.0± 0.3	-	-	-

Table 1. Fitting results of the TG responses of CdSe QDs adsorbed onto nanostructured TiO_2 films measured in air and in Na_2S solution (hole acceptor) as well as their "difference response" as shown in Fig. 3 with a double exponential decay equation (eq. (2)). τ_1 and τ_2 are time constants; A_1, A_2 and y_0 are constants (Shen et al., 2010a).

2.5 Electron injection from CdSe QDs to TiO_2 nanostructured electrode

To investigate the electron transfer rate from CdSe QDs to TiO_2 electrodes, Shen and co-workers measured the TG kinetics of CdSe QDs adsorbed on both nanostructured TiO_2 electrodes and glass substrates under the same deposition conditions (Shen et al., 2008).

Figure 4 shows the TG kintics (pump beam wavelength: 388 nm) of CdSe QDs adsorbed on a TiO$_2$ nanostructured electrode and on a glass substrate (CBD method). The average diameter of the CdSe QDs was about 5.5 nm for both kinds of substrates. The TG kinetics were fitted to a double exponential decay function (Eq. (3)) using a least-squares fitting method, convoluting with a 1ps Gaussian representing the laser pulse.

$$y = A_1 e^{-t/\tau_1} + A_2 e^{-t/\tau_2} \tag{3}$$

where A_1 and A_2 are constants, and τ_1 and τ_2 are the time constants of the two decay processes, respectively.

The fast decay time constants τ_1 were obtained to be 2.3 ps for both the two kinds of substrates. The slow decay time constants τ_2 were obtained to be approximately 140 ps and 570 ps for the CdSe QDs adsorbed on the TiO$_2$ electrode and on the glass substrate, respectively. It was found that τ_1 was the same for both kinds of substrates. However, the value of τ_2 for the CdSe QDs adsorbed on the nanostructured TiO$_2$ electrode was much smaller than that on the glass substrate. Under the experimental conditions (pump intensity < 2 μJ/pulse in this case), the waveforms of the responses overlapped each other very well when they were normalized at the peak intensity, as mentioned above. Therefore, it is reasonable to assume that the two decay processes in the CdSe QDs adsorbed on these two kinds of electrodes were due to one-body recombination, such as trapping, decay into intrinsic states and/or a transfer process. Because the fast decay time constant is independent of the substrates and it is known that hole trapping time is much faster than that of electron relaxation (Underwood et al., 2001; Braun et al., 2002), we believe that the fast decay process with a time scale of 2-3 ps mainly corresponded to a decrease in the photoexcited hole carrier density due to trapping at the CdSe QD surface/interface states or relaxation into the intrinsic QD states, as mentioned earlier.

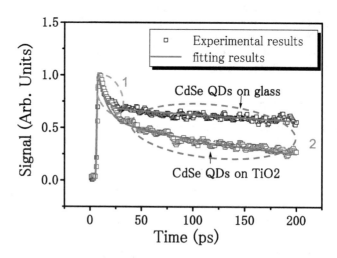

Fig. 4. TG kinetics of the CdSe QDs (average diameter: 5.5 nm) adsorbed on a nanostructured TiO$_2$ electrode and on a glass substrate under same adsorption conditions (Shen et al., 2008).

This result is in good agreement with the experimental results obtained with a femtosecond fluorescence up-conversion technique and transient absorption measurements (Underwood et al., 2001). The slow decay process was considered to reflect the photoexcited electron relaxation processes. For the glass substrate, no electron transfer from the QDs to the glass substrate could occur. Thus, the slow decay process only corresponded to the relaxation of photoexcited electrons in the CdSe QDs, which was mostly due to trapping at the CdSe QD surfaces. For the nanostructured TiO_2 substrate, the CdSe QDs were adsorbed onto the surfaces of the TiO_2 nanoparticles and electron transfer from the CdSe QDs to the TiO_2 nanoparticles could occur, which had been confirmed by the photocurrent measurements (Shen et al., 2004a, 2004b). As mentioned above, the TG signal due to the change in the refractive index $\Delta n(t)$ of the TiO_2 nanoparticles resulted from the injected free electrons from the CdSe QDs would be very small and can be ignored. Therefore, the slow decay process of the CdSe QDs adsorbed on the nanostructrued TiO_2 film mostly reflected the decrease of the photoexcited electron densities in the CdSe QDs due to both electron trapping at the CdSe QD surfaces/interfaces and electron transfer from the CdSe QDs to the TiO_2 nanoparticles. This is the reason why the τ_2 obtained for the CdSe QDs adsorbed on the nanostructured TiO_2 electrode was much smaller than that obtained for the CdSe QDs adsorbed on the glass substrate. For the CdSe QDs on the nanostructured TiO_2 electrode, the decay rate k_2 ($k_2 = 1/\tau_2$) can be expressed as

$$k_2 = k_r + k_{et} \tag{4}$$

where k_r is the intrinsic decay rate (mostly trapping) in CdSe QDs and k_{et} is the electron-transfer rate from CdSe QDs to TiO_2. The intrinsic decay time of electrons in the CdSe QDs adsorbed on the nanostructured TiO_2 electrode was assumed to be the same as that in the CdSe QDs adsorbed on the glass substrate. In this way, we can estimate the electron transfer rate from the CdSe QDs into the nanostructured TiO_2 electrode to be approximately 5.6×10^9 s^{-1} using the values of τ_2 obtained from the CdSe QDs adsorbed on these two kinds of substrates. This result is very close to the electron transfer rates from CdSe QDs to TiO_2 substrates measured recently using a femtosecond transient absorption (TA) technique by Kamat and co-workers (Robel et al., 2007).

2.6 Changes of carrier dynamics in CdSe QDs adsorbed onto TiO_2 nanostructured electrodes versus adsorption conditions

The photoexcited carrier dynamics of CdSe QDs, including the electron injection rates to the TiO_2 electrodes, depends greatly on the size of the CdSe QDs, the adsorption method (mode of attachment), the adsorption conditons such as adsorption time, and the properties of the TiO_2 electrodes such as crystal structure. In the following, two examples will be introduced.

The first example is the carrier dynamics of CdSe QDs adsorbed onto TiO_2 electrodes with SILAR method (Guijarro et al., 2010a). Figure 5 shows the normalized TG kinetics of CdSe adsorbed TiO_2 electrodes with different SILAR cycles and an example of the fitting result of the TG kinetics with eq. (3) to determine the parameters A_1, A_2, τ_1 and τ_2 shown in Fig. 5. Figure 6 shows the dependence of τ_1 and τ_2 on the number of SILAR cycles. As shown in Fig. 6, τ_1 increases from 2.8 to 6.3 ps and τ_2 increases from 83 to 320 ps upon increasing the number of SILAR cycles from 2 to 9-15. With the SILAR cycle increasing, both the average QD size and the number of QDs increase. Therefore, for a low number of cycles, all the CdSe QDs would be in direct contact with the TiO_2 electrode, favouring electron injection.

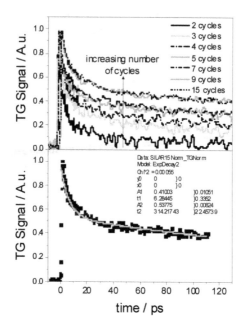

Fig. 5. Transient grating kinetics of CdSe QD adsorbed TiO₂ electrodes with different SILAR cycles (a) and an example of the fitting of the TG kinetics (15 cycles) to a double exponential decay (eq. (3)) (b) (Guijarro et al., 2010a).

However, for a large number of deposition cycles, a large fraction of CdSe QDs would not be in direct contact with the TiO₂ electrode, resulting in an increase in electron injection time. As discussed in depth by Guijarro and co-workers (Guijarro et al., 2010a), in this case, the fast decay component was attributed to direct electron transfer from the first layer of deposited CdSe QDs besides the hole trapping, and the slow one to electron injection into the TiO₂ from CdSe QDs in the outer layers (and trapping at surface and interfacial QD/QD states). For the fast decay component, the increase in QD average sizes resulted from the increase of SILAR cycles likely plays a key role in determining the values of τ_1. The results were in qualitative agreement with those reported by Kamat and co-workers (Robel et al., 2007) showing that electron injection into TiO₂ from CdSe QDs becomes faster as the QD size gets smaller. It was considered to be due to an increase in the driving force as the CdSe conduction band shifts toward higher energies. A plot of τ_1 versus the number of SILAR cycle numbers (Fig. 6(a)) shows approximately a linear dependence. For the slow decay component, the electron relaxation time τ_2 also increases upon increasing the number of SILAR cycles (Fig. 6(b)), which can be explained as follows. The increase in the number of SILAR cycles results in not only an average increase of the CdSe quantum dot size but also an increase in the average distance of the CdSe QDs to the oxide particle. Correspondingly, the concentration of QDs not in direct contact with the TiO₂ would also grow. This leads to an increase in the average time needed for transferring the photoexcited electrons from the QDs to the oxide. Therefore, the values of A_1 and A_2 mainly reflected the relative contribution of the fast electron transfer (from the first QD layer in direct contact with the TiO₂) and the slow one (from the QDs in the outer layers), respectively. A_1 decreases from

0.84 to 0.41 while A_2 increases from 0.22 to 0.54 as the SILAR cycle increases from 2 to 9-15. For a low number of SILAR cycles, A_1 is large, indicating the predominance of electron injection from QDs in direct contact with the TiO_2. As the SILAR cycle increases, A_2 increases because the contribution of the slower electron transfer through CdSe/CdSe interfaces becomes more and more important.

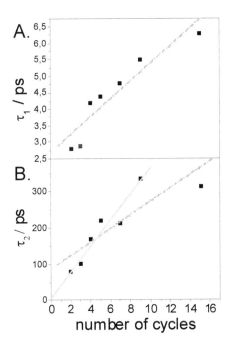

Fig. 6. Dependence of τ_1 (A) and τ_2 (B) on the number of SILAR cycles. Dashed lines correspond to linear fittings including all data, whereas the solid line shows a linear fitting forced to pass through the origin without considering the point corresponding to 15 SILAR cycles (Guijarro et al., 2010a).

The second example is the carrier dynamics of CdSe QDs adsorbed onto TiO_2 electrodes with linker-assisted (LA) method (Guijarro et al., 2010b). Figure 7 shows the TG kinetics for TiO_2 nanostrucuted electrodes modified by linker-adsorbed (using cysteine, p-mercaptobenzoic acid (MBA) and mercaptopropionic acid (MPA)) CdSe QDs (average size: 3.5 nm), together with the TG kinetics of the CdSe QD solution used for the adsorption. The TG kinetics of the LA adsorption samples were not sensitive to adsorption time. It indicates that the specific interaction between the thiol group (-SH) of the three kinds of linkers and the QD leads to homogeneous adsorption, in which all the QDs are adsorbed directly on the TiO_2 surface with no aggregation. As shown in Fig. 7, strong dependence of the decay of the TG kinetics on the linker nature was observed. This result suggests that, not only the length, but also other factors such as the dipole moment, the redox properties or the electronic structure of the linkers may play a role in the carrier dynamics. The TG kinetics of the LA adosprtion samples were fitted very well to eq. (3), and electron transfer rate constants were calculated according to eq. (4) by assuming that the intrinsic decay rate constant in adsorbed TiO_2 electrodes is the same as that in the CdSe colloidal dispersion (Table 2). The values of the electron transfer rate constant from CdSe to TiO_2 via MPA are in good agreement with the results obtained by Kamat's group (Robel et al., 2007), ranging from 1.0×10^9 to 2×10^{11} s^{-1}. With respect to the effect of QD size, it was found that the smaller the QD, the faster the electron injection. Very interestingly, a direct correalation between the ultrafast carrier dynamics and the incident photon current conversion efficiency (IPCE) values measured in the absence of electron acceptors in solution for CdSe QDSCs was found. (Guijarro et al., 2010b).

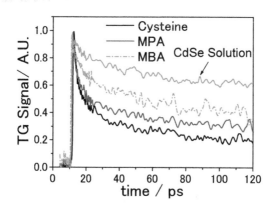

Fig. 7. TG kinetics for CdSe-sensitized TiO_2 electrodes using cysteine, MPA and MBA as linkers. All the curves are normalized to the maximum value. The TG kinetics corresponding to the CdSe solution used for LA adsorption is also given for comparison (Guijarro et al., 2010b).

Sample	A_1	τ_1 (ps)	A_2	τ_2 (ps)	$k_{1,et} \times 10^{-11}$ (s^{-1})	$k_{2,et} \times 10^{-9}$ (s^{-1})
CdSe solution	0.18 ± 0.01	8.8 ± 0.8	0.77 ± 0.01	506 ± 45		
LA (cysteine)	0.46 ± 0.01	3.9 ± 0.3	0.41 ± 0.01	127 ± 7	1.4 ± 0.3	5.9 ± 0.6
LA (MPA)	0.50 ± 0.01	4.2 ± 0.2	0.46 ± 0.01	204 ± 13	1.2 ± 0.2	2.9 ± 0.5
LA (MBA)	0.38 ± 0.02	7.7 ± 0.3	0.77 ± 0.01	245 ± 24	0.2 ± 0.1	2.1 ± 0.6

Table 2. Fitting parameters of TG kinetics (Fig. 7) of CdSe solution and LA adsorption samples (eq. (3)) and calculated electron transfer rate constants (eq. (4)) (Guijarro et al., 2010b).

2.7 Effect of surface modification on the ultrafast carrier dynamics and photovoltaic properties of CdSe QD sensitized TiO₂ solar cells

As mentioned above, the energy conversion efficiency of QDSCs is still less than 5% at present (Mora-Sero', et al., 2010; Zhang et al., 2011) and more studies are necessary for improving the photovoltaic properties of QDSCs. The poor photovoltaic performance of QDSCs usually results from three kinds of recombination. The first one is the recombination of the injected electrons in TiO_2 electrodes with electrolyte redox species. The second one is the recombination of photoexcited electrons in the QDs with electrolyte redox species. The third one is the recombination of photoexcited electrons and holes through surface and/or interface defects. To improve the photovoltaic performance, surface passivation should be carried out. In the previous work (Shen et al., 2008; Diguna et al., 2007a), it was found that ZnS surface coating on the CdSe QDs could imporve the photovoltaic properties of CdSe QDSCs significantly. In addition to ZnS, the authors also tried surface mosifications with Zn^{2+} and S^{2-} surface adsorption. The effects of the three kinds of surface modification (ZnS, Zn^{2+}, and S^{2-}) on the photovoltaic performances and ultrafast carrier dynamics of CdSe QDSCs have been investigated.

The three kinds of surface modifications for CdSe QD adsorbed TiO$_2$ nanostructured electrodes (24 h adsorption at 10 °C using CBD method) were carried out as follows: (1) the samples were coated with ZnS by alternately dipping them two times in 0.1 M Zn(CH$_2$COO) and 0.1 M Na$_2$S aqueous solutions for 1 minute for each dip; (2) the samples were adsorbed with Zn^{2+} by dipping them in 0.1 M Zn(CH$_2$COO) for 6 min; (3) the samples were adsorbed with S^{2-} by dipping them in 0.1 M Na$_2$S for 6 min. For characterization of the photovoltaic performances, solar cells were assembled using a CdSe QD sensitized TiO$_2$ electrode as the working electrode and a Cu$_2$S film as the counter electrode (Hodes et al., 1980). Polysulfide electrolyte (1 M S and 1 M Na$_2$S solution) was used as the regenerative redox couple (Shen et al., 2008b). The active area of the cells was 0.25 cm^2. The photovoltaic characteristics of the solar cells were measured using a solar simulator (Peccell Technologies, Inc.) at one sun (AM1.5, 100 mW/cm^2). For the TG measurements, the pump and the probe wavelengths are 530 nm and 775 nm, respectively.

Figure 8 shows the photocurrent density-photovoltage curves of CdSe QDSCs, in which the QD surfaces were not modified and modified with Zn^{2+}, S^{2-}, and ZnS, respectively. The photovoltaic properties of short circuit current density, open circuit voltage and fill factor (FF) were improved after modifying the surface with ZnS or Zn^{2+}, especially the short circuit current density was improved significantly (Table 3). As a result, the energy conversion efficiencies of the samples with ZnS or Zn^{2+} surface modification increased as many as 3 times compared to the sample without surface modification. The improvement of the photovoltaic performances by means of the ZnS or Zn^{2+} surface modification may be due to (1) the decrease of surface states of TiO$_2$ and CdSe QDs and (2) the formation of a potential barrier at the CdSe QD/electrolyte and TiO$_2$/electrolyte interfaces. However, there is little effect of the modification with S^{2-} on the photovoltaic properties. So, maybe Se^{2-} is rich on the CdSe QD surfaces.

Figure 9 shows the TG kinetics of the CdSe QD adsorbed TiO$_2$ electrodes before and after Zn^{2+} and ZnS surface modification. From the TG kinetics, two decay processes were observed. The fast and slow decay processes were mostly attributed to photoexcited hole trapping and electron transfer dynamics, as discussed earlier. For the sample before and after Zn^{2+} surface modification, the fast decay times were 4.5 and 2 ps, and the slow decay times were 53 ps and 25 ps, respectively. It indicates that the hole and electron relaxation became faster after the surface modification. Similar results were also obtained after the ZnS and S^{2-} surface modifications as shown in Fig. 10. The faster electron transfer due to the surface modification may result from the decrease of the surface trapping states. Further studies on the mechanism of the effects of the surface modifications on the photovoltaic performances and the carrier dynamics are in progress now.

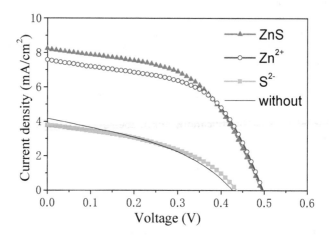

Fig. 8. Photocurrent density-photovoltage characteristics of CdSe QD-sensitized solar cells without and with surface modifications.

Sample	J_{sc} (mA/cm²)	V_{oc} (V)	FF	η (%)
TiO₂/CdSe	4.2	0.43	0.39	0.69
TiO₂/CdSe/ZnS	8.2	0.49	0.53	2.1
TiO₂/CdSe/Zn²⁺	7.6	0.50	0.53	2.1
TiO₂/CdSe/S²⁻	3.8	0.44	0.43	0.82

Table 3. Photovoltaic properties of CdSe QDSCs before and after surface modifications with Zn^{2+}, S^{2-}, and ZnS, respectively. J_{sc} is the short circuit current density, V_{oc} is the open circuit voltage, FF is the fill factor, and η is the energy conversion efficiency.

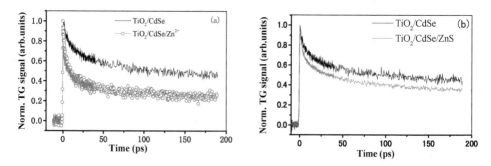

Fig. 9. Transient grating kinetics of CdSe QD adsorbed TiO₂ electrode before and after the surface modification with Zn^{2+} (a) and ZnS (b).

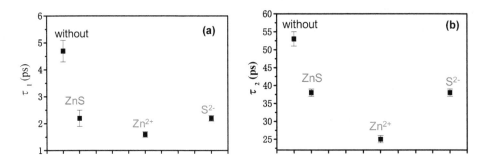

Fig. 10. Fast (τ_1) (a) and slow (τ_2) (b) decay times of the transient grating kinetics for CdSe QD adsorbed TiO$_2$ electrodes before and after the surface modifications with ZnS, Zn^{2+} and S^{2-}.

3. Conclusion

In summary, the photoexcited carrier dynamics in CdSe QDs adsorbed onto TiO$_2$ nanostructured electrodes have been characterized by using the improved TG technique. It has been demonstrated that both photoexcited electron and hole dynamics can be detected by using this improved TG technique. By comparing the TG responses measured in air and in a Na$_2$S solution (hole acceptor), the dynamics of photoexcited electrons and holes in the CdSe QDs has been sucssesfully separated from each other. It was found that charge separation in the CdSe QDs occurred over a very fast time scale from a few hundreds of fs in the Na$_2$S solution via hole transfer to S^{2-} ions to a few ps in air via hole trapping. On the other hand, the electron dynamics in the CdSe QDs, including trapping and injection to the metal oxide electrodes, depends greatly on the QD size, and adsorption methods (such as CBD, SILAR, DA and LA adsorption methods) and conditons (such as adsorption time and SIALR cycle number). In addition, surface modifications such as ZnS coating and adsorption with Zn^{2+} have been demonstrated to improve the photovoltaic properties and have a great influence on the ultrafast carrier dynamics of CdSe QDSCs. Detailed studies on the correlations between the carrier dynamics and the photovoltaic properties in QDSCs are in progress now.

4. Acknowledgment

Part of this research was supported by JST PRESTO program, Grant in Aid for Scientific Research (No. 21310073) from the Ministry of Education, Sports, Science and Technology of the Japanese Government. The authors would like to thank Dr. Kenji Katayama and Dr. Tsuguo Sawada for their help in the setup of TG equipment and also thank Mr. Yasumasa Ayuzawa for his help in some experiments. The authors are grateful to Prof. Juan Bisquert, Prof. Roberto Go´mez, Mr. Ne´stor Guijarro, Dr. Sixto Gime´nez, and Dr. Iva´n Mora-Sero´ for their collaboration and help in the research.

5. References

Adachi, M.; Murata, Y.; Okada, I.; &Yoshikawa, S. (2003), *J. Electrochem. Soc.*, Vol. 150, G488.
Bawendi, M. G.; Kortan, A. R.; Steigerwals, M.; & Brus, L. E. (1989), *J. Chem. Phys.*, Vol. 91, p. 7282.

Blackburn, J. L.; Selmarten, D. C. & Nozik, A. J. (2003); *J. Phys. Chem. B*, Vol. 107, p. 14154.
Blackburn, J. L.; Selmarten, D. C.; Ellingson, R. J.; Jones, M.; Micic, O. & Nozik, A. J. (2005); *J. Phys. Chem. B*, Vol. 109, p. 2625.
Braun, M.; Link, S.; Burda C.; & El-Sayed, M. A. (2002), *Chem. Phys. Lett.*, Vol. 361, p. 446.
Chiba, Y.; Islam, A.; Watanabe, Y.; Koyama, R.; Koide, N.; & Han, L. (2006), *Jpn. J. Appl. Phys.* Vol. 43, p. L638.
Diguna, L. J.; Shen, Q.; Kobayashi, J.; & Toyoda, T (2007a), *Appl. Phys. Lett.*, Vol. 91, p. 023116.
Diguna, L.J., Shen, Q., Sato, A., Katayama, K., Sawada, T., Toyoda, T. (2007b), *Mater. Sci. Eng. C*, Vol. 27, p. 1514.
Eichler, H.J.; Gunter P. & Pohl, D. W. (1986). *Laser-induced Dynamic Gratings.*, Springer, Berlin.
Giménez, S; Mora-Seró, I; Macor, L.; Guijarro, N.; Lala-Villarreal, T.; Gómez, R.; Diguna, L.;Shen, Q.; Toyoda, T; & Bisquert, J (2009), *Nanotechnology*, Vol. 20, p. 295204.
Glorieux, C.; Nelson, K. A.; Hinze, G. & Fayer, M. D. (2002), *J. Chem. Phys.*, Vol. 116, p. 3384.
Gorer, S.; & Hodes, G. (1994), *J. Phys. Chem.*, Vol. 98, p. 5338.
Grätzel, M. (2003), *J. Photochem. Photobiol. C: Photochem. Rev.*, Vol. 4, p. 145.
Guijarro, N.; Lana-Villarreal, T.; Mora-Sero´, I.; Bisquert, J.; Go´mez, R. (2009), *J. Phys. Chem. C*, Vol. 113, p. 4208.
Guijarro, N.; Lana-Villarreal, T.; Shen, Q.; Toyoda, T.& Gómez, R. (2010a), *J. Phys. Chem. C*, Vol. 114, p. 21928.
Guijarro, N.; Shen, Q.; Giménez, S.; Mora-Seró, I.; Bisquert, J.; Lana-Villarreal, T.; Toyoda, T. & Gómez, R. (2010b), *J. Phys. Chem. C*, Vol. 114, p. 22352.
Hanna, M. C. & Nozik, A. J. (2006), *J. Appl. Phys.*, Vol. 100, p. 074510.
Hodes, G; Manassen, J; & Cahen, D. (1980), *J. Electrochem. Soc.*, Vol. 127, p. 544.
Hodes, G. (2008), *J. Phys. Chem. C*, Vol. 112, p. 17778.
Kashiski, J. J.; Gomez-Jahn, L. A.; Faran, K. Y.; Gracewski, S. M. & Miller, R. J. D. (1989), *J. Chem. Phys.*, Vol. 90, p. 1253.
Kamat , P. V. (2008), *J Phys Chem C*, Vol. 112, p. 18737.
Katayama, K.; Yamaguchi, M.; & Sawada, T. (2003), *Appl. Phys. Lett.*, Vol. 82, p. 2775.
Kim, J.; Meuer, C.; Bimberg, D. & Eisenstein, G. (2009), *Appl. Phys. Lett.*, Vol. 94, p. 041112.
Klimov, V. I. (2006), *J. Phys. Chem. B*, Vol. 110, p. 16827.
Law, M.; Greene, L. E.; Johnson, J. C.; Saykally, R. & Yang, P. (2005), *Nat. Mater.*, Vol. 7, p. 1133.
Lee, Y-L.; & Lo, Y-S. (2009), *Adv. Func. Mater.*, Vol. 19, p. 604.
L. Min, L. & Miller, R. J. D. (1990), *Appl. Phys. Lett.*, Vol. 56, p. 524.
Miyata, R.; Kimura, Y. & Terazima, M. (2002), *Chem. Phys. Lett.*, Vol. 365, p. 406.
Mora-Seró, I; Giménez, S; Fabregat-Santiago, F.; Gómez, R.; Shen, Q.; Toyoda, T; & Bisquert, J. (2009), *Acc. Chem. Res.*, Vol. 42, 1848.
Mora-Seró, I & Bisquert, J. (2010), *J Phys Chem Lett*, Vol. 1, p. 3046.
Niitsoo, O.; Sarkar, S. K.; Pejoux, P.; Rühle, S.; Cahen, D.; & Hodes, G., *J. Photochem. Photobiol. A*, Vol. 182, 306.
Nishimura, S.; Abrams, N.; Lewis, B. A.; Halaoui, L. I.; Mallouk, T. E.; Benkstein, K. D.; Lagemaat, J. & Frank, A. K. (2003), *J. Am. Chem. Soc.*, Vol. 125, p. 6306.

Nozik, A. J. (2002), *Physica E*, Vol. 14, p. 16827.

Nozik, A. J. (2008), *Chem. Phys. Lett.* , Vol. 457, p. 3.

Nozik, A. J.; Beard, M. C.; Luther, J. M.; Law, M.; Ellingson, R. J. & Johnson, J. C. (2010), *Chem. Rev.*, Vol. 110, p. 6873.

Ohmor, T.; Kimura, Y.; Hirota, N. & Terazima, M. (2001), *Phys. Chem. Chem. Phys.*, Vol. 3, p. 3994.

O'Regan, B.; & Grätzel (1991), *Nature*, Vol. 353, p. 737.

Park, B.-W.; Inoue, T.; Ogomi, Y.; Miyamoto, A; Fujita, S; Pandey, S. S.&Hayase, S. (2011) *Appl. Phys. Express*, Vol. 4, p. 012301.

Paulose, M.; Sharkar, K.; Yoriya, S.; Prakasam, H. E.; Varghese, O. K.; Mor, G. K.; Latempa, T. A.; Fitzgerald, A. & Grimes, C. A. (2006), *J. Phys Chem. B*, Vol. 110, p. 16179.

Peter, L. M.; Riley, D. J.; Tull, E. J. & Wijayanta, K. G. U. (2002), *Chem. Commun.*, Vol. 2002, p. 1030.

Plass, R.; Pelet, S.; Krueger, J.; Gratzel, M. & Bach, U. (2002), *J. Phys. Chem. B*, Vol. 106, p. 7578.

Polo, A. S.; Itokatu, M. K. & Iha, N. Y. M. (2004), *Coord. Chem. Rev.*, Vol. 248, p. 1343.

Rajesh, R. J. & Bisht, P. B. (2002), *Chem. Phys. Lett.*, Vol. 357, p. 420.

Robel, I.; Subramanian, V.; Kuno, M. & Kamat, P. V. (2006), *J. Am. Chem. Soc.*, Vol. 128, p. 2385.

Robel, I.; Kuno, M. & Kamat, P. V. (2007), *J. Am. Chem. Soc.*, Vol. 129, p. 4136.

Schaller, R. D. & Klimov, V. I. (2004), *Phys. Rev. Lett.*, Vol. 92., P. 186601.

Shen, Q.; & Toyoda, T (2003), *Thin Solid Films*, Vol. 438-439, p. 167.

Shen, Q.; & Toyoda, T (2004a), *Jpn. J. Appl. Phys.*, Vol. 43, p. 2946.

Shen, Q.; Arae, D.; & Toyoda, T. (2004b), *J. Photochem. Photobiol. A*, Vol. 164, p. 75

Shen, Q.; Katayama, K.; Yamaguchi, M.; Sawada, T.& Toyoda, T. (2005), *Thin Solid Films*, Vol. 486, p. 15.

Shen, Q.; Katayama, K.; Sawada, T.; Yamaguchi, M.& Toyoda, T. (2006a), *Jpn. J. Appl. Phys.*, Vol. 45, p. 5569.

Shen, Q.; Sato, T.; Hashimoto, M.; Chen, C. C. & Toyoda, T. (2006b), *Thin Solid Films*, Vol. 499, p. 299.

Shen, Q.; Yanai, M.; Katayama, K.; Sawada, T.& Toyoda, T.(2007), *Chem. Phys. Lett.*,Vol. 442, p. 89.

Shen, Q.; Katayama, K.; Sawada, T.& Toyoda, T. (2008a), *Thin Solid Films*, Vol. 516, p. 5927.

Shen, Q.; Kobayashi, J.; Doguna, L. J.; & Toyoda, T. (2008b), *J. Appl. Phys.*, Vol. 103, p. 084304.

Shen, Q.; Ayuzawa, Y.; Katayama, K.; Sawada, T.; & Toyoda, T. (2010a), *Appl. Phys. Lett.*, Vol. 97, p. 263113.

Shen, Q.; Yamada, A.; Tamura, S. & Toyoda, T. (2010b), *Appl. Phys. Lett.*, Vol. 97, p. 23107.

Song, M. Y.; Ahn, Y. R.; Jo, S. M.; Kim, D. Y. & Ahn, J. P. (2005), *Appl. Phys. Lett.*, Vol. 87, p. 113113.

Terazima, M.; Nogami, Y. & Tominaga, T. (2000), *Chem. Phys. Lett.*, Vol. 332, p. 503.

Toyoda, T.; Saikusa, K. & Shen, Q. (1999), *Jpn. J. Appl. Phys.*, Vol. 38, p. 3185.

Toyoda, T.; Sato, J. & Shen, Q. (2003), *Rev. Sci. Instrum.* , Vol. 74, p. 297.

Vogel, R.; Pohl, K. & Weller, H. (1990), *Chem. Phys. Lett.*, Vol. 174, p. 241.

Tvrdy, K.; Frantsuzov, P. A.; Kamat, P. V. (2011), *Proceedings of the National Academy of Sciences of the United States of America*, Vol. 108, p. 29.

Vogel, R.; Hoyer, P. & Weller, H. (1994), *J. Phys. Chem.*, Vol. 98, p. 3183.

Yamaguchi, M.; Katayama, K.; & Sawada, T. (2003), *Chem. Phys. Lett.*, Vol. 377, p. 589.

Underwood, D. F.; Kippeny, T.; & Rosenthal, S. J. (2001), *Eur. Phys, J. D*, Vol. 16, p. 241.

Yan, Y. I.; Li, Y.; Qian, X. F.; Zhu, Z. K. (2003), *Mater. Sci. Eng. B*, Vol. 103, p. 202.

Yang, S. M.; Huang, C. H.; Zhai, J.; Wang, Z. S.; & Liang, L. (2002), *J. Mater. Chem.*, Vol. 12, p. 1459.

Yu, P. R.; Zhu, K.; Norman, A. G.; Ferrere, S.; Frank, A. J. & Nozik, A. J. (2006), *J. Phys. Chem. B*, Vol. 110, p. 25451.

Zhang, Q. X.; Guo, X. Z.; Huang, X. M.; Huang, S. Q.; Li, D. M.; Luo, Y. H.; Shen, Q.; Toyoda, T. & Meng, Q. B. (2011), *Phys. Chem. Chem. Phys.*, Vol. 13, p. 4659.

3

Progress in Organic Photovoltaic Fibers Research

Ayse Bedeloglu
Dokuz Eylül University,
Turkey

1. Introduction

Energy management including production, distribution and usage of energy is an important issue, which determines internal and external policy and economical situation of countries. For generating electrical energy, use of traditional energy sources in particular fossil based fuels through long ages, caused environmental problems, in recent years. Renewable energy technologies using power of wind, sun, water, etc. can be remedies to hinder negative effects of pollution, emissions of carbon dioxide and irreversible climate change problem, which it caused. Photovoltaic technology, which converts photons of the sun into electrical energy by using semiconductors, is one of the most environmental friendly sources of renewable energy (Dennler et al., 2006a). Solar cells are used in many different fields such as in solar lambs and calculators, on roofs and windows of buildings, satellites and space craft, textile structures (fibers, fabrics and garments) and accessories (bags and suitcases).

In addition, there is an increasing interest in organic electronics from a wide range of science disciplines in which researchers search for novel, efficient and functional materials and structures. Organic materials based optoelectronic devices such as organic photovoltaics (organic solar cells), organic light emitting diodes and organic photo detectors (Curran et al., 2009) are desirable in many applications due to interesting features of organic materials such as cost advantage and flexibility. Production of electrical energy, which is necessary in both industrial and human daily life by converting sunlight using organic solar cells (organic photovoltaic technology) via easy and inexpensive techniques is also very interesting (Günes et al., 2007).

A photovoltaic textile, which is formed by combining a textile structure with a solar cell, and on which carries physical properties of textile and working principle of solar cell together, can generate electricity for powering different electrical devices. Photovoltaic fiber providing more compatibility to textiles in terms of flexibility and lightness owing to its thin and polymer-based structure may be used in a wide variety of applications such as tents, jackets, soldier uniforms and marine fabrics. This review is organized as follows: In the first section, an overview of photovoltaic technology, smart textiles and photovoltaic textiles will be presented. In the second section, a general introduction to organic solar cells and organic semi conductors, features, the working principle, manufacturing techniques, and characterization of organic solar cells as well as polymer based organic solar cells and studies about nanofibers and flexible solar cells will be given. In the third part, recent studies about photovoltaic fiber researches, production methods, and materials used and application areas will be recounted. Finally, suggestions on future studies and the conclusions will be given.

1.1 Photovoltaic technology

"Photovoltaic" is a marriage of two words: "photo", which means light, and "voltaic", which means electricity. Electrical energy produced by solar cells is one of the most promising sustainable alternative energy and can provide energy demand of the world, in the future (Green, 2005). Today, silicon based solar cells having the highest power conversion efficiency are dominated in the market; however they have still high production costs. Electricity generation by solar cells is still more expensive than that of fossil fuels due to materials and manufacturing processes used in solar cells and installation costs. However, photovoltaic technology, compared to traditional energy production technologies, have interesting features such as using endless and abundant source of sun's energy, direct, environmental friendly and noiseless energy generation without the need of additional generators, customization according to requirements, having low maintenance costs and portable modules producing power ranging between milliwatt to megawatts even in remote areas, which make it unique (Dennler et al., 2006a). A photovoltaic system can convert sun light into electricity on both sunny and cloudy days (European PhotoVoltaic Industry Association (EPIA), 2009). The worldwide cumulative photovoltaic power installed reached about 23GW, in the beginning of 2010 and produces about 25TWh of electricity on a yearly basis (European PhotoVoltaic Industry Association (EPIA), 2010).

The electricity produced by solar cells can be utilized in many applications such as cooling, heating, lighting, charging of batteries and providing power for different electrical devices (Curran et al., 2009). Solar cells using silicon wafers are classified as first generation technology having high areal production costs and moderate efficiency. Thin-film solar cells using Amorphous silicon (a-Si), Cadmium telluride (CdTe) and Copper indium gallium (di)selenide (CIGS) as second generation technology have advantages such as increased size of the unit of manufacturing and reduction in material costs. However, this technology has modest efficiency beside these advantages compared to first generation technology. Therefore, third generation technology concept has been developed to eliminate disadvantages of earlier photovoltaic technologies (Green, 2005). There are two approaches in third generation photovoltaic technology. The first one aims to achieve very high efficiencies and second one tries to achieve cost per watt balance via moderate efficiency at low cost. Therefore, this uses inexpensive semiconductor materials and solutions at low temperature manufacturing processes. The third generation photovoltaics use various technologies and grouped under organic solar cells (Dennler et al., 2006a).

1.2 Smart textiles

Humankind has always been inspired to mimic intelligence of nature to create novel materials and structures with fascinating functions. Over the last decades, in industrial and daily life, paralleling to growth in world population and advancements in science and technology, human requirements have changed and begun to diverge from each other. Therefore, different functional products have emerged according to expectations and requirements of human kind. One of these, intelligent materials, can coordinate their characteristic behavior according to changes of external or internal stimulus (chemical, mechanical, thermal, magnetic, electrical and so on) as in biological systems and have different functions owing to their unique molecular structure (Mattila, 2006; Tani et al.,

1998). Intelligent materials and structures can sense and react and more, adapt it and perform a function of changes (Takagi, 1990; Tao, 2001).Intelligent material systems consist of three parts: a sensor, a processor and an actuator. Intelligent materials can provide advancements in many fields of science for energy generation, medical treatments, and engineering applications and so on.

There are also many application areas for interactive textiles, which use intelligent materials such as shape memory alloys or polymers, phase change materials, conductive materials and etc. Intelligent textiles are defined as structures that are capable of sensing external and internal stimuli and respond or adapt to them in a pre-specified way. Knowledge from different scientific fields (biotechnology, microelectronics, nanotechnology and so on) is required for intelligent textile research (Mattila, 2006). Intelligent textiles find uses in many applications ranging from space to healthcare and entertainment.

Power supply by using discrete batteries is an important obstacle towards functionality of intelligent textiles. Besides, flexibility, comfort and durability are other parameters concerned to manufacture consistent products (Coyle & Diamond, 2010). Flexible solar cells (Schubert & Werner, 2006), micro fuel cells (Gunter et al., 2007; enfucell, 2011), power generation by body motion and body heat (Beeby, 2010; Starner, 1996) can be some alternatives to the traditional batteries. Photovoltaic fibers and textiles can overcome this power supply problem since they use the working principle of solar cells.

1.3 Photovoltaic textiles

Small electronic devices such as personal digital assistants, mobile phones, mp3 players, and notebook computers, usually use traditional batteries of which energy is used up in a short time. Integration of flexible solar cells into apparels and fabrics, which cells are positioned in/on the textile, can provide required electrical energy for these portable electronic devices (Schubert & Werner, 2006). Photovoltaic textiles can be formed by integrating solar cells into textile structure or making textile structure itself from photovoltaic materials. Photovoltaic textile research needs cooperation of different sciences consisting of textile, electronics, physics and chemistry. Incorporation of solar cells with fibers and textiles that are flexible can extend the applications of photovoltaics from military and space applications to lighting and providing power for consumer electronics of humankind in daily usage.

Textile based solar cells are also named as photovoltaic textiles, solar textiles, energy harvesting textiles, solar powered textiles in the literature. Photovoltaic textiles, which are high value added intelligent products, and, which have a large application area, can benefit textile industry by increasing its competitiveness with long term development.

Power conversion efficiency and price properties beside the flexibility, weight, comfort, durability and washability properties of the products are also important from a customer point of view. Position of the flexible solar cells on fabric is also important to take efficient irradiation from the light source. Places of needed wires, controllers and batteries, which have to be lightweight under the cloth, are needed to be concerned to develop viable photovoltaic textiles (Schubert & Werner, 2006).

In recent years, there has been an increase in studies about developing photovoltaic fibers which can take charge in different textile and clothing applications. An active photovoltaic fiber, which is produced by using advanced design and suitable materials, and, which consist of adequate smooth layers, efficiency and stability, is capable of forming a flexible fabric by suitable knitting or weaving techniques, or integrating as a yarn into a cloth to generate power for electronic devices by converting sunlight (DeCristofano, 2009)

Fiber based photovoltaics take the advantage of being flexible and lightweight. Integration of photovoltaic fibers into fabrics and clothes is easy to manufacture wearable technology products. Small surface of a fiber also provide large area photoactive surfaces in the case of fabric, so higher power conversion efficiency can be obtained.

Traditional solar cells using silicon based semiconductors are generally rigid and are not suitable to be used with textiles. The thin film solar cells based on inorganic semiconductors can be made flexible and however they are more suitable for patching onto fabrics (Schubert & Werner, 2006).

Inexpensive electricity production can be achieved, when both low-cost and high efficient manufacturing of photovoltaic cells are achieved. A potential alternative approach to conventional rigid solar cells is organic solar cells, which can be coated on both rigid and flexible substrates using easy processing techniques. In addition, the polymer based organic solar cells can be used to produce fully flexible photovoltaic textiles easily, in any scale, from fibers to fabrics and using low-cost methods.

2. Organic photovoltaic technology

2.1 Organic semiconductors

Organic semiconductors, which are generally considered as intrinsic wide band gap semiconductors (band gap>1.4 eV), have many advantages to be used in solar cells. For example, organic semiconductors of which electronic band gap can be engineered by chemical synthesis with low-cost (Günes et al., 2007) have generally high absorption coefficients.

Organic semiconductors consist of different chemical structures (Nunzi, 2002) including polymers, oligomers, dendrimers, dyes, pigments, liquid crystals (Yilmaz Canli et al., 2010) etc. In carbon-based semiconductors, conductivity is obtained by conjugation, which single and double bonds between the carbon atoms alternate (Pope & Swenberg, 1999).

Conjugated organics are challenging materials for solar cells owing to their semiconducting and light absorbing features. As a compound of organic solar cells, organic semiconductors can be processed by thermal evaporation techniques or solution based coating or printing techniques at low temperatures (Deibel & Dyakonov, 2010).

2.2 Organic solar cells

As a promising renewable energy source, organic photovoltaics have attracted attention during the last decades resulting in significant progress in cell efficiency exceeded 5% (AM1.5, 1000 W/m²) (Green et al., 2010) in the conventional bulk heterojunction solar cell architecture consisting of a polymer donor and fullerene acceptor blend. Organic solar cells achieving photovoltaic energy conversion by organic semiconductor or conductor are compatible with flexible substrates like textiles for use in novel application areas.

Photovoltaic effect, production of electricity by converting photons of the sunlight, occurs in an organic solar cell by the following steps (Nunzi, 2002): Absorption of photons of the light in the solar cell and exciton (electron-hole pair) creation; separation of charges and carriers generation from exciton dissociation; transport and then collection of charges by respective electrodes (Günes et al., 2007; Nunzi, 2002)

There are some approaches such as using conjugated polymers (Antonradis et al., 1994) and their blends (Granström et al., 1998; Halls et al., 1995; Yu & Heeger, 1995), small molecules (Tang, 1986; Wöhrle & Meissner, 1991) polymer-small molecule bilayers (Jenekhe & Yi, 2000;

Breeze et al., 2002) and their blends (Tang, 1986; Shaheen et al., 2001; Dittmer et al., 2000) or combinations of inorganic-organic materials (O`Reagan & Graetzel, 1991; Greenham et al., 1996; Günes et al., 2008; to develop organic solar cells (Güneş & Sariçiftçi, 2007). Mostly, two concepts are considered in organic solar cell researches: first one, (Krebs, 2009a) which is the most successful is using conjugated polymers (Fig. 1) with fullerene derivatives by solution based techniques and second one is cooperating small molecular materials (as donor and acceptor) by thermal evaporation techniques (Deibel & Dyakonov, 2010).

A conventional organic solar cell (Fig. 2) device is based on the following layer sequence: a semi-transparent conductive bottom electrode (indium tin oxide (ITO)) or a thin metal layer), a poly(3,4-ethylenedioxythiophene:poly(styrene sulfonic acid) (PEDOT:PSS) layer facilitating the hole injection and surface smoothness, an organic photoactive layer (most common poly(3- hexylthiophene):[6,6]-*phenyl-C61-butyric acid* methyl ester (P3HT:PCBM)) to absorb the light and a metal electrode (Aluminum, Al and Calcium, Ca) with a low work function to collect charges on the top of the device (Brabec et al., 2001a; Brabec et al., 2001b; Padinger et al., 2003). To form a good contact between the active layer and metal layer, an electron transporting layer (i.e. Lithium Fluoride, LiF) is also used (Brabec et al., 2002).

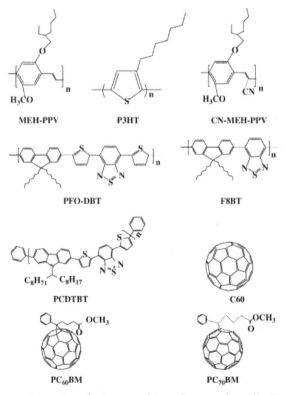

Fig. 1. Example of organic semiconductors used in polymer solar cells. Reprinted from Solar Energy Materials and Solar Cells, 94, Cai, W.; Gong, X. & Cao, Y. Polymer solar cells: Recent development and possible routes for improvement in the performance, 114–127, Copyright (2010), with permission from Elsevier.

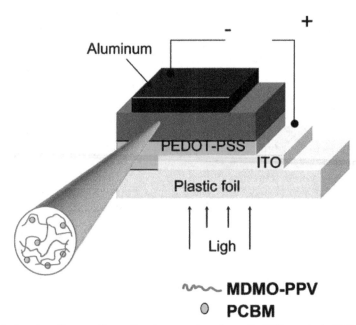

Fig. 2. Bulk heterojunction configuration in organic solar cells (Günes et al., 2007)

ITO is the most commonly used transparent electrode due to its good transparency in the visible range and good electrical conductivity (Zou et al., 2010). However, ITO, which exhibits poor mechanical properties on polymer based substrates, has limited conductivity for fabricating large area solar cells and needs complicated techniques, which tend to increase the cost of the solar cells (Zou et al., 2010). Indium availability is also limited. To alleviate limitations arised from ITO, alternative materials are needed to replace transparent conducting electrode. There are some approaches such as using carbon nanotubes (CNTs) (Rowell et al., 2006; Glatthaar et al., 2005; (Celik) Bedeloglu et al., 2011; Dresselhaus et al., 2001), graphenes (Eda et al., 2008), different conductive polymers (i.e. PEDOT:PSS and its mixtures (Ouyang et al., 2005; Kushto et al., 2005; Huang et al., 2006; Ahlswede et al., 2008; Zhou et al., 2008), metallic grids (Tvingstedt & Inganäs, 2007; Kang et al., 2008), nanowires (Lee et al., 2008) for potential candidates to substitute ITO layer and to perform as hole collecting electrode. In particular, CNTs have a wide variety of application area due to their unique features in terms of thermal, mechanical and electrical properties (Ajayan, 1999; Baughman et al., 2002). A nanotube has a diameter of a few nanometers and from a few nanometers to centimeters in length. Carbon nanotubes can be classified into two groups according to the number of combinations that form their walls: Single-walled nanotubes (SWNTs) and multi-walled nanotubes (MWNTs) (Wang et al., 2009) Recently, CNTs are used in solar cells and can substitute ITO as a transparent electrode in organic solar cells (Rowell et al., 2006; Glatthaar et al., 2005; (Celik) Bedeloglu et al., 2011; Dresselhaus et al., 2001).

In the organic solar cell, the photoactive layer, light absorbing layer, is formed by combination of electron donor (p) and an electron accepting (n) materials (Deibel & Dyakonov, 2010) C_{60}, its derivatives and Perylen pigments are mostly used as electron accepting materials. Also, phtalocyanines, porphyrins, poly(3- hexylthiophene) (P3HT) and poly[2-methoxy-5-(2'-ethyl-hexyloxy)-1,4-phenylene vinylene] (MEH-PPV) are good donors (Nunzi,2002). Most of the time, evaporation step is indispensable in the manufacturing of conventional organic photovoltaic devices, but this process tends to increase the cost of the cell. Besides, ITO-PEDOT:PSS interface (de Jong et al., 2000) and Al top electrode (Do et al., 1994) is known to be quite unstable, which limits the lifetime of organic solar cells. Organic solar cells have low charge transport and a mismatch between exciton diffusion length and organic layer thickness. While the efficient absorption of light is provided by organic film based on P3HT:PCBM having a thickness over 250 nm (Liu et al., 2007a), exciton diffusion length of which surpassing causes exciton recombination, is about 10-20 nm in polymer based and in organic semiconductors (Nunzi, 2002). Although, organic solar cells have lower power conversion efficiency (~5%) than inorganic traditional solar cells (for crystalline silicon based solar cells ~25% in laboratory conditions); their cost and processing parameters are favorable (Deibel & Dyakonov, 2010).

In order to avoid the limitations of organic semiconductor and to improve the power conversion efficiency of organic solar cells, several approaches such as optical concepts, different device configurations such as inverted layer architecture, multijunction solar cells, novel materials with lower band gap (Park et al., 2009; Chen et al., 2009; Huo et al., 2009; Coffin et al., 2009), wider absorption ranges, higher dielectric constants and higher charge carrier mobility are some approaches are studied in last few years (Deibel & Dyakonov, 2010). Reversing the nature of charge collection in organic solar cells using a less air sensitive high work function metal (Ag, Au) (de Jong et al., 2000; Do et al., 1994; Liu et al., 2007a; Park et al., 2009; Chen et al., 2009; Huo et al., 2009; Coffin et al., 2009; Wong et al., 2006) as hole collecting electrode at the back contact and a metal oxide (TiO$_x$, ZnO) as hole blocking barrier and electron selective contact at the ITO interface to block the oxidation (Hau et al., 2008) is a beneficial approach to avoid from low power conversion efficiency, which limited absorption in solar spectrum causes. In particular, non-clorinated solvents are more appropriate for high volume manufacturing with low cost. Besides, active layer can be protected by use of metal oxides (i.e. vanadium oxide and cesium carbonate), which are used as buffer layer of inverted polymer solar cells (Li et al., 1997). However, there is still a trade-off between stability and photovoltaic performance in inverted solar cells (Hsieh et al., 2010).

Organic solar cells have many advantages such as potential to be semi-transparent, manufacture on both large or small areas compatible with mass production and low-cost, production possibility with continuous coating and printing processes on lightweight and flexible substrates (for example textiles), ecological and low-temperature production possibilities (Dennler et al., 2006a; Dennler & Sariciftci, 2005).

Polymer heterojunction organic solar cells have attracted much attention because of their potential applications in large area, flexible and low-cost devices (Park et al., 2009; Yu et al., 1995; Chang et al., 2009; Dridi et al., 2008; Peet et al., 2009; Oey et al., 2006; Yu et al., 2008). Polymer based thin films on flexible and non-flexible substrates can be achieved by various printing techniques (see Fig. 3) (screen printing, inkjet printing, offset printing, flexo printing and so on), solution based coating techniques (dipping, spin coating, doctor blading, spray coating, slot-die coating and so on), and electrospinning (Krebs, 2009a).

Roll to roll, reel to reel, process is suitable for solar cells, which are on long flexible substrates (polymeric substrates and thin metal foils), and, which can be wound on a roll

(Krebs, 2009a). Various coating and printing techniques including knife-over-edge coating and slot die coating can be used for manufacturing flexible solar cells. The most appropriate processes for flexible photovoltaics should be free from Indium, toxic solvents and chemicals, and should have solution based manufacturing steps (coating and printing techniques), which results an environmentally recyclable product (Hoppe & Sariciftci, 2006). The studies about improving polymer solar cells (Günes et al., 2007; Hoppe & Sariciftci, 2006; Bundgaard & Krebs, 2007; Jørgensen et al., 2008; Thompson & Frechet, 2008; Tromholt et al., 2009) include developments in material properties and manufacturing techniques.

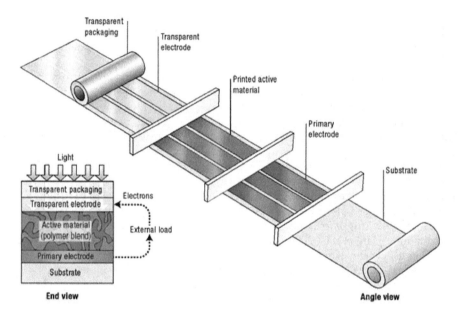

Fig. 3. Schematic description of printing process used for manufacturing of polymer-based photovoltaic cells. Reprinted by permission from Macmillan Publishers Ltd: Nature Photonics, *Gaudiana, R. & Brabec*, C. (2008). Organic materials: Fantastic plastic. Vol.2, pp.287- 289, copyright (2008).

2.3 Characterization of organic solar cells

Characterization of organic solar cells performed by measuring efficiencies in the dark and under an illumination intensity of 1000 W/m² (global AM1.5 spectrum) at 25°C (IEC 60904-3: 2008, ASTM G-173-03 global) (Green et al., 2010). Generally a solar simulator is used as illumination source for simulating AM1.5 conditions. Air Mass (AM) is a measure of how much atmosphere sunlight must travel through to reach the earth's surface. AM1.5 means that the sun is at an angle about 48° (Benanti & Venkataraman, 2006). A graph (Fig. 4) on which shows the current-voltage characteristics in the dark and under an illumination, gives significant information about photovoltaic performance and photoelectrical behavior of the cells (Nunzi, 2002).

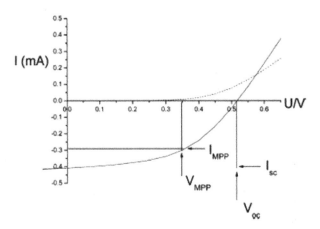

Fig. 4. Current-voltage (I-V) curves of an organic solar cell (dark, - - -; illuminated, -). The characteristic intersections with the abscissa and ordinate are the open circuit voltage (Voc) and the short circuit current (Isc), respectively. The largest power output (Pmax) is determined by the point where the product of voltage and current is maximized. Division of Pmax by the product of Isc and Voc yields the fill factor FF (Günes et al., 2007).

The overall efficiency of a solar cell can be expressed as follows :

$$\eta = \frac{P_{max}}{P_{in}} = \frac{I_{sc}\,V_{oc}\,FF}{P_{in}} \tag{1}$$

Here, maximum power point is the point on the I-V curve where maximum power (Pmax) is produced. The photovoltaic power conversion efficiency (η) of a solar cell is defined as the ratio of power output to power input. I_{sc} is the short-circuit current, which flows through the cell when applied voltage is zero, under illumination. Under an external load, current will always be less than max. current value. V_{oc}, the open-circuit voltage is the voltage when no current is flowing, under illumination. When current flows, voltage will be less than max. voltage value. FF, fill factor is the ratio of max. power output to the external the short-circuit current and open-circuit voltage values. FF is given by following formula:

$$FF = \frac{P_{max}}{I_{sc}\,V_{oc}} = \frac{I_{mpp}\,V_{mpp}}{I_{sc}\,V_{oc}} \tag{2}$$

where I_{mpp} and V_{mpp} represent the current and the voltage at the maximum power point (Pmax) in the four quadrant, respectively (Nunzi,2002; Benanti & Venkataraman, 2006).
The incident photon to collected electron (IPCE) or external quantum efficiency (EQE) under monochromatic lightning at a wavelength λ includes losses by reflection and transmission (Benanti & Venkataraman, 2006) and gives the ratio of collected charge carriers per incident photons (Dennler et al., 2006a):

$$IPCE = \frac{1240\,I_{sc}}{\lambda\,P_{in}} \tag{3}$$

2.4 Flexible organic solar cells

Solar cells generally developed on rigid substrates like glass and suffer from heavy, fragile and inflexible devices. However, stiff substrates limit usage, storage and transport of photovoltaic devices. Therefore, a big interest from both industrial and academic sides has been observed for research and development of flexible (foldable or rollable) solar cells, recently. Organic solar cells, easy scalable and suitable to roll-to-roll production with low-cost have potential to be used with flexible substrates such as textiles and fibers. Materials used in organic solar cells are also capable of producing lightweight photovoltaics. Polymer based substrates, which are used to replace rigid substrates and which have adequate flexibility are required to have mechanically and chemically stable, while organic solar cell manufacturing processes continue. Optimum substrate should have some features such as resistance to effects of chemical materials, water and air and also, mechanical robustness, low coefficients of thermal expansion, anti-permeability and smooth surface properties and so on.

Polyethylene terephtalate (PET), ITO coated PET, Poly(ethylene naphthalate) (PEN), Polyimide (PI), Kapton and Polyethersulphone (PES) are used as substrates to develop flexible solar cells.

Polyethylene terephtalate (PET) based fibers, which melt about 260°C show good stability to UV light and most of the chemicals and exhibit good mechanical properties including flexibility and comfort ability both in fiber and fabric form (Mather & Wilson, 2006). However, PET foils are often used as substrate of the flexible photovoltaics (Breeze, et al., 2002; Aernouts et al., 2004; Winther-Jensen & Krebs, 2006; Krebs, 2009b). Manufacturing of photovoltaics using PET substrates is suitable for reel to reel production, which reduces material and production costs. PET foils are preferable materials for solar cells due to their price, mechanical flexibility and easy availability comparing to other substrates. However, use of PET foils is limited because it melts about 140 °C (Zimmermann et al., 2007; Krebs, 2009a; Krebs, 2009b). Thermocleavable materials are used for the preparation of very stable solar cells (Liu et al., 2004; Krebs & Spanggaard 2005; Krebs & Norrman, 2007; Krebs, 2008; Bjerring et al., 2008; Krebs et al., 2008) but, these materials are required to heat to a temperature of around 200°C to achieve insolubility, which is a limitation to use of PET foils in conventional methods. However, PET substrates can be used with thermocleavable materials thanks to longer processing time (Krebs, 2009b).

Spin coating and screen printing techniques are used to coat highly conductive PEDOT:PSS dispersions onto flexible PET substrates as anode, which also improves an application of a metallic silver(Ag)-grid deposited between foil and electrode (Aernouts et al., 2004). A silk screen printing procedure (Winther-Jensen & Krebs, 2006) can be applied to develop PEDOT electrode with surface resistances down to 20 Ω^{-1} on flexible PET substrates. Researchers obtained 0.7 V open circuit voltage, 1 mA/cm^2 short circuit current and 0.2%, efficiency under simulated sun light (AM1.5 at 1000 W m^{-2}) with an active area of 4.2 cm^2 based on MEH-PPV:PCBM mixture and Al counter electrode.

In recent years, carbon nanotubes have found a wide variety of applications in photovoltaics. Films of SWNT networks can be printed on PET foils to get flexible transparent conducting electrodes. The well dispersed and stable solutions of SWNTs can be produced as electrode of flexible polymer based solar cells with various methods, which are inexpensive, scalable to large areas, and allows for the transfer of the film to a variety of surfaces. Such a flexible photovoltaic device configuration (Rowell et al., 2006) (see Fig. 5) (PET/SWNTs/PEDOT:PSS/P3HT:PCBM/Al) gave 2.5% efficiency of which efficiency is very close to conventional ITO coated glass based rigid solar cells.

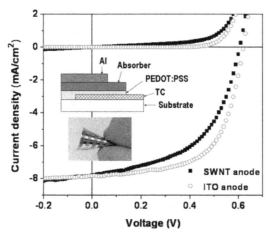

Fig. 5. Current density-voltage characteristics of P3HT:PCBM devices under AM1.5G conditions using ITO on glass (open circles) and flexible SWNTs on PET (solid squares) as the anodes, respectively. Inset: Schematic of device and photograph of the highly flexible cell using SWNTs on PET. Reprinted with permission from Rowell, M.W.; Topinka, M.A.; McGehee, M.D.; Prall, H.J.; Dennler, G.; Sariciftci, N.S.; Hu, L. & Gruner, G. (2006). Organic Solar Cells with Carbon Nanotube Network Electrodes. *Applied Physics Letters*, Vol.88, pp. 233506. Copyright 2006, American Institute of Physics.

An inverted layer sequence (see Fig. 6) was used (Zimmermann et al., 2007) in an ITO-free wrap through approach of which device configuration included PET/ AL-Ti/ Absorber (P3HT:PCBM)/ PEDOT:PSS/Au layer sequence. Thermal evaporation, e-beam evaporation and spin coating techniques were used for device fabrication on flexible substrate. Researchers obtained a power conversion efficiency of 1.1% (under $1000W/m^2$ AM 1.5) from the device with additional serial circuitry, which employed top illumination by avoiding the use of ITO.

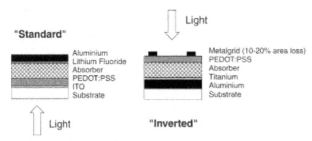

Fig. 6. Comparison of the widely used layer sequence on ITO/PEDOT:PSS electrode (left) and inverted layer sequence (right), where the ITO is replaced by a metal grid for small area devices. Reprinted from Sol. Energy Mater. Sol. Cells, 91, Zimmermann, B.; Glatthaar, M.; Niggemann, M.; Riede, M. K.; Hinsch, A. & Gombert, A., ITO-free wrap through organic solar cells–A module concept for cost- efficient reel-to-reel production., 374- 378, Copyright (2007), with permission from Elsevier.

The commercially available ITO-coated PET foils are used mostly in studies about flexible organic solar cells. PET layer is as polymeric substrate and ITO is the transparent conducting electrode of photovoltaic device.

Researchers (Brabec et al., 1999) performed efficiency and stability studies on large area (6 cm x 6 cm) flexible solar cells based on MDMO-PPV and PCBM materials and compared them with small area devices. Thin films were produced on two different substrates including ITO coated glass substrates and ITO coated PET foils with different active areas. The overall conversion efficiency of the flexible plastic solar cell is calculated with app. 1,2 % and a filling factor FF - 0.35 under monochromatic illumination (488 nm) with 10 mW/cm². It is possible to produce organic solar cells on flexible substrates without loosing efficiency, whereas fullerene bulk heterojunctions was still limited by charge transport. Al-Ibrahim et al. (Al-Ibrahim et al., 2005) developed photovoltaic devices based on P3HT and PCBM materials on ITO coated polyester foils with an active cell area of 0.5×0.5 cm² with the following photovoltaic device configuration: PET/ITO/PEDOT:PSS/P3HT:PCBM/Al. Device parameters without any special postproduction treatment were obtained as: V_{OC} = 600 mV, I_{SC} = 6.61 mA/cm², FF=0.39 and η=1.54% under irradiation with white light (AM1.5, 100mW/cm²). These results were hopeful for device up-scaling and development of processing technologies for reel to reel production of flexible organic photovoltaic devices.

Different oligothiophene materials are used to develop (Liu et al., 2008) flexible organic photovoltaic devices on ITO-coated PET films. The organic layers (5-formyl-2,2':5',2":5",2""-quaterthiophene (4T-CHO), 5-formyl-2,2':5', 2":5",2"":5"',2""-quinquethiophene (5T-CHO) and 3,4,9,10-perylenetetracarboxylic dianhydride (PTCDA)) were deposited by vacuum deposition. While the PET-ITO/4T-CHO/PTCDA/Al device showed an open circuit voltage (V_{oc}) of 1.56 V and a photoelectric conversion efficiency of 0.77%, the PET-ITO/5T-CHO/PTCDA/Al device exhibited a V_{oc} of 1.70 V and photoelectric conversion efficiency of 0.84%. Stakhira et al. (2008) fabricated an organic solar cell consisting of an ITO/PEDOT:PSS/ pentacene (Pc)/Al multilayer structure on flexible PET substrate coated with conductive ITO layer. PEDOT:PSS/Pc and Al contact were formed by electron beam deposition technique. The photovoltaic effect was measured with open circuit voltage of ~0.5 V, short circuit current of 0.6 1A and fill factor 0.2. Researchers (Blankenburg et al., 2009) used continuous reel-to-reel (R2R) slot die coating process to develop polymer based solar cells on plastic foils with adjustable coating thicknesses. Transparent conducting and photoactive layers were prepared with good reproducibility and promising power conversion efficiencies (0.5–1% (1.7% as maximum value)).

Krebs et al. (2007) fabricated organic solar cells on ITO coated PET substrates. Active layer consisted of MEH-PPV was coated by screen-printing method and an optional layer of fullerene (C_{60}) and the final Al electrode were applied by vacuum coatings. Thirteen individual solar cells with an active area of 7.2 cm² were connected in series. In the simple geometry ITO/MEH-PPV/Al the module gave a V_{oc} of 10.5 V, an I_{sc} of 5 A, a FF of 13% and an efficiency (η) of 0.00001% under AM1.5 illumination with an incident light intensity. A geometry (ITO/MEH-PPV/C_{60}/Al) employing a sublimed layer of C_{60} improved Voc, Isc, FF and η to 3.6V, 178 A, 19% and 0.0002%, respectively. The results of roll-to-roll coated flexible large-area polymer solar-cell modules (eight serially connected stripes), which was performed in 18 different laboratories in Northern America, Europe and Middle East, were presented in another study (Krebs et al., 2009c). In all steps, roll-to-roll processing was employed. A zinc oxide nanoparticle layer, P3HT-PCBM and PEDOT:PSS layers were coated onto ITO coated PET by a modified slot-die coating procedure, respectively. ZnO as buffer layer has high electron mobility compared to titanium oxide (Yip et al., 2008) and so, can be ideal electron selective contact layer in polymer solar cells (Hau et al., 2008). The devices were completed by screen-printing silver paste and lamination of PET protective layer on top. In another study of Krebs (2009c) they prepared polymer solar cell module using all-solution processing on ITO coated PET substrates. zinc oxide nanoparticles (ZnO-nps) were applied using either knife-over- edge coating or slot-die coating. A mixture of the thermocleavable poly-(3-(2-methylhexan-2-yl)- oxy-carbonyldithiophene)(P3MHOCT) and ZnO-nps was applied by a modified slot-die coating procedure as second layer. The third layer was patterned into stripes and juxtaposed with the ITO layer. The fourth layer comprised screen-printed or slot-die-coated PEDOT:PSS and the fifth and the final layer comprised a screen-printed or slot-die-coated silver electrode. Coating ITO onto the PET substrate by sputtering process in a vacuum, cost of ITO and thermal disadvantage of PET foils (temperatures only up to 140°C) were some implications of the research. Also, efficient inverted polymer solar cell fabricated by roll-to-roll (R2R) process could be obtained in terms of both power conversion efficiency and operational stability. Maximum 1.7% efficiency for the active area of the full module was obtained from eight serially connected cells (Krebs et al., 2009a). They (Krebs et al., 2009b) showed the versatility of the polymer solar cell technology with abstract forms for the active area, a flexible substrate, processing entirely from solution, complete processing in air using commonly available screen printing, and finally, simple mechanical encapsulation using a flexible packaging material and electrical contacting post-production using crimped contacts. Following two different devices were developed:

PET/ITO/ZnO/P3CT/ZnO/PEDOT:PSS/Ag paste/Cold laminated PET with acrylic resin and

PET/ITO/ZnO/P3CT/PCBM/ZnO/PEDOT:PSS/Ag paste/Cold laminated PET with acrylic resin

Poly(ethylene naphthalate) (PEN), has higher glass transition temperature than PET and this provides potential post-treatment of devices (Dennler et al., 2006b). However, shrinkage is seen in the material and so, subsequent processes will be problematic (Krebs, 2009b). PEN substrates (Dennler et al., 2006b) were used to develop flexible solar cells and were coated with ultra-high barrier multilayer coatings (Fig. 7). Shelf lifetime of conjugated polymer:fullerene

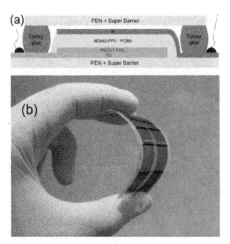

Fig. 7. (a) Cross-sectional view of the conjugated polymer:fullerene solar cells investigated here; (b) picture of a bent device. Reprinted from Thin Solid Films, 511–512, Dennler, G.; Lungenschmied, C.; Neugebauer, H.; Sariciftci, N. S.; Latreche, M.; Czeremuszkin, G. & Wertheimer, M. R., A new encapsulation solution for flexible organic solar cells, 349–353, Copyright (2006), with permission from Elsevier.

solar cells fabricated on PEN substrates and encapsulated with flexible, transparent PEN-based ultra-high barrier material entirely fabricated by plasma enhanced chemical vapor deposition (PECVD) was studied. ITO bottom electrodes were sputtered through a mask onto flexible substrates and so, good adhesion and ~60 Ω/square sheet resistance was obtained. The complete device provided a shelf lifetime of more than 3000h. Lungenschmied et al. (2007) also studied interconnected organic solar cell modules on flexible ultrahigh barrier foils (Fig. 8). Flexible solar cell modules had 11 cm² total active area and reached 0.5% overall powerconversion efficiency under AM1.5 conditions. ITO bottom electrode was structured by deposition through a shadow mask directly onto substrate and a sheet resistance of approximately 60 Ω/square was obtained. PEDOT:PSS and P3HT: PCBM were coated using the doctor blade technique. Al top electrode was thermally evaporated using a shadow mask.

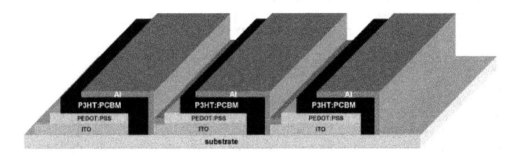

Fig. 8. Serial connection of organic solar cells. Reprinted from Sol. Energy Mater. Sol. Cells, 91, Lungenschmied, C.; Dennler, G.; Neugebauer, H. ; Sariciftci, N. S. ; Glatthaar, M. ; Meyer, T. & Meyer, A., Flexible, long-lived, large-area, organic solar cells, 379–384, Copyright (2007), with permission from Elsevier.

A roll-to-roll process enables fabrication of polymer solar cells with many layers on flexible substrates. Inverted solar cell designs (Krebs, 2009b) can be used on both transparent and non-transparent flexible substrates. Silver nanoparticles on PEN were developed as bottom electrode. ZNO-nps from solution, P3HT-PCBM as active layer and PEDOT:PSS as hole transporting layers were coated, respectively, using slot-die coating. The last electrode was applied by screen printing of a grid structure that allowed for transmission of 80% of the light. The devices were tested under simulated sunlight (1000Wm⁻², AM1.5G) and gave 0.3% of power conversion efficiency for the active layer. The illumination of the device is through the top electrode enabling the use of non-transparent substrates. The poor optical transmission in PEDOT:PSS-silver grid electrode caused a decrease in performance.

Polyimide (PI) films, which show high glass transition temperatures, low surface roughnesses, low coefficients of thermal expansion, and high chemical resistance under manufacturing conditions, are suitable for fabrication of flexible electronics. Inverted polymer solar cells were studied on PI substrates (Hsiao et al., 2009). Surface-nickelized polyimide films (NiPI films) as cathodes (back contact electrode) and high-conductivity PEDOT:PSS films as anodes were coated using solution based processes (see Fig. 9). The resulting FF of 0.43 was lower than that of standard devices. However, this ITO-free inverted polymer solar cells exhibited high performance, with the power conversion efficiency reaching 2.4% under AM 1.5 illumination (100mWcm⁻²).

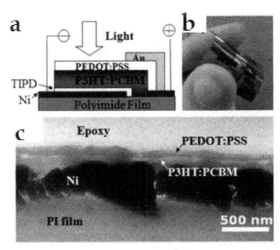

Fig. 9. (a) Architecture of an inverted PSC featuring an inverted sequence on NiPI as the back contact electrode. (b) Optical image of an inverted PSC on NiPI. (c) TEM cross-sectional image of an inverted PSC on NiPI. Scale bars, 500 nm. Reprinted from Org.Electron, 10, Hsiao, Y. S.; Chen, C. P. ; Chao, C. H. & Whang, W. T., All-solution-processed inverted polymer solar cells on granular surface-nickelized polyimide, 551-561, Copyright (2009), with permission from Elsevier.

Kapton®, which was synthesized by polymerizing an aromatic dianhydride with an aromatic diamine, has good chemical and thermal resistance (>400 °C). Kapton® polyimide films can be used in a variety of electrical and electronic uses such as wire and cable tapes, substrates for printed circuit boards, and magnetic and pressure-sensitive tapes (matweb, 2010). Guillen and Herrero (2003) developed both bottom and top electrodes onto polyimide sheets (Kapton KJ) to be used in applications of lightweight and flexible thin film photovoltaic devices. ITO as the frontal electrical contact and Mo, Cr and Ni layers as the back electrical connections were prepared and then compared with conventional electrodes on glass substrates. ITO deposited polyimide sheets showed similar optical transmittance and higher electrical conductivity than ITO coated glass substrates. Mo, Cr and Ni coated polyimide sheets showed similar structure and electrical conductivity to Mo, Cr and Ni coated glass substrates without bending or adhesion failure.

A commercially available polyimide foil (Kapton), which was overlayed with copper, was used as the substrate of polymer solar cell in a roll-to-roll process that does not involve ITO (Krebs, 2009d). Titanium metal was sputtered onto the kapton/copper layer in the vacuum and both the monolithic substrate and back electrode for the devices were obtained. PEDOT:PSS and the active layer were slot-die coated onto the kapton (25 µm) /Cu/Ti foil, respectively. A front electrode, a protective layer and finally a silver grid was applied by screen printing technique. Vacuum coating step was the current limitation of the device.

Polyethersulphone (PES), which is related to polyetheretherketone and polyetherimide, is used as thermoplastic substrate and has high glass transition temperature (Tg ~223°C). PES was used as substrate to fabricate small molecule organic solar cells, which have single heterojunction structure, and, which use PEDOT:PSS anodes possessing low sheet resistance (~450 Ω/ℝ). High conductivity PEDOT:PSS layers were prepatterned using photolithographic technique and spin cast onto fully flexible thermoplastic PES-based substrates having %90 optical transmission. Both organic solar cells, which have plastic and glass based substrates, and, which use a hole transport material, 4,4-bis[N-(1-naphthyl)-N-phenyl-amino]biphenyl (α-NPD) and C_{60} bilayer structure, exhibited high carrier mobilities and high V_{oc}=0.85V (AM1.5, 97 mW/cm^2) (Kushto et al., 2005).

2.5 Solar cell integrated textiles

Among the photovoltaic technologies, organic solar cells are the most suitable ones to textile structures in terms of favorable features such as flexibility, lightness, cost-effectiveness and usage performance. Studies about photovoltaic textiles consider two main approaches: First, solar cell is formed elsewhere and then, photovoltaic structure is integrated in/onto textiles using various techniques, i.e. patching. Second, solar cell is formed in fiber or textile form. So, it can be used as fiber itself or can form textile structures, which are partly or completely photovoltaic. Shelf lifetime, cost and efficiency of organic solar cells are still important issues for also photovoltaic fibers and textiles to be overcome before commercialization.

Utilizing flexible solar cells with textiles can open many application fields for photovoltaic textiles such as electronic textiles besides powering movable electronic devices. Solar cell integrated bags, jackets and dresses are some of the recent applications of polymer based solar cells. For example, in study of Krebs et al. (2006) incorporation of polymer based organic solar cells into textile structures were performed by two ways: In first one, PET substrate was coated with ITO, MEH-PPV, C_{60} and Al, respectively. Then, device was laminated using PET. In second one, PE layer was laminated onto textile substrate. Then, by applying PEDOT, active material and final electrode, respectively, device was completed. Completed devices were integrated into clothes (Fig. 10).

Fig. 10. An example of patterned polymer solar cells on a PET substrate incorporated into clothing by sewing through the polymer solar cell foil using an ordinary sewing machine. Connections between cells were made with copper wire that could also be sewn into the garment. The solar cells were incorporated into a dress and a belt. Design by Tine Hertz Reprinted from Sol. Energy Mater. Sol. Cells, 90, Krebs F.C.; Biancardo M.; Jensen B.W.; Spanggard H. & Alstrup J., Strategies for incorporation of polymer photovoltaics into garments and textiles, 1058-1067, Copyright (2006), with permission from Elsevier.

2.6 Studies about polymer nanofibers for solar cells

There are several studies about developing conductive polymer nanofibers used to fabricate solar cells. Various methods such as self-assembly (Merlo & Frisbie, 2003), polymerization in nanoporous templates (Martin, 1999), dip-pen nano-lithography (Noy et al., 2002), and electrospinning (Babel et al., 2005; Wutticharoenmongkol et al., 2005; Madhugiri; 2003) techniques are used to produce conductive polymer nanowires and nanofibers. Nanofibers having ultrafine diameters provide some advantages including mechanical performance, very large surface area to volume ration and flexibility to be used in solar cells (Chuangchote et al., 2008a).

Since morphology of the active layer in organic solar cells plays an important role to obtain high power conversion efficiencies, many researchers focus on developing P3HT nanofibers for optimized morphologies (Berson et al., 2007; Li et al., 2008; Moulé & Meerholz, 2008). Nanofibers can be deposited onto both conventional glass-based substrates flexible polymer based substrates, which have low glass transition temperature (Bertho et al., 2009).

A fabrication method (Berson et al., 2007) was presented to produce highly concentrated solutions of P3HT nanofibers and to form highly efficient active layers after mixing these with a molecular acceptor (PCBM), easily. A maximum PCE of 3.6% (AM1.5, 100 mWcm^{-2}) has been achieved without any thermal post-treatment with the optimum composition:75 wt% nanofibers and 25 wt% disorganized P3HT. Manufacturing processes were appropriate to be used with flexible substrates at room temperatures. Bertho et al. (Bertho et al., 2009) demonstrated that the fiber content of the P3HT-fiber:PCBM casting solution can be easily controlled by changing the solution temperature. Optimal solar cell efficiency was obtained when the solution temperature was 45 °C and the fiber content was 42%. Fiber content in the solution effected the photovoltaic performances of cells.

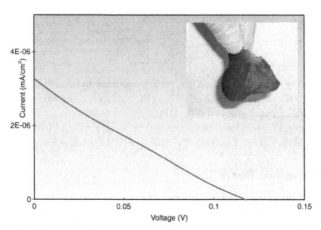

Fig. 11. Jsc–V graph of the P3HT/PCBM based solar cloth measured under 1 Sun conditions. Inset shows a picture of the solar cloth fabricated using electrospinning. Reprinted from *Materials Letters*, 64, Sundarrajan, S.; Murugan, R.; Nair, A. S. & Ramakrishna, S., 2369 -2372., Copyright (2010), with permission from Elsevier.

Electrospinning technique (Chuangchote et al., 2008b) is also used to prepare photoactive layers of polymer-based organic solar cells without thermal post-treatment step. Electrospun MEH-PPV nanofibers were obtained after polyvinylpyrrolidone (PVP) was removed from as-spun MEH-PPV/PVP fibers. A ribbon-like structure aligned with wrinkled surface in fiber direction was gained. Bulk heterojunction organic solar cells were manufactured by using the electrospun MEH-PPV nanofibers with a suitable acceptor. Chuangchote et al. produced ultrafine MEH-PPV/PVP composite fibers (average diameters ranged from 43 nm to 1.7 mm) by electrospinning of blended polymer solutions in mixed solvent of chlorobenzene and methanol under the various conditions.

Recently, a photovoltaic fabric (Sundarrajan et al., 2010) based on P3HT and PCBM materials were developed. The non-woven organic solar cloth was formed by co-electrospinning of two materials: the core-shell nanofibers as the core and PVP as the shell. The efficiency of the fiber-based solar cloth was obtained as 8.7×10^{-8} due to processing conditions and thickness of structure (Fig. 11-12). However, this is an novel and improvable approach to develop photovoltaic fabrics for smart textiles.

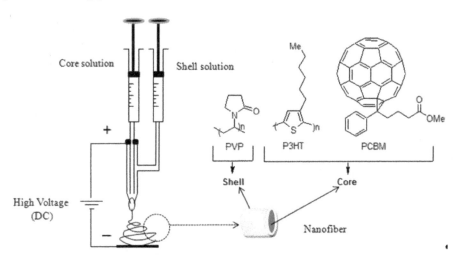

Fig. 12. Schematic diagram of core-shell electrospinning set-up used in this study: direct current voltage at 18 KV, the flow rate of P3HT/PCBM in chloroform/toluene (3:1 ratio, as core) and PVP in chloroform/ethanol (1:1 ratio, shell) was set at 1.3 mL/h and 0.8 mL/h, Respectively. Reprinted from *Materials Letters*, 64, Sundarrajan, S.; Murugan, R.; Nair, A. S. & Ramakrishna, S., 2369 -2372., Copyright (2010), with permission from Elsevier

3. Organic photovoltaic fibers

In recent years, attention on fibrous and flexible optoelectronic structures is increased in both scientific and industrial areas in terms of lightweight, low-cost and large scale production possibilities. Photovoltaic fibers, cost effective and scalable way of solar energy harvesting, work with the principle of solar cell, which produces electricity by converting photons of the sun. Although solar cells made from silicon and other inorganic materials are far more efficient for powering devices than organic solar cells, they are still too expensive to be used in widespread and longterm applications. In studies of fiber-based solar cells, which are incorporated in textiles, organic semiconductors that are naturally flexible and light-weight, are ideal candidates compared to conventional inorganic semiconductors.

For developing optimum photovoltaic textile, choice of the fiber type, which determines UV resistance and maximum processing temperature for photovoltaics and textile production methods (Mather & Wilson, 2006) need to be considered.
In recent years, there are several studies about photovoltaic fibers based on polycrystalline silicon (Kuraseko et al., 2006), dye sensitized solar cells (Fan et al., 2008; Ramier et al., 2008; Toivola et al., 2009) and organic solar cells (Bedeloglu et al., 2009, 2010a, 2010b, 2010c, 2011; Curran et al., 2006; Curran et al., 2008; Curran et al., 2009; Lee et al., 2009; Liu et al., 2007a; Liu et al., 2007b; O'Connor et al., 2008; Zhou et al., 2009; Zou et al., 2010). Protection of liquid electrolyte in DSSCs is problematic causing leakage and loss of performance. However, solid type DSSCs suffer from cracking due to low elongation and bending properties. The organic solar cells based fibers still suffer from low power conversion efficiency and stability. However, organic materials are very suitable to develop flexible photovoltaic fibers with low-cost and in large scale (Bedeloglu et al., 2009; DeCristofano, 2008).
The fiber geometry due to circular cross-section and cylindrical structure brings advantages in real usage conditions. Contrast to planar solar cells, absorption and current generation results in a greater power generation, which can be kept constant during illumination owing to its symmetric structure. A photovoltaic fiber has very thin coatings (about a few hundred nanometers). Therefore, a photovoltaic fabric made from this fiber will be much lighter than that of other thin film technologies or laminated fabric (Li et al., 2010a).
Organic photovoltaic fibers have been produced in different thicknesses and lengths, using different techniques and materials in previous studies. In order to develop fiber based solar cells, mainly solution based coating techniques were applied to develop polymer based electrodes and light absorbing layers. However, deposition techniques in a vacuum were used to develop a photovoltaic fiber formation, too.
Current studies about fiber shaped organic photovoltaics used different substrate materials such as optical fibers (Do et al., 1994), polyimide coated silica fibers (O'Connor et al., 2008), PP fibers and tapes (Bedeloglu et al., 2009, 2010a, 2010b, 2010c, 2011) and stainless steel wires (Lee et al., 2009).
In order to fabricate photovoltaic fiber with low-cost and high production rate, an approach is using a drawing a metal or metalized polymer based fiber core through a melt containing a blend of photosensitive polymer. A conductor can also be applied parallel to the axis of the photoactive fiber core (Shtein & Forrest, 2008).
In optical fiber concept, photovoltaic fiber takes the light and transmitted down the fiber by working as an optical can. The fiber shaped photovoltaics approach can reduce the disadvantage of organic solar cells, which is trade-off between exciton diffusion length and the photoactive film thickness in conjugated polymers based solar cells, by forming the solar cell around the fiber (Li et al., 2010b).

3.1 Device structures

Organic solar cell materials are generally coated around the fibers concentrically in an order in photovoltaic fibers, as in planar solar cells. The Substrate, active layer and conductive electrodes do their own duties. Recent studies about photovoltaic fibers can be classified in two groups: First one is interested with photovoltaic fibers that were illuminated from outside as in photovoltaic textiles, second one is the study of illuminated from inside the photovoltaic fiber (Zou et al., 2010).
For the outside illuminated photovoltaic fibers, different device sequences and manufacturing techniques were used. A fiber-shaped, ITO-free organic solar cell using small molecular

organic compounds was demonstrated by Shtein and co-workers (O'Connor et al., 2008). Light was entered the cell through a semitransparent outer electrode in the fiber-based photovoltaic cell. Concentric thin films of Mg/Mg:Au/Au/CuPc/C$_{60}$/Alq$_3$/Mg:Ag/Ag were deposited onto rotated polyimide coated silica fibers having 0.48 mm diameter by thermal evaporation technique in a vacuum (see Fig. 13). The cell exhibited 0.5% power conversion efficiency, which was much less dependent on variations in illumination angle. However, coated fiber length was limited by the experimental deposition chamber geometry.

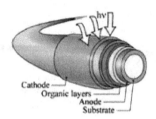

Fig. 13. A flexible polyimide coated silica fiber substrate device, with the layers deposited concentrically around the fiber workers. Reprinted with permission from O'Connor, B.; Pipe, K. P. & Shtein, M. (2008). Fiber based organic photovoltaic devices. *Appl. Phys. Lett.*, vol. 92, pp. 193306-1–193306-3. Copyright 2008, American Institute of Physics.

Bedeloglu et al. developed flexible photovoltaic devices (Bedeloglu et al., 2009, 2010a, 2010b, 2010c, 2011) to manufacture textile based photovoltaic tape and fiber by modifying planar organic solar cell sequence. The non-transparent and non-conductive polymeric materials (PP tapes and fibers) were used as substrate and dip coating and thermal evaporation technique were used to coat active layer and top electrode, respectively. Devices gave moderate efficiencies in photovoltaic tape (PP/Ag/PEDOT:PSS/P3HT:PCBM/LiF/Al) and in photovoltaic fiber (PP/PEDOT:PSS/P3HT:PCBM/LiF/Al) (see Fig. 14). Light entered the photovoltaic structure from the outer semi-transparent cathode (10 nm LiF/Al). Obtained structures that were very flexible and lightweight were hopeful for further studies using textile fibers.

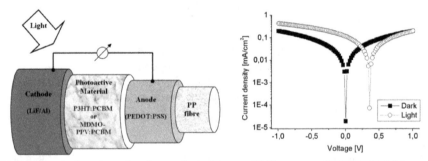

Fig. 14. Schematic drawing of a photovoltaic fiber and I–V curves of P3HT:PCBM -based photovoltaic fibers, lighting through the cathode direction. The final, definitive version of this paper has been published in < Textile Research Journal>, 80/11/July/2010 by <<SAGE Publications Ltd.>>/<<SAGE Publications, Inc.>>, All rights reserved. ©.

Flexible photovoltaic wires based on organic materials can also be produced to be used in a broad range of applications including smart textiles (Lee et al., 2009). In the study, a stainless steel wire used as primary electrode was coated with TiO$_x$, P3HT and PC$_{61}$BM,

PEDOT ·PSS materials as electron transport layer, active layer and hole transport layer, respectively (Fig. 15). Another wire as secondary electrode was wrapped around the coated primary wire with a rotating stage similar to commercial wire winding operations. In the best cell, the short circuit current density was 11.9 mA/cm² resulting 3.87% power conversion efficiency.

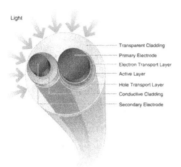

Fig. 15. Schematic of a complete fiber showing the potential for shadowing by the secondary electrode. From Lee, M. R.; Eckert, R. D. ; Forberich, K. ; Dennler, G.; Brabec, C. J. & Gaudiana, R. A. (2009). Solar power wires based on organic photovoltaic materials. *Science*, Vol. 324, pp. 232–235. Reprinted with permission from AAAS.

Many researchers considered photovoltaic fiber design for different function from an optical perspective to capture or trap more light. An optical design was investigated (Curran et al., 2006) to increase the efficiency of photovoltaic device by directing the incident light into the photoactive layer using optical fibers. Prepared fibers are worked up into bundle to confine the light in the device. Polymer based organic solar cell materials are used to develop an optical fiber-based waveguide design (Liu et al., 2007a). P3HT:PCBM is commonly used composite material to form active layer. Carroll and co-workers added top electrode (Al) to only one side of the fiber and tested the photovoltaic fibers under standard illumination at the cleaved end of the fibers. Optical loss into the fiber based solar cell increased as the fiber diameter decreased (See Fig. 16) and increasing efficiency was obtained by the smaller diameter photovoltaic fibers. In their other study (Liu et al., 2007b), performances of the photovoltaic fibers were compared as a function of incident angle of illumination (varied from 0° – 45°) on the cleaved face of the fiber. 1/3 of the circumference was coated with thick outer electrode (LiF/Al) due to fibers having small diameter. Photovoltaic performance of the devices was dependent on fiber diameter and the angle of the incidence light onto the cleaved fiber face.

Using an optical fiber having 400 μm in diameter, microconcentrator cell (Curran et al., 2008) was fabricated to develop an efficient method of light capturing for the optical concentration by using a mathematical based model to pinpoint how to concentrate light within the microconcentrator cell. Behaviour of light between the fiber entrance and active semiconductor layer was investigated. The fiber-based photovoltaic cell, which was a solar collector that utilized internal reflector to confine light into an organic absorber, collected nearly 80% of the incoming photons as current, at ~3 kOhms.cm (Zhou et al., 2009). Li et al. (2010) developed a mathematical model that was also supported by experimental results, for light transmission, absorption and loss in fiber-based organic solar cells using ray tracing and optical path iteration. A patent was developed about photovoltaic devices having fiber structure and their applications (Curran et al., 2009). A tube-based photovoltaic structure was developed to capture optical energy effectively within the absorbing layer without reflective losses at the front and rear surfaces of the devices (Li et al., 2010b). That architecture was enabled that the absorption range of a given polymer (P3HT:PCBM) can be broaden by producing power from band edge absorption.

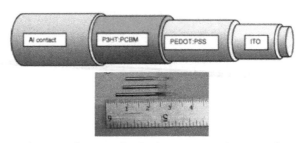

Fig. 16. (a) Schematic diagram showing the device structure (we note that a 0.5nm LiF layer is added below the metal contact but not shown), and (g) optical micrographs of the finished fibers. Reprinted with permission from Liu, J. W.; Namboothiry, M. A. G. & Carroll, D. L. (2007). Fiber-based architectures for organic photovoltaics. *Appl. Phys. Lett.*, Vol. 90, pp. 063501-1–063501-3. Copyright 2007, American Institute of Physics.

4. Conclusions

Polymer solar cells carry various advantages, which are suitable to flexible and fiber-shaped solar cells. However, optimum thickness for photovoltaic coatings and adequate smoothness for the surface of each layer (substrate, photoactive layer and electrodes) are required to obtain higher power conversion efficiencies and to prevent the short-circuiting in the conventional and flexible devices. Suitable coating techniques and materials for developing photovoltaic effect on flexible polymer based textile fibers are also needed not to damage photovoltaic fiber formation in continuous or discontinuous process stages. Many studies still continue for improving stability and efficiency of photovoltaic devices.

Flexible solar cells can expand the applications of photovoltaics into different areas such as textiles, membranes and so on. Photovoltaic fibers can form different textile structures and also can be embedded into fabrics forming many architectural formations for powering portable electronic devices in remote areas. However, optimal photovoltaic fiber architecture and the suitable manufacturing processes to produce it are still in development stage. More studies are required to design and perform for a working photovoltaic fiber.

A viable photovoltaic fiber that is efficient and have resistance to traditional textile manufacturing processes, which are formed from some consecutive dry and wet applications, and, which damage to textile structure, will open new application fields to concepts of smart textiles and smart fabrics.

5. References

Aernouts, T.; Vanlaeke, P.; Geens, W.; Poortmans, J.; Heremans, P.; Borghs, S.; Mertens, R.; Andriessen,R. & Leenders, L. (2004). Printable anodes for flexible organic solar cell modules. *Thin Solid Films*, Vol.22, pp.451-452, ISSN: 0040-6090.

Ajayan, P. M. (1999). Nanotubes from Carbon, *Chem. Rev.* Vol.99, pp.1787- 1800, ISSN: 1520-6890.

Ahlswede, E.; Muhleisen, W.; bin Moh Wahi, M.W.; Hanisch, J. & Powalla, M. (2008). Highly efficient organic solar cells with printable low-cost transparent contacts. *Appl. Phys. Lett.* Vol.92, pp.143307, ISSN: 1077-3118.

Al-Ibrahim, M.; Roth, H.-K.; Zhokhavets, U.; Gobsch, G. & Sensfuss S. (2005) Flexible large area polymer solar cells based on poly(3-hexylthiophene)/fullerene. *Solar Energy Materials and Solar Cells*, Vol.85, No.1, pp. 13-20, ISSN 0927-0248.

Antoniadis, H.; Hsieh, B. R.; Abkowitz, M. A.; Stolka, M., & Jenekhe, S. A. (1994). Photovoltaic and photoconductive properties of aluminum/poly(p-phenylene vinylene) interfaces. *Synthetic Metals*, Vol. 62, pp. 265-271, ISSN: 0379-6779.

Babel, A.; Li, D.; Xia, Y. & Jenekhe, S. A. (2005). Electrospun nanofibers of blends of conjugated polymers: Morphology, optical properties, and field-effect transistors. *Macromolecules*, Vol. 38, pp.4705- 4711, ISSN: 1520-5835.

Baughman, R. H.; Zakhidov, A. A. & de Heer, W. A. (2002). Carbon Nanotubes – The Route Towards Applications. *Science*, Vol.297, pp.787-792, ISSN: 0036-8075 (print), 1095-9203 (online).

Bedeloglu, A.; Demir, A.; Bozkurt, Y. & Sariciftci, N.S. (2009). A flexible textile structure based on polymeric photovoltaics using transparent cathode. *Synthetic. Metals*, Vol.159, pp.2043–2047, ISSN: 0379-6779.

Bedeloglu, A.; Demir, A.; Bozkurt, Y.& Sariciftci, N.S. (2010a).A Photovoltaic Fibre Design for. Smart Textiles. *Textile Research Journal*, Vol.80, No.11, pp.1065-1074, eISSN: 1746-7748, ISSN: 0040-5175.

Bedeloglu, A.; Koeppe, R.; Demir, A.; Bozkurt, Y. & Sariciftci, N.S. (2010b). Development of energy generating photovoltaic textile structures for smart applications. *Fibers and Polymers*, Vol.11, No.3, pp.378-383, ISSN: 1229-9197 (print version), 1875-0052 (electronic version).

Bedeloglu, A.; Demir, A.; Bozkurt, Y. & Sariciftci, N.S. (2010c). Photovoltaic properties of polymer based organic solar cells adapted for non-transparent substrates. *Renewable Energy*, Vol.35, No.10, pp.2301-2306, ISSN: 0960-1481.

Bedeloglu, A.; Jimenez, P.; Demir,A.; Bozkurt, Y.; Maser, W. K., & Sariciftci, NS. (2011). Photovoltaic textile structure using polyaniline/carbon nanotube composite materials. *The Journal of The Textile Institute*, Vol. 102, No. 10, pp. 857–862, ISSN:1754-2340 (electronic) 0040-5000 (paper).

Beeby, S.P. (2010). Energy Harvesting Materials for Smart Fabrics and Interactive Textiles. *http://gow.epsrc.ac.uk/ViewGrant.aspx?GrantRef=EP/I005323/1*

Benanti, T. L. & Venkataraman, D. (2006). Organic Solar Cells: An Overview Focusing on Active Layer Morphology. *Photosynthesis Research*, Vol. 87, pp.73-81, ISSN: 0166-8595 (print version), 1573-5079 (electronic version).

Berson, S.; De Bettignies, R.; Bailly, S. & Guillerez, S. (2007). Poly(3-hexylthiophene) Fibers for Photovoltaic Applications. *Adv. Funct. Mater.* , Vol.17, pp. 1377-1384, Online ISSN: 1616-3028.

Bertho, S.; Oosterbaan, W.; Vrindts, V.; D'Haen, J.; Cleij, T.; Lutsen, L.; Manca, J. & Vanderzande, D. (2009). Controlling the morphology of nanofiber-P3HT:PCBM blends for organic bulk heterojunction solar cells. *Organic Electronics*, Vol. 10, No.7, pp.1248-1251, ISSN: 1566-1199.

Bjerring, M.; Nielsen, J. S.; Siu, A.; Nielsen, N. C. & Krebs, F. C. (2008). An explanation for the high stability of polycarboxythiophenes in photovoltaic devices – A solid-state NMR dipolar recoupling study *Sol. Energy Mater. Sol.Cells.,*Vol. 92,pp. 772– 784, ISSN 0927-0248.

Blankenburg, L.; Schultheis, K.; Schache, H.; Sensfuss, S. & Schrodner, M. (2009). Reel-to-reel wet coating as an efficient up-scaling technique for the production of bulk heterojunction polymer solar cells. *Sol. Energy Mater. Sol. Cells,* Vol.93, pp.476, ISSN: 0927-0248.

Brabec, C.J.; Padinger, F.; Hummelen, JC; Janssen RAJ. & Sariciftci, NS. (1999). Realization of Large Area Flexible Fullerene - Conjugated Polymer Photocells: A Route to Plastic Solar Cells. *Synthetic Metals,* Vol. 102, pp. 861-864, ISSN: 0379-6779.

Brabec, C. J.; Shaheen S. E.; Winder C. & Sariciftci, N. S. (2002). Effect of LiF/metal electrodes on the performance of plastic solar cells. *Appl. Phys. Lett.,* Vol.80, pp.1288–1290, Print: ISSN 0003-6951, Online: ISSN 1077-3118.

Brabec, C.; Sariciftci, N. & J. Hummelen, (2001a). Plastic Solar Cells. *Adv. Funct. Mater.* Vol.11, No.1, pp.15-26, Online ISSN: 1616-3028.

Brabec, C.; Shaheen, S.; Fromherz, T.; Padinger, F.; Hummelen, J.; Dhanabalan, A.; Janssen, R. & Sariciftci, N.S. (2001b). Organic Photovoltaic Devices produced from Conjugated Polymer /Methanofullerene Bulk Heterojunctions. *Synth. Met.* Vol.121, pp.1517-1520, ISSN: 0379-6779.

Breeze, A. J.; Salomon, A.; Ginley, D.S.; Gregg, B. A.; Tillmann, H. & Hoerhold, H. H. (2002). Polymer - perylene diimide heterojunction solar cells. *Appl. Phys. Lett.,* Vol. 81, pp.3085–3087, ISSN: 1077-3118.

Bundgaard, E. & Krebs, F. C. (2007). Low band gap polymers for organic photovoltaics. *Sol. Energy Mater. Sol. Cells,* vol. 91, pp.954– 985, ISSN 0927-0248.

Cai, W.; Gong, X. & Cao, Y. (2010). Polymer solar cells: Recent development and possible routes for improvement in the performance. *Solar Energy Materials and Solar Cells,* Vol.94, No.2, pp.114–127, ISSN 0927-0248.

Chang Y T.; Hsu S L.; Su M. H. & Wei K.H. (2009). Intramolecular Donor–Acceptor Regioregular Poly(hexylphenanthrenyl-imidazole thiophene) Exhibits Enhanced Hole Mobility for Heterojunction Solar Cell Applications. *Adv. Mater.,* Vol.21, pp.2093-2097, Online ISSN: 1521-4095.

Chen, H. Y.; Hou, J. H.; Zhang, S. Q.; Liang, Y. Y.; Yang, G. W.; Yang, Y.; Yu, L. P.; Wu, Y. & Li, G. (2009). Polymer solar cells with enhanced open-circuit voltage and efficiency. *Nat. Photonics,*Vol. 3, pp.649-653, ISSN: 1749-4885, EISSN: 1749-4893.

Chuangchote, S.; Sagawa, T. & Yoshikawa, S. (2008a) Fiber-Based Organic Photovoltaic Cells. *Mater. Res. Soc. Symp. Proc.,* 1149E, 1149-QQ11-04.

Chuangchote, S.; Sagawa, T. & Yoshikawa S. (2008b). Electrospun Conductive Polymer Nanofibers from Blended Polymer Solution. *Japanese Journal of Applied Physics,* Vol.47, No.1, pp.787-793, *ISSN* (electronic): 1347-4065.

Coffin, R. C.; Peet, J.; Rogers, J. & Bazan, G. C. (2009). Streamlined microwave-assisted preparation of narrow-bandgap conjugated polymers for high-performance bulk heterojunction solar cells. *Nat. Chem.* Vol.1, pp.657 – 661, *ISSN* : 1755-4330; EISSN : 1755-4349.

Coyle, S.& Diamond, D. (2010). Smart nanotextiles: materials and their application. In: *Encyclopedia* of Materials: Science and Technology *Elsevier* pp. 1-5. ISBN 978-0-08-043152-9.

Curran, S.A.; Carroll, D.L. & Dewald, L. (2009). Fiber Photovoltaic Devices and Applications Thereof, Pub.No.: US2009/0301565 A1.

Curran, S.; Gutin, D. & Dewald, J. (2006). The cascade solar cell, *2006 SPIE – The International Society for Optical Engineering*, 10.1117/2.1200608.0324 , pp. 1-2.

Curran, S.; Talla, J.; Dias, S. & Dewald, J. (2008). Microconcentrator photovoltaic cell (the m-C cell): Modeling the optimum method of capturing light in an organic fiber based photovoltaic cell. *J. Appl. Phys.*, Vol. 104, pp. 064305-1–064305-6, *ISSN* (electronic): 1089-7550, *ISSN* (printed): 0021-8979.

De Jong, M. P.; van Ijzendoorn, L. J. & de Voigt, M. J. A. (2000). Stability of the interface between indium-tin-oxide and poly(3,4-ethylenedioxythiophene)/poly(styrenesulfonate) in polymer light-emitting diodes. *Appl. Phys. Lett.* Vol. 77, pp.2255-2257, ISSN: 1077-3118.

DeCristofano, B.S. (2009). Photovoltaic fibers for smart textiles, RTO-MP-SET-150.

Deibel, C. & Dyakonov, V. (2010). Polymer-fullerene bulk heterojunction solar cells. *Rep. Prog. Phys.* Vol.73, No.096401, pp. 1-39, *ISSN*: 0034-4885.

Dennler, G.; Lungenschmied, C.; Neugebauer, H.; Sariciftci, N. S.; Latreche, M.; Czeremuszkin, G. & Wertheimer, M. R. (*2006* b). A new encapsulation solution for flexible organic solar cells. *Thin Solid Films,*Vol. 511–512, pp.349–353, ISSN: 0040-6090.

Dennler, G. & Sariciftci, N.S. (*2005*). Flexible conjugated polymer-based plastic solar cells: From basic to applications. *Proceedings of the IEEE*, Vol.96, No 8, pp.1429- 1439, *ISSN*: 0018-9219.

Dennler, G.; Sariciftci, N.S. & Brabec, C.J. (2006a). Conjugated Polymer-Based Organic Solar Cells, In: *Semiconducting Polymers: Chemistry, Physics and Engineering* Hadziioannou, G., Malliaras, G. G.,Vol.1, pp.455-519, WILEY-VCH Verlag GmbH & Co. KGaA, *ISBN*: 3527295070, Weinheim.

Do, M.; Han, E. M.; Nidome, Y.; Fujihira, M.; Kanno, T.; Yoshida, S.; Maeda, A. & Ikushima, A. J., (1994). Observation of degradation processes of Al electrodes in organic electroluminescence devices by electroluminescence microscopy, atomic force microscopy, scanning electron microscopy, and Auger electron spectroscopy. *J. Appl. Phys.* Vol. 76, pp.5118-5121, *ISSN* (electronic): 1089-7550, *ISSN* (printed): 0021-8979.

Dresselhaus, M.S.; Dresselhaus, G. & Avouris, P. (2001). Carbon nanotubes: Synthesis, structure, properties and applications. Springer, *ISBN*: 3-54041-086-4, Berlin.

Dittmer, J. J.; Marseglia, E. A. & Friend, R. H. (2000). Electron Trapping in Dye/ Polymer Blend Photovoltaic Cells. *Adv. Mat.*, Vol. 12, pp.1270-1274, Online ISSN: 1521-4095.

Dridi, C.; Barlier, V.; Chaabane, H.; Davenas, J. & Ouada, H.B. (2008). Investigation of exciton photodissociation, charge transport and photovoltaic response of poly(N-vinyl carbazole):TiO_2 nanocomposites for solar cell applications. *Nanotechnology.* Vol.19, pp.375201–375211, ISSN :0957-4484 (Print), 1361-6528 (Online).

Eda, G.; Lin, Y. Y.; Miller, S.; Chen, C. W.; Su, W. F. & Chhowalla, M. (2008). Transparent and Conducting Electrodes for Organic Electronics from Reduced Graphene Oxide. *App Phys. Lett. Vol.92*, pp.233305-1–233305-3, ISSN: 1077-3118.

Enfucell (2011). http://www.enfucell.com/products-and-technology.

European PhotoVoltaic Industry Association (EPIA), (2009). Photovoltaic energy, Electricity from sun, http: www.epia.org

European PhotoVoltaic Industry Association (EPIA) (2010). Global Market Outlook for Photovoltaics until 2014, http: www.epia.org

Fan, X.; Chu, Z.; Chen, L.; Zhang, C.; Wang, F.; Tang, Y.; Sun, J. & Zou, D. (2008).Fibrous flexible solid-type dye-sensitized solar cells without transparent conducting oxide, *Appl. Phys. Lett.* Vol.92, pp. 113510-1 - 113510-3, ISSN: 1077-3118.

Glatthaar, M.; Niggemann, M.; Zimmermann, B.; Lewer, P.; Riede, M.; Hinsch, A. and Luther, J. (2005). Organic solar cells using inverted layer sequence. *Thin Solid Films,* Vol. 491, pp.298-300, ISSN: 0040-6090.

Granström, M.; Petritsch, K.; Arias, A.C.; Lux, A.; Andersson, M.R. & Friend, R.H. (1998). Laminated fabrication of polymeric photovoltaic diodes. *Nature,* Vol. 395, pp.257-260, ISSN: 0028-0836, EISSN: 1476-4687.

Green, M. A.; Emery, K.; Hishikawa, Y. & Warta, W. *(2010).* Solar cell efficiency tables (version 36). *Prog. Photovolt: Res. Appl.,* Vol.*18,* pp.*346–* 352, Online ISSN: 1099-159X.

Green, M.A. (2005) Third Generation Photovoltaics, Advanced Solar Energy Conversion, Springer-Verlag Berlin Heidelberg, ISSN 1437-0379.

Greenham, N. C.; Peng, X. & Alivisatos, A.P. (1996). Charge separation and transport in conjugated-polymer/semiconductor-nanocrystal composites studied by photoluminescence quenching and photoconductivity. *Phys. Rev. B,* Vol. 54, pp. 17628-17637, ISSN 1098-0121 Print, 1550-235X Online.

Guillen C. & Herrero, J. (2003). Electrical contacts on polyimide substrates for flexible thin film photovoltaic devices. *Thin Solid Films,* Vol.431–432, pp.403-406, ISSN: 0040-6090.

Gunter, J.C.; Hodge, R. C. & Mcgrady, K.A. (2007). Micro-scale fuel cell fibers and textile structures there from, United States Patent Application 20070071975.

Gunes, S.; Marjanovic, N.; Nedeljkovic J. M. &. Sariciftci, N.S. (2008). Photovoltaic characterization of hybrid solar cells using surface modified TiO2 nanoparticles and poly(3-hexyl)thiophene. *Nanotechnology,* Vol.19, pp.424009-1 – 424009-5, ISSN :0957-4484 (Print), 1361-6528 (Online).

Gunes, S.; Neugebauer H. & Sariciftci, N.S. (2007). Conjugated polymer-based organic solar cells, *Chem. Rev.* Vol.107, pp.1324-1338, *ISSN:* 0009-2665.

Gunes, S. &. Sariciftci, N.S. (2007). An Overview Of Organic Solar Cells. *Journal of Engineering and Natural Sciences,* Vol.25, No.1, pp.1-16.

Halls, J.J.M.; Walsh, C.A.; Greenham, N.C.; Marseglia, E.A.; Friends, R.H.; Moratti,S.C. & Holmes, A.B. (1995). *Nature,* Vol. 78, pp.451, ISSN: 0028-0836, EISSN: 1476-4687.

Hau, S. K.; Yip, H. L.; Baek, N. S. ; Zou, J.; O'Malley, K. & Jen, A. (2008). Air-stable inverted flexible polymer solar cells using zinc oxide nanoparticles as an electron selective layer. *Appl. Phys. Lett.,* Vol.92, No.25, pp.253301-1 -253301-3, ISSN: 1077-3118.

Hoppe, H. & Sariciftci, N. S. (2006). Morphology of polymer/fullerene bulk heterojunction solar cells. *J. Mater. Chem.* Vol.16, pp.45– 61, *ISSN:* 0959-9428.

Hsiao, Y. S.; Chen, C. P. ; Chao, C. H. & Whang, W. T. (2009). All-solution-processed inverted polymer solar cells on granular surface-nickelized polyimide. *Org. Electron.* Vol.10, No.4, pp.551-561, *ISSN:* 1566-1199.

Hsieh, C.-H.; Cheng, Y.-J.; Li, P.-J. ; Chen, C.-H. ; Dubosc, M.; Liang, R.-M. & Hsu, C.-S. (2010). Highly Efficient and Stable Inverted Polymer Solar Cells Integrated with a Cross-Linked Fullerene Material as an Interlayer. *Journal of the American Chemical Society,*Vol.132, No.13, pp.4887-4893, *ISSN:* 0002-7863.

Huang, J.; Wang, X.; Kim, Y.; deMello, A.J.; Bradley, D.D.C. & deMello, J.C. (2006). High efficiency flexible ITO-free polymer/fullerene photodiodes. *Phys. Chem. Chem.Phys.,* Vol.8, pp.3904–3908, *ISSN:* 1463-9076.

Huo, L. J.; Hou, J. H.; Chen, H. Y.; Zhang, S. Q.; Jiang, Y.; Chen, T. L. & Yang, Y. (2009). Bandgap and Molecular Level Control of the Low-Bandgap Polymers Based on 3,6-Dithiophen-2-yl-2,5-dihydropyrrolo[3,4-c]pyrrole-1,4-dione toward Highly Efficient Polymer Solar Cells, *Macromolecules,* Vol.42, pp.6564–6571, , ISSN: 1520-5835.

Jenekhe, S. A. & Yi, S. (2000). *Efficient photovoltaic cells* from semiconducting polymer heterojunctions. *App. Pyhsics Lett.,* Vol. 77, pp.2635-2637, *ISSN:* 0003-6951.

Jørgensen, M.; Norrman, K. & Krebs, F. C. (2008). Stability/degradation of polymer solar cells. *Sol. Energy Mater. Sol. Cells,* Vol.92, pp.686– 714, ISSN 0927-0248.

Kang, M.-G.; Kim, M.-S.; Kim, J. & Guo, L. J. (2008). Organic solar cells using nanoimprinted transparent metal electrode. *Adv. Mater.* Vol.20, pp.4408–4413, Online ISSN: 1521-4095.

Krebs, F. C. (2009a). Fabrication and processing of polymer solar cells: a review of printing and coating techniques. *Sol. Eng. Mater. Sol. Cells* 93, 394–412, ISSN 0927-0248.

Krebs, F. C. (2009 b). All solution roll-to-roll processed polymer solar cells free from indium tin-oxide and vacuum coating steps. *Org. Electron.* Vol.10, No.5, pp.761–768, , *ISSN:* 1566-1199.

Krebs, F. C. (2009 c). Polymer solar cell modules prepared using roll-to-roll methods: Knife over-edge coating, slot-die coating and screen printing," Sol. Energy Mater. Sol. Cells 93(4), 465–475, ISSN 0927-0248.

Krebs, F. C. (2009 d). Roll-to-roll fabrication of monolithic large area polymer solar cells free from indium-tin-oxide. *Sol. Energy Mater. Sol. Cells* Vol.93, No.9, pp.1636–1641, ISSN 0927-0248.

Krebs, F. C.; Gevorgyan, S. A. & Alstrup, J. (2009 a). A roll-to-roll process to flexible polymer solar cells: model studies, manufacture and operational stability studies, *J. Mater. Chem.* Vol.19, No.30, pp.5442–5451, *ISSN:* 0959-9428.

Krebs, F. C.; Jorgensen, M.; Norrman, K.; Hagemann, O. , Alstrup, J.; Nielsen, T. D.; Fyenbo, J.; Larsen, K. & Kristensen, J. (2009 b). A complete process for production of flexible large area polymer solar cells entirely using screen printing—First public demonstration. *Sol. Energy Mater. Sol. Cells,* Vol.93, No.4, pp.422–441, ISSN 0927-0248.

Krebs F.C.; Biancardo M.; Jensen B.W.; Spanggard H. & Alstrup J. (2006). Strategies for incorporation of polymer photovoltaics into garments and textiles, *Sol Energy Mater Sol Cells,* Vol.90, pp.1058-1067, ISSN 0927-0248.

Krebs, F.; Spanggard, H.; Kjar, T.; Biancardo, M. & Alstrup, J. (2007).Large Area Plastic Solar Cell Modules. *Mater. Sci. Eng. B,* Vol.138, No.2, pp.106–111, *ISSN:* 0921-5107.

Krebs, F.C.; Gevorgyan, S.A.; Gholamkhass, B.; Holdcroft, S.; Schlenker, C.; Thompson, M.E.; Thompson, B.C.; Olson, D.; Ginley, D.S.; Shaheen, S.E.; Alshareef, H.N.; Murphy, J.W.; Youngblood, W.J.; Heston, N.C.; Reynolds, J.R.; Jia, S.J.; Laird, D.; Tuladhar, S.M.; Dane, J.G.A.; Atienzar, P.; Nelson, J.; Kroonl, J.M.; Wienk, M.M.;.

Janssen, R.A.J.; Tvingstedt, K. ; Zhang, F.L.; Andersson, M.; Inganas, O.; Lira-Cantu, M. ; Bettignies, R.; Guillerez, S.; Aernouts, T.; Cheyns, D.; Lutsen, L.;Zimmermann, B.; Wurfel, U.; Niggemann, M.; Schleiermacher, H.F.; Liska, P.; Gratzel, M.; Lianos, P.; Katz, E.A.; Lohwasser, W. & Jannon, B. (2009c). A round robin study of flexible large-area roll-to-roll processed polymer solar cell modules, *Sol. Energy Mater. Sol. Cells*, Vol.93, pp.1968-1977, ISSN 0927-0248.

Krebs, F. C. & Spanggaard, H. (2005). Significant Improvement of Polymer Solar Cell Stability. *Chem. Mater.*, Vol.17, pp.5235- 5237, *ISSN:* 0897-4756.

Krebs, F. C. & Norrman, K. (2007). Analysis of the failure mechanism for a stable organic *photovoltaic* during 10000 h of testing. *Prog. Photovoltaics: Res. Appl.*, Vol.15, pp.697-712, Online ISSN: 1099-159X.

Krebs, F. C. (2008). *Sol. Energy Mater. Sol. Cells*, Vol. 92, pp.715- 726, ISSN 0927-0248.

Krebs, F. C.; Thomann, Y.; Thomann, R. & Andreasen, J. W. (2008). A *simple nanostructured polymer/ZnO hybrid solar* cell-preparation and operation in air. *Nanotechnology* Vol.19, pp.424013-1 -424013-12. , ISSN :0957-4484 (Print), 1361-6528 (Online).

Kuraseko, H.; Nakamura, T. ; Toda, S. ; Koaizawa, H. ; Jia, H. & Kondo, M. (2006). Development of flexible fiber-type poly-Si solar cell, *IEEE 4th World Conf. Photovolt. Energy Convers.* Vol.2, pp.1380-1383.

Kushto, G.P.; Kim, W. & Kafafi, Z.H. (2005). Flexible organic photovoltaics using conducting polymer electrodes, *Appl. Phys. Lett.*, Vol.86, pp.093502-1 -093502-3, ISSN: 1077-3118.

Lee, J.-Y.; Connor, S. T.; Cui, Y. & Peumans P. (*2008*). Solution-Processed Metal Nanowire Mesh Transparent Electrodes. *Nano Lett.* Vol.*8*, pp.689-692, *ISSN* (printed): 1530-6984.

Lee, M. R.; Eckert, R. D. ; Forberich, K. ; Dennler, G.; Brabec, C. J. & Gaudiana, R. A. (2009). Solar power wires based on organic photovoltaic materials. *Science*, Vol. 324, pp. 232-235, ISSN: 0036-8075 (print), 1095-9203 (online).

Li, F.; Tang, H.; Anderegg, J. & Shinar, J. (1997). Fabrication and electroluminescence of double-layered organic light-emitting diodes with the Al$_2$O$_3$/Al cathode. *Appl. Phys. Lett.* Vol.70, pp.1233- 1235, ISSN: 1077-3118.

Li, L.; Lu, G. & Yang, X. (2008). Improving performance of polymer photovoltaic devices using an annealing-free approach *via* construction of ordered aggregates in solution, *J. Mater. Chem.* Vol.18, pp.1984-1990, *ISSN:* 0959-9428.

Li, Y.; Nie, W. ; Liu, J.; Partridge, A. & Carroll, D. L. (2010a). The Optics of Organic Photovoltaics: Fiber-Based Devices, *IEEE J. Sel. Top. Quantum Electron.* Vol.99, pp.1-11.

Li ,Y.; Peterson, E.D.; Huang, H.; Wang, M.; Xue, D.; Nie, W.; Zhou, W., & Carroll, D.L. (2010b). Tube-based geometries for organic photovoltaics. *Appl. Phys. Lett.*, Vol. 96, pp.243505-1 -.243505-3, ISSN: 1077-3118.

Lin, Y.Y.; Wang, D.Y.; Yen, H. C.; Chen, H. L.; Chen, C. C.; Chen, C. M.; Tang, C. Y. & Chen, C. W. (2009). Extended red light harvesting in a poly(3-hexylthiophene)/iron disulfide nanocrystal hybrid solar cell. *Nanotechnology*, Vol.20, pp. 405207-1 -405207-5, ISSN :0957-4484 (Print), 1361-6528 (Online).

Liu, J. W.; Namboothiry, M. A. G. & Carroll, D. L. (2007a). Fiber-based architectures for organic photovoltaics. *Appl. Phys. Lett.*, Vol. 90, pp. 063501-1-063501-3, ISSN: 1077-3118.

Liu, J. W.; Namboothiry, M. A. G. & Carroll, D. L. (2007 b). Optical geometries for fiber-based organic photovoltaics. *Appl. Phys. Lett.*, Vol. 90, pp. 133515-1–133515-3, ISSN: 1077-3118.

Liu, J. S.; Kadnikova, E. N.; Liu, Y. X.; McGehee, M. D. & Fréchet, J. M. J. (2004). Polythiophene containing thermally removable solubilizing groups enhances the interface and the performance of polymer-titania hybrid solar cells. *J. Am. Chem.Soc.* Vol.126, pp. 9486– 9487, *ISSN*: 0002-7863.

Liu, P.; Chen, T.H.; Pan, W.Z.; Huang, M.S.; Deng, W.J.; Mai, Y.L.; Luan, A.B. (2008). Flexible organic photovoltaic devices based on oligothiophene derivatives. *Chinese Chemical Letters.* Vol.19, No.2, pp.207–210, *ISSN*: 1001-8417.

Lungenschmied, C.; Dennler, G.; Neugebauer, H. ; Sariciftci, N. S. ; Glatthaar, M. ; Meyer, T. & Meyer, A. (2007). Flexible, long-lived, large-area, organic solar cells. *Sol. Energy Mater. Sol. Cells, Vol.* 91, No.5, pp.379–384, ISSN 0927-0248.

Madhugiri, S.; Dalton, A.; Gutierrez, J.; Ferraris, J. P. & Balkus, Jr. K. J. (2003). Electrospun MEH-PPV/SBA-15 Composite Nanofibers Using a Dual Syringe Method. *J. Am. Chem.Soc.*, Vol.125, pp.14531-14538, , *ISSN*: 0002-7863.

Martin, C. R. (1994). Nanomaterials: A Membrane-Based Synthetic Approach. *Science,* Vol.266, pp.1961-1966, ISSN: 0036-8075 (print), 1095-9203 (online).

Mather, R. R. & Wilson, J., (2006). Solar textiles: production and distribution of electricity coming from solar radiation. In Mattila, H. (Eds.), Intelligent textiles and clothing, Woodhead Publishing Limited, first ed., ISBN: 1 84569 005 2, England.

Mattila, H. (Eds.), (2006). Intelligent textiles and clothing, Woodhead Publishing Limited, first ed., ISBN: 1 84569 005 2, England.

Matweb, www.matweb.com/search/datasheettext.aspx?matguid=4827771a785e44ed9a0d400dba931354

Merlo, J. A. & Frisbie, C. D. (2003). Field effect conductance of conducting polymer nanofibers. *J. Polym. Sci. Part B: Polym. Phys.*, Vol. 41, pp.2674-2680, *ISSN* (printed): 0887-6266.

Moule, A.J.& Meerholz, K. (2008). Controlling Morphology in Polymer–Fullerene Mixtures, *Adv. Mater.* Vol.20, pp. 240-245, Online ISSN: 1521-4095.

Noy, A.; Miller, A. E. ; Klare, J. E.; Weeks, B. L.; Woods, B. W. & DeYoreo, J. J. (2002). Fabrication and imaging of luminescent nanostructures and nanowires using dip-pen nanolithography. *Nano Lett.* Vol.2, pp. 109–112, *ISSN* (printed): 1530-6984.

Nunzi, J.-M. (2002). Organic photovoltaic materials and devices, *C. R. Physique.* Vol.3, pp.523–542, ISSN: 1631-0705.

O'Connor, B.; Pipe, K. P. & Shtein, M. (2008). Fiber based organic photovoltaic devices. *Appl. Phys. Lett.*, vol. 92, pp. 193306-1–193306-3, ISSN: 1077-3118.

O'Reagan, B. & Graetzel, M. (1991). A low-cost, high-efficiency solar cell based on dye-sensitized colloidal TiO_2 films. *Nature*, Vol. 353, pp. 737- 740, ISSN: 0028-0836, EISSN: 1476-4687.

Oey, C. C.; Djurisic, A.B.; Wang, H.; Man, K. K. Y.; Chan,W. K.; Xie, M.H.; Leung, Y. H.; Pandey, A.; Nunzi, J. M. & Chui, P. C. (2006). Polymer-TiO2 solar cells: TiO2 interconnected network for improved cell performance. *Nanotechnology*, Vol. 17, pp.706- 713, ISSN :0957-4484 (Print), 1361-6528 (Online).

Ouyang, J.; Chu, C.-W.; Chen, F.-C.; Xu, Q.,& Yang, Y. (2005). High-conductivity poly (3,4-ethylenedioxythiophene): poly(styrene sulfonate) film and its application in polymer optoelectronic devices, *Adv. Funct. Mater.* Vol.15, pp.203-208, Online ISSN: 1616-3028.

Padinger, F.; Rittberger, R. & Sariciftci, N.S. (2003). Effect of postproduction treatment on plastic solar cells. *Adv. Funct. Mater.* Vol.13, pp.1-4, . Online ISSN: 1616-3028.

Park, S. H., Roy, A., Beaupre, S., Cho, S., Coates, N., Moon, J. S., Moses, D., Leclerc, M., Lee, K. & Heeger, A. J. (2009). Bulk heterojunction solar cells with internal quantum efficiency approaching 100%, *Nat. Photonics,*Vol.3, pp.297-302, *ISSN*:1749-4885 (Print).

Peet, J.; Senatore, M. L.; Heeger, A. J. & Bazan, G. C. (2009). The Role of Processing in the Fabrication and Optimization of Plastic Solar Cells. *Adv. Mater.* Vol.21, pp.1521-1527, Online ISSN: 1521-4095.

Pope, M. & Swenberg, C.E. (1999). Electronic Processes in Organic Crystals and Polymers 2nd edn., Oxford University Press, *ISBN*-10: 0195129636, New York.

Ramier, J.; Plummer, C.; Leterrier, Y.; Manson, J.; Eckert, B. & Gaudiana, R. (2008). Mechanical integrity of dye-sensitized photovoltaic fibers, *Renewable Energy*, Vol.33, pp. 314-319, *ISSN: 0960-1481*.

Rowell, M.W.; Topinka, M.A.; McGehee, M.D.; Prall, H.J.; Dennler, G.; Sariciftci, N.S.; Hu, L. & Gruner, G. (2006). Organic Solar Cells with Carbon Nanotube Network Electrodes. *Applied Physics Letters*, Vol.88, pp. 233506.-1 -233506.-3, ISSN: 1077-3118.

Gaudiana, R. & Brabec, C. (2008). Organic materials: Fantastic plastic. *Nature Photonics*, Vol.2, pp.287- 289, , ISSN: 1749-4885, EISSN: 1749-4893.

Schubert, M. B. & Werner, H. J. (2006). Flexible solar cells for clothing. *Materials Today*, Vol.9, No.6, pp. 42-50, *ISSN*: 1369-7021

Shaheen, S.; Brabec, C. J.; Sariciftci, N. S.; Padinger, F.; Fromherz, T. & Hummelen, J. C. (2001). 2.5% Efficient Organic Plastic Solar Cells. *App. Phys. Lett.*, Vol. 78, pp.841., ISSN: 1077-3118.

Starner, T. (1996). Human Powered Wearable Computing. *IBM Systems J*, Vol.35, pp. 618-629.

Shtein, M. & Forrest S.R. (2008). Organic devices having a fiber structure, US2008/0025681A1

Stakhira, P.; Cherpak, V. ; Aksimentyeva, O.; Hotra, Z. ; Tsizh B.; Volynyuk, D. & Bordun, I. (2008). Properties of flexible heterojunction based on ITO/poly(3,4-ethylenedioxythiophene): poly(styrenesulfonate)/pentacene/Al. *J Non-Cryst Solids*, Vol.354 pp. 4491–4493, *ISSN*: 0022-3093.

Sundarrajan, S.; Murugan,, R.; Nair, A. S. & Ramakrishna, S. (2010). Fabrication of P3HT/PCBM solar cloth by electrospinning technique. *Materials Letters*, Vol.64, No.21, pp.2369-2372, *ISSN*: 0167-577X.

Takagi, T. (1990). A concept of Intelligent Materials. *J. of Intell. Mater. Syst. And Struct.* Vol.1, pp.149, *ISSN*:1045-389X (Print).

Tang, C.W. (1986). Two Layer Organic Photovoltaic Cell. *Appl. Phys. Lett.*, Vol. 48, pp.183-185, ISSN: 1077-3118.

Tani, J.; Takagi, T. & Qiu, J. (1998). Intelligent material systems: Application of functional material. *Appl. Mech. Rev.* Vol.51. *ISSN*: 0003-6900.

Tao, X. (2001). Smart Fibres, Fabrics and Clothing: Fundamentals and Applications, Woodhead Publishing Ltd., *ISBN*: 1 85573 546 6.

Thompson, B. C. & Fréchet, J. M. J. (2008). Organic photovoltaics—Polymer- fullerene composite solar cells. Angew. Chem., Int. Ed.Vol. 47, pp.58– 77, Online *ISSN*: 1521-3773.

Toivola, M. , Ferenets, M.; Lund, P. and Harlin, A. (2009) Photovoltaic fiber. *Thin Solid Films*, Vol.517, No.8, pp.2799-2802, , ISSN: 0040-6090.

Tromholt, T., Gevorgyan, S.A. ; Jorgensen, M.; Krebs, F.C. & Sylvester-Hvid, K. O. (2009). Thermocleavable Materials for Polymer Solar Cells with High Open Circuit Voltage—A Comparative Study. *ACS Appl. Mater. Interfaces*, Vol.1, No.12, pp.2768–2777, *ISSN*:1944-8244 (Print).

Tvingstedt, K. & Inganas, O. (2007). Electrode grids for ITO-free organic photovoltaic devices, *Adv. Mater.*, Vol.19, pp.2893–2897, Online ISSN: 1521-4095.

Wöhrle, D. & Meissner, D. (1991). Organic Solar Cells. *Adv. Mat.*, Vol.3, pp.129- 137, *ISSN*: 1022-6680.

Wang, X.; Li, Q.; Xie, J.; Jin, Z.; Wang, J.; Li, Y.; Jiang, K.; Fan, S. (2009). Fabrication of Ultralong and Electrically Uniform Single-Walled Carbon Nanotubes on Clean Substrates. *Nano Letters* Vol.9, No.9, pp.3137–3141, *ISSN* (printed): 1530-6984.

Winther-Jensen, B. & Krebs, F.C. (2006) High-conductivity large-area semi-transparent electrodes for polymer photovoltaics by silk screen printing and vapour- phase deposition. *Sol. Energy Mater. Sol. Cells*, Vol.90, pp.123–132, ISSN 0927-0248.

Wong, K. W.; Yip, H. L.; Luo. Y.; Wong, K. Y.; Laua, W. M.; Low, K. H.; Chow,H. F.; Gao, Z. Q.; Yeung,W. L. & Chang, C. C. (2006). Blocking reactions between indium-tin oxide and poly (3,4-ethylene dioxythiophene):poly(styrene sulphonate) with a self-assembly monolayer. *Appl. Phys. Lett.*, Vol.80, pp.2788-2790, ISSN: 1077-3118.

Wutticharoenmongkol, P.; Supaphol, P.; Srikhirin, T.; Kerdcharoen, T. & Osotchan, T. (2005). Electrospinning of polystyrene/poly(2-methoxy-5-(2'-ethylhexyloxy)-1, 4-phenylene vinylene) blends. *J. Polym. Sci., B, Polym . Phys.*, Vol.43, pp.1881-1891, Online *ISSN*: 1099-0488.

Yilmaz Canli, N.; Gunes, S.; Pivrikas, A.; Fuchsbauer, A.; Sinwel, D.; Sariciftci, N.S.; Yasa O. & Bilgin-Eran, B. (2010). Chiral (S)-5-octyloxy-2-[{4-(2-methylbuthoxy)-phenylimino}-methyl]-phenol liquid crystalline compound as additive into polymer solar cells, *Sol. Energy Mater. Sol. Cells*, Vol.94, pp.1089–1099, ISSN 0927-0248.

Yip, L.; Hau, S. K.; Baek, N. S.; Jen, A. (2008). Self-assembled monolayer modified ZnO/metal bilayer cathodes for polymer/fullerene bulk-heterojunction solar cells. *Appl. Phys. Lett.*, Vol.92, No.19, pp.193313-1 - 193313-3, ISSN: 1077-3118.

Yu, G. & Heeger, A. J. (1995). Charge separation and photovoltaic conversion in polymer composites with internal donor/acceptor heterojunctions. *J. App. Physics*, Vol. 78, pp.4510-4515, *ISSN* (printed): 0021-8979.

Yu, B. Y.; Tsai, A.; Tsai, S. P.; Wong, K. T.; Yang, Y.; Chu, C. W. & Shyue, J. J. (2008). Efficient inverted solar cells using TiO2 nanotube arrays. *Nanotechnology*, Vol.19, pp.255202, ISSN :0957-4484 (Print), 1361-6528 (Online).

Yu, G.; Gao, J.; Hummelen, J. C.; Wudl, F. & Heeger, A. J. (1995). Polymer Photovoltaic Cells: Enhanced Efficiencies via a Network of Internal Donor-Acceptor Heterojunctions. *Science*, Vol.270, pp. 1789-1791, ISSN: 0036-8075 (print), 1095-9203 (online).

Zhou, Y.; Zhang, F.; Tvingstedt, K.; Barrau, S.; Li, F.; Tian, W. & Inganas, O. (2008). Investigation on polymer anode design for flexible polymer solar cells. *Appl. Phys. Lett.*,Vol. 92, pp.233308-1- .233308-3,ISSN: 1077-3118.

Zhou, W.; Nie, W. ; Li, Y.; Liu, J. & Carroll, D. L. (2009). Fabrication considerations in fiber based organic photovoltaics. Proceedings of SPIE Vol. 7416, pp. 741613-1- 741613-5, ISBN: 9780819477064.

Zimmermann, B.; Glatthaar, M.; Niggemann, M.; Riede, M. K.; Hinsch, A. & Gombert, A. (2007). ITO-free wrap through organic solar cells–A module concept for cost efficient reel-to-reel production. *Sol. Energy Mater. Sol. Cells*, Vol. 91, No.5, pp.374–378, ISSN 0927-0248.

Zou, D.; Wang, D.; Chu, Z.; Lv, Z.; & Fan, X. (2010). Fiber-shaped flexible solar cells. *Coordination Chemistry Reviews*, Vol. 254, pp.1169–1178, ISSN 0010-8545.

Zou, J.; Yip, H.-L.; Hau, S. K. & Jen, A. K.-Y. (2010). Metal grid/conducting polymer hybrid transparent electrode for inverted polymer solar cells. *Applied Physics* Letters, Vol.96, pp.203301-1- 203301-3, ISSN: 1077-3118.

4

Transparent Conducting Polymer/Nitride Semiconductor Heterojunction Solar Cells

Nobuyuki Matsuki[1,3], Yoshitaka Nakano[2], Yoshihiro Irokawa[1],
Mickael Lozac'h[1] and Masatomo Sumiya[1]
[1]National Institute for Materials Science, Namiki, Tsukuba, Ibarak,
[2]Institute of Science and Technology Research,
Chubu University, Matsumoto, Kasugai, Aichi,
[3]Department of Electrical and Electronic Engineering,
Faculty of Engineering, Gifu University, Yanagido, Gifu,
Japan

1. Introduction

Energy supplies that depend on fossil fuels evoke significant concern about the future depletion of those resources and the emission of carbon dioxide and sulfidizing gas, which are believed to cause environmental problems including climate change and acid precipitation (Solomon et al., 2007). Solar cells, which convert sunlight directly to electric power, are one of the most promising devices for a clean and enduring energy source. The standard energy-weighted power density of sunlight, which is defined as air mass 1.5, is $1 kW/m^2$ under clear and sunny weather conditions (Myers et al., 2000). The maximum available amount of sunlight is usually lower than the value described above due to the weather and the total hours of sunlight in the region.

Thus, the first important aim for developing a solar cell is to derive the highest possible photovoltaic conversion efficiency from the utilized materials and structure. When a solar cell with a single bandgap, E_g, is exposed to the solar spectrum, a photon with less energy than E_g does not contribute to the cell output. Therefore, a multilayer structure comprising a variety of bandgaps is effective for the collection of photons in a wide range of the solar spectrum.

The current (2010) best research-cell efficiencies of typical solar cells are as follows (Green, 2010): crystalline Si (25.0%), multicrystalline Si (20.4%), crystalline GaAs (26.4%), CuInGaSe (19.4%), CdTe (16.7%), amorphous Si (10.1%), dye-sensitized polymers (10.4%), and organic polymers (5.15%). In addition to these, there have been a number of studies focused on developing "third-generation photovoltaics" with ultra-high conversion efficiencies at a low cost (Green, 2001). More recently, after the discovery of the wide band gap range of 0.65–3.4 eV in $In_xGa_{1-x}N$, this material is considered to be one of the most promising candidates for third-generation photovoltaic cells.

Aiming at developing multijunction solar cells based on III-nitrides, we have focused on the potential of a transparent conductive polymer (TCP) as a UV-transparent window layer for the cell instead of adopting the conventional all-inorganic p-i-n structure. In this chapter, we describe the concept and experimental results of the development of TCP/nitride semiconductor heterojunction solar cells. In addition, prospects for their further development are discussed.

2. Basic concepts

2.1 Background

In 2002, an epochal report on the E_g of InN was published; the E_g, which had been believed to be 2.0 eV for many years, was found to be less than 1.0 eV by photoluminescence characterization (Matsuoka et al., 2002). Subsequent investigations verified that the correct E_g is 0.7 eV (Wu et al., 2003). This fact immediately impelled III-nitride-researchers to consider applying III-nitrides to solar cells because $In_xGa_{1-x}N$, which is the III-nitride compound obtained from InN (E_g = 0.7 eV) and GaN (E_g = 3.4 eV), is a direct transition semiconductor that would widely cover the solar spectrum. Furthermore, the strong Piezo-electric-field that forms in III-nitride semiconductors, which is a critical problem for optical emission devices due to the suppression of carrier recombination (Takeuchi, 1998), will be more advantageous to photovoltaic devices in which carrier separation is necessary. There have been reports on the theoretical predictions of the conversion efficiency of In_xGa_{1-x} N solar cells that suggest that the maximum conversion efficiency of In_xGa_{1-x} N solar cells will reach 35-40% (Hamzaoui, 2005; Zhang, 2008). Experimental results of In_xGa_{1-x} N-based solar cells have been also reported (Chen, 2008; Zheng, 2008; Dahal, 2009; Kuwahara, 2010). Although the potential conversion efficiency of $In_xGa_{1-x}N$ solar cells is promisingly high, the highest one so far obtained through an InGaN/InGaN superlattice structure remains as low as 2.5% (Kuwahara, 2011).

The challenges for the development of high efficiency InGaN solar cells are mainly attributed to the necessity for: (1) a conductive crystalline substrate to grow high quality nitride layers in order to reduce series resistance, (2) a high quality film growth technique to reduce carrier recombination, (3) high-efficiency p-type doping, and (4) a novel cell design that allows absorption in a wide range of the solar spectrum and efficient collection of the photo-generated carrier.

Our research has targeted issues (3) and (4) above by introducing a novel Schottky contact consisting of a transparent conducting polymer/nitride semiconductor heterojunction. In this section, the advantages of the polymer/nitride semiconductor heterojunction are described in comparison with those of a conventional nitride p-n homojunction. In addition, the optical and electrical properties of the transparent conducting polymers are shown.

2.2 Issues with solar cell window layer

Figure 1 shows a schematic structure of the In_xGa_{1-x} N-based solar cell that exhibits 2.5% conversion efficiency (Kuwahara, 2011). Due to the low doping efficiency and activity of Mg in p-type III nitride semiconductors, the In_xGa_{1-x} N-based solar cell requires a highly conductive front layer on top of the p-type layer to collect the photo-generated carriers.

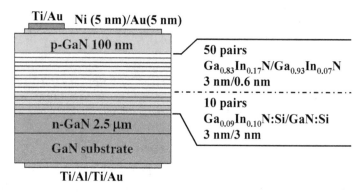

Fig. 1. Schematic of In_xGa_{1-x} N-based solar cell exhibiting 2.5% conversion efficiency (Kuwahara, 2011).

In Figure 1, the electrode on the window side consists of a Ni/Au semitransparent thin film similar to that in the conventional III-nitride-based photoelectric devices. Despite the transparency of the Ni/Au thin-film being as low as 67%, this material is utilized because it forms good ohmic contact with the III-nitride semiconducting layer (Song et al., 2010). With the aim of increasing the transparency of the window-side electrode, indium tin oxide (ITO) was applied to a III-nitride light-emitting diode (LED) (Shim et al, 2001; Chang et al., 2003). In the same study, although the light emitting intensity in the ITO/GaN LED was enhanced compared with that of a Ni/Au/GaN LED under the same current density, the lifetime of the device was significantly shortened due to the heat generated by the high contact resistance between ITO and GaN. Thus, ITO is not a suitable alternative candidate for the metal semitransparent layer unless the contact resistance problem is solved. The low optical transparency and/or the high contact resistance of the front conductive layer are a critical disadvantage for solar cell applications; therefore, new materials that can overcome these issues are highly desirable.

2.3 Conducting polymers as electrodes

Recently, the electronic properties of conducting polymers have been significantly improved and they have been extensively applied in various electric devices (Heeger, 2001).

The study of polymers began with the accidental discovery of vinyl chloride by H. V. Regnault (1835). Thereafter, various kinds of polymers were found and industrialized including ebonite (1851), celluloid (1856), bakelite (1907), polyvinyl chloride (1926), polyethylene (1898; 1933), nylon (1935), etc. Polymers show good electrical insulating properties due to the lack of free electrons; therefore, they have been extensively applied as electrical insulators. However, in 1963, D. E. Weiss and his colleagues discovered that polypyrrole became electrically conductive by doping it with iodine (Bolto et al., 1963). In 1968, H. Shirakawa and his colleagues accidentally discovered a fabrication process for thin-film polyacetylene. In 1975, A. G. MacDiamid noticed the metallic-colored thin-film polyacetylene when he visited Shirakawa's laboratory. Thereafter, collaborative works by A. Heeger, A. G. MacDiamid, and H. Shirakawa began and soon they found a remarkable effect that the electrical conductivity of the polyacetylene thin-film increased over seven orders of magnitude, from 3.2×10^{-6} to 3.8×10^2 Ω^{-1} cm^{-1}, with iodine doping (Shirakawa et al., 1977). Since these early studies, various sorts of π-conjugated polymer thin films have been produced and efforts to improve their conductivity have been made.

We briefly describe the origin of conductivity in degenerate π-conjugated polymers below (Heeger, 2001). In degenerate π-conjugated polymers, stable charge-neutral-unpaired-electrons called solitons exist due to defects at the counterturned connection of the molecular chain. When the materials are doped with acceptor ions like I_2, the acceptor ion abstracts an electron from the soliton; then the neutral soliton turns into a positively-charged soliton while I_2 becomes I_3^-. If the density of the positively-charged solitons is low, the positively-charged soliton tends to pair with a neutral soliton to form a polaron. The polaron is mobile along the polymer chain, thus it behaves as a positive charge. However, the mobility of the polaron is quite low due to the effect of Coulomb attraction induced by the counterion (I_3^-). The Coulomb attraction is reduced by increasing the density of the counterions, which block the electric field. Thus, a high doping concentration of up to ~20% is required to gain high conductivity of over 10^2 Ω^{-1} cm^{-1}. Typical conducting polymers that have high conductivity are fabricated based on polyacetylene (PA), polythiophene (PT), polypyrrole (PPy), polyethylenedioxythiophene (PEDOT), and polyaniline (PANI) (Heeger, 2001).

2.4 Transparent conducting polymers as Schottky contacts

Among the various kinds of conducting polymers, we have focused primarily on polyaniline (PANI) and poly(ethylenedioxythiophene)-polystyrene sulfonate (PEDOT:PSS) because of their high conductivity (~1000 Ω^{-1} cm^{-1}) and high optical transparency (>80%) (Lee et al., 2006; Ha, 2004). Conducting polymers with high optical transparency are known as transparent conducting polymers (TCPs). PANI and PEDOT:PSS also have the advantage in a high workfunction of 5.2–5.3 eV (Brown, 1999; Jang, 2008). This workfunction value is comparable to that of Ni (5.1 eV) and Au (5.2 eV). The high workfunction properties of PANI and PEDOT:PSS make them feasible candidates as hole injection layers in polymer light emitting devices (Jang, 2008). If we assume that a heterojunction consists of a metallic layer and an n-type semiconductor, it is expected that electric barrier, or Schottky barrier, will form at the metal-semiconductor interface. The ideal Schottky barrier height, ϕ_B, is given by following equation (Schottky, 1939; Mott, 1939):

$$q\phi_B = q(\phi_m - \chi) \tag{1}$$

where q is the unit electronic charge, ϕ_m is the workfunction of the metallic material, and χ is the electron affinity of the semiconductor. In general, the experimentally observed Schottky barrier is modified due to the influence of image-force surface states of the semiconductor and/or the dipole effect (Tung, 2001; Kampen, 2006). Nevertheless, the ideal Schottky barrier height estimated from Eq. (1) is still useful to evaluate the potential barrier formation. There have been precedential reports on heterojunctions consisting of TCPs and inorganic monocrystalline semiconductors including: sulfonated-PANI/n-type Si (Wang et al., 2007; da Silva et al., 2009), PEDOT:PSS/SrTiO$_3$:Nb (Yamaura et al., 2003), and PEDOT:PSS/ZnO (Nakano et al., 2008). The $(\phi_m - \chi)$ values of these TCP/semiconductor heterojunctions, and those of PEDOT:PSS or AlN with III-nitrides including AlN, GaN and InN, are summarized in Table 1.

| TCP | | Semiconductor | | $\phi_m - \chi$ | |
material	ϕ_m (eV)	material	χ (eV)	(eV)	References
PANI	5.3[a), b)]	n-type Si	4.05[c), d)]	1.25	a) Brown et al., 1999;
		AlN	0.25[e)]	5.05	b) Jang st al., 2008
		GaN	3.3[f)]	2.0	c) Wang et al, 2007
		InN	5.7[g)]	-0.4	d) da Silva et al., 2009
PEDOT:PSS	5.2[a), b)]	SrTiO$_3$:Nb	4.1[h)]	1.1	e) Grabowski et al., 2001
		ZnO	4.3[i)]	0.7	f) Wu et al., 1999
		AlN	0.25[e)]	4.95	g) Wu et al., 2004
		GaN	3.3[f)]	1.9	h) Yamaura, 2003
		InN	5.7[g)]	-0.5	i) Nakano et al., 2008

Table 1. Summary of workfunction barrier height properties of TCP/inorganic semiconductor heterojunction.

The theoretical Schottky barrier height ($\phi_m - \chi$) is considerably high for AlN and GaN. Thus, it was expected that combinations of these TCPs and III-nitrides would exhibit high-quality Schottky contact properties. When light is irradiated on the Schottky contact, the hole-electron pairs that are photo-generated in the depletion region of the semiconductor are separated due to the strong electric field. As a result, the carriers can be collected as a photocurrent. This suggests that the TCP Schottky contact can be a novel window layer for III-nitride solar cells as an alternative to a p-type layer. Based on this, we began to study transparent conducting polymer/nitride semiconductor heterojunction solar cells.

3. Fabrication processes

3.1 Sample preparation for optical transmittance, workfunction, and conductivity characterizations

Synthetic silica plates (500 μm thick) were utilized as the substrates to prepare samples for characterization to determine their optical transmittance, workfunction, and conductivity. A conductive polymer-dispersed solution of PEDOT:PSS (Clevios PH500, H. C. Starck; without dimethyl sulfoxide dopant) or PANI (ORMECON - Nissan Chemical Industries, Ltd.) was utilized to form the transparent conductive polymer films on the substrate. The same fabrication process was applied to both the PEDOT:PSS and PANI samples. The procedure was as follows:

1. The substrate (2 × 2 cm²) was cleaned using ethanol and acetone for 5 min each in an ultrasonic cleaning bath at ambient temperature.
2. The cleaned substrate was set in a spin coater (MIKASA Ltd., 1H-D7), and the polymer-dispersed solution were dropped onto the substrate using a dropper.
3. The substrate was spun at a 4000 rpm rotating speed for 30 s.
4. The drop and spin procedures were repeated 4 times in total to obtain a sufficient thickness.
5. The coated sample was baked in air at 130 °C on an electric hotplate for 15 min.

The resulting PEDOT:PSS and PANI film thicknesses were measured using a surface profilometer (Dektak 6M) and were found to be 420 and 170 nm, respectively. In the spin-coat process, we applied the same conditions to both the PEDOT:PSS and PANI samples. Their thicknesses unintentionally differed due to differences in the viscosities of their source solutions.

In order to measure the conductivity, a coplanar electrode was fabricated by adding Ag paste to the TCP/synthetic silica plate sample. The electrode-gap width were both 3.3 mm and the lengths were 10.7 and 11.2 mm, respectively, for the PEDOT:PSS and PANI samples.

3.2 Transparent conducting polymer/nitride semiconductor heterojunction solar cells

We fabricated a TCP/III-nitride heterojunction solar cell structure by employing PEDOT:PSS or PANI for the TCP layer and epitaxial GaN (epi.-GaN) for the III-nitride layer (Matsuki et al, 2009, 2010, 2011). Silicon-doped gallium nitride (GaN) was grown on a sapphire (0001) substrate (sapp (0001)) surface by typical metal-organic vapor-phase epitaxy (MOVPE). Ammonia and trimethylgallium were used as the N and Ga sources, respectively. Nitrogen was used as the carrier gas. An undoped buffer GaN layer with a thickness of 1 μm was deposited, followed by the growth of a 2 μm thick Silicon-doped layer. The carrier concentration and electron mobility of the GaN film was determined to be 6.3×10^{17} cm^{-3} and 360 cm^2/V·s, respectively, by Hall measurement.

The PEDOT:PSS or PANI thin film was formed on the epi.-GaN surface using the same process described in section 3.1. Then, in order to fabricate isolated cells, the TCP film was divided into several ~3–9 mm^2 square-shaped sections using a scratching tool. Finally, an ohmic contact for the GaN layer was made by soldering indium metal onto the area from which the TCP layer was removed. Figure 2 shows the schematic structure of the fabricated TCP/epi.-GaN heterojunction solar cell.

4. Characterization methods

4.1 Photoemission electron spectroscopy for workfunction determination

The workfunctions of the TCPs were determined using photoemission electron spectroscopy. The photoemission electron yield Y is expressed as follows Kane (1962):

$$Y = \alpha(h\nu - E_t)^n \qquad (2)$$

where α is a proportional constant, h is Planck's constant, ν is the frequency, E_t is the threshold energy, and the value of n ranges from 1 to 5/2 depending on the system. For metallic materials, an n value of 2 is recommended, and the E_t is consistent with the

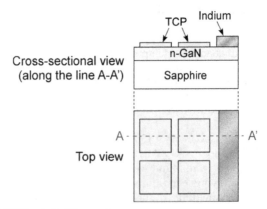

Fig. 2. Schematic of TCP/epi.-GaN sample.

photoelectric workfunction. Thus, if we modify Eq. (2) to the following form (Eq. (2)'), we can determine E_t by extrapolating the linear portion of a $Y^{1/2}$ vs. hv plot:

$$Y^{1/2} = \alpha(hv - E_t) \,. \tag{2}'$$

We employed a photoemission yield spectrometer (AC-3, Riken Keiki Co., Ltd.) to determine the workfunctions of the TCPs. The sample, which was prepared as described in section 3.1, was installed in the spectrometer and the photoemission yield was measured in air.

4.2 Evaluation of current-voltage characteristics
The diode (rectifying) and photovoltaic characteristics were evaluated using an electronic measurement system consisting of an electrometer and a light source. It is necessary for the diode characterization to cover a wide current range from ~10^{-11} to ~10^{-1} A to estimate the Schottky barrier height (SBH) based on the saturation current of the thermionic emission theory (Crowell, 1965). Thus, for the evaluation of the diode characteristics, we employed a high-precision electrometer with a built-in voltage source (Keithley 6487) and performed the measurement under dark conditions. The sample was put on a measurement stage and probe needles were connected to the indium and TCP parts. A xenon-arc light source (HX-504/Q, Wacom Electric Co., Ltd.) was utilized for the evaluation of the photovoltaic characteristics. The light passed though an AM1.5 filter (Bunko Keiki Co., Ltd) and guided onto the TCP side by an aluminum mirror. The values for the source voltage and measured current were acquired by a computer through a GPIB-USB device (National Instruments Co. ltd.).

4.3 Capacitance measurements
The depletion layer width and built-in potential of the GaN layer in the TCP/GaN heterojunction solar cell were estimated using a capacitance measurement setup. A solartron 1255B frequency response analyzer was utilized for the measurement. The sample was set on a sample stage, which was in a vacuum chamber to avoid any influences from light and humidity.

5. Experimental results and discussion

5.1 Conductivity, transparency, and workfunction of polyaniline and PEDOT:PSS
The electrical conductivity was evaluated using a current-voltage (I-V) measurement setup under dark conditions. The conductivities estimated from the result of the I-V measurements were 3.4×10^2 S/cm and 5.7×10^{-1} S/cm for PANI and PEDOT:PSS, respectively.

The optical transmittance was evaluated using a UV-visible-near-infrared spectrophotometer (UV-3150, Shimadzu Co., Ltd.). Figure 3(a) shows the optical transmittance spectra of the PEDOT:PSS and PANI films. Both of the films exhibited transmittance greater than 80% within the wavelength region between 250 and 1500 nm. This is superior to conventional transparent contact materials such as transparent conductive oxides or semi-transparent metals (Kim et al., 2002; Satoh et al., 2007), which exhibit significant drops in transparency particularly near the UV region, as seen in Figure 3.

The workfunctions of the TCPs were estimated using an ultraviolet photoelectron emission spectrometer (AC-3, Riken Keiki Co., Ltd.). Figure 3(b) depicts the photoelectron emission spectra of the PEDOT:PSS and PANI films. The spectra consist of two parts: one with a constant slope and another that linearly increases against the photon energy. The workfunction of PEDOT:PSS and PANI were found to be 5.3 and 5.2 eV, respectively, from Figure 3(b) by assuming that the threshold energy for photoelectron emission is located at the intersection point of the two straight lines that are fitted to the constant-slope and linearly-increasing-slope regions of the plots. These workfunction values show good agreement with those reported previously (Brown et al., 1999; Jang et al., 2008).

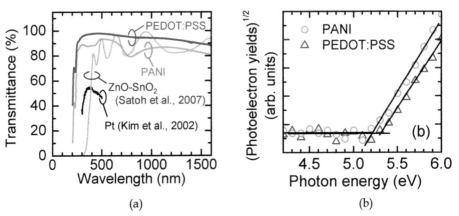

(a)   (b)

Fig. 3. (a) Optical transmittance spectra of PEDOT:PSS and PANI films. The transmittance spectra of ZnO-SnO$_2$ (Satoh et al., 2007) and semi-transparent Pt (Kim et al., 2002) thin films are also shown for comparison. (b) Photoelectron emission yield spectra of PEDOT:PSS and PANI films.

5.2 Diode characteristics of transparent conducting polymer/nitride structures

Figure 4(a) shows the current density-voltage (J-V) characteristics of the TCP/epi.-GaN samples. The diode ideality factor, n, and the SBH, ϕ_B, were evaluated by fitting the theoretical values obtained using the following equation based on the thermionic emission theory (Crowell, 1965):

$$J = A^* T^2 \exp\left(-\frac{\phi_B}{kT}\right) \cdot \left[\exp\left(\frac{qV}{nkT}\right) - 1\right] \tag{3}$$

where q is the electronic charge, A^* represents the effective Richardson constant, which is defined as $A^* \equiv 4\pi m_e^* k^2 / h^3$ (26.4 A/(cm^2·K^2) for GaN), T is the absolute temperature, k is the Boltzmann constant, V is the applied bias, m^* is the effective electron mass (0.2 m_e for GaN), and h is Planck's constant. The n and ϕ_B values derived using the J-V characteristics were 3.0 and 0.90 eV, respectively, for PEDOT:PSS/epi.-GaN, and 1.2 and 0.97 eV, respectively, for PANI/epi.-GaN. The low reverse leakage current, which ranged between 10^{-8} and 10^{-9} A/cm^2 at a reverse bias voltage of -3 V, indicates that the TCP/epi.-GaN heterojunctions had a Schottky contact property comparable to that exhibited by conventional metal Schottky contacts.

The depletion width, W_D, in the n-type GaN of the TCP/epi.-GaN heterojunction is expressed by

$$W_D = \sqrt{\frac{2\varepsilon_S\varepsilon_0}{qN_D}\left(V_{Built-in} - V\right)} \qquad (4)$$

where ε_S is the relative dielectric constant of GaN and equals 8.9 (Wu, 2009), ε_0 is the vacuum dielectric constant, $V_{built-in}$ is the built-in voltage formed in GaN, V is the bias voltage, and N_D is the donor concentration. The space charge, Q_{SC}, in the depletion layer is given by $Q_{SC} = qN_DW_D$, thus, the depletion layer capacitance C_D is obtained by

$$C_D = \frac{|\partial Q_{SC}|}{\partial V} = \sqrt{\frac{q\varepsilon_S\varepsilon_0 N_D}{2(V_{built-in} - V)}} . \qquad (5)$$

Equation (5) can also be written in the following form:

$$\frac{1}{C_D^2} = \frac{2(V_{built-in} - V)}{q\varepsilon_S\varepsilon_0 N_D} . \qquad (5)'$$

Equation (5)' suggests that if $1/C_D^2$ exhibits linear plots against V, $V_{built-in}$ can be obtained at the V-intercept of extrapolated fit-line of the plots. Figure 4(b) shows the plot of $1/C_D^2$ as a function of the applied voltage. The frequencies used for the capacitance measurements were 100 Hz and 1 KHz for the PANI/epi.-GaN and PEDOT:PSS/epi.-GaN heterojunctions, respectively. The frequency for measurement was chosen within a range that was sufficiently lower than the cut-off frequency, which is described in section 5.4. In Figure 4(b), both the data sets are linear and straight lines were successfully fitted to the data. The determined diode characteristics of the TCP/epi.-GaN heterojunction determined from the J-V characteristics and capacitance measurements are summarized in Table 2. The observed barrier height was comparable to that obtained by conventional metal Schottky contacts (Tracy et al., 2003). In the case of the conventional metal Schottky contacts, elaborate surface cleaning processes and moderate metal deposition in ultra-high-vacuum conditions are required to attain good Schottky contact with a ϕ_B of more than 1 eV. It is worth noting that the good Schottky contact properties in the TCP/epi.-GaN heterojunction were achieved with convenient spin coating of a water-dispersed TCP solution onto the GaN layer in air at ambient temperature.

The observed ϕ_B of the TCPs were much lower than expected from the energy difference $\phi_m - \chi$. There are various possibilities for the lower barrier heights including the Schottky effect, which is caused by the electronic mirror force, interface dipole effect, surface defects of GaN, inhomogeneous workfunctions in the TCP film, and/or residual contamination (Sze, 1981; Kampen, 2006). However, the major candidates for the modification of the barrier height have been discussed and are still controversial even in conventional metal/semiconductor Schottky heterojunctions (Tung, 2001). Further detailed investigation is required to determine which effects dominate in lowering the barrier in the TCP/epi.-GaN heterojunction.

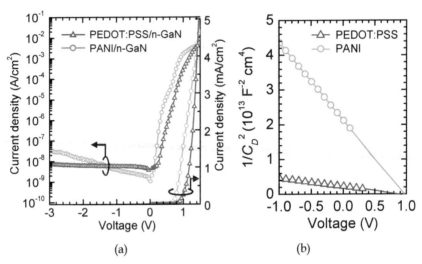

(a)   (b)

Fig. 4. (a) J-V characteristics and (b) Capacitance-voltage plots of TCP/GaN heterojunction solar cells.

	Polymer thickness (nm)	Schottky contact area (mm²)	J-V		C-V	
			n	ϕ_B (eV)	W_D (nm)	$V_{Built-in}$ (V)
PANI	170	7.1	1.2	0.97	39	0.94
PEDOT:PSS	420	3.0	3.0	0.90	40	0.95

Table 2. Diode characteristics of PEDOT:PSS/epi.-GaN (0001) and PANI/epi.-GaN.

5.3 Photovoltaic characteristics of transparent conducting polymer/nitride semiconductor heterojunction solar cells

Figure 5(a) shows the photovoltaic characteristics (J-V measurements under AM1.5 light irradiation) of the PANI/epi.-GaN and PEDOT:PSS/epi.-GaN samples. Table 3 represents a summary of the resulting photovoltaic and resistivity characteristics, which include open-circuit voltage (V_{OC}), short-circuit current density (J_{SC}), maximum output power (P_{max}), fill factor (FF), shunt resistivity, and series resistivity. Note that the V_{OC} exhibited high values (>0.5 V), which was much higher than the photovoltage observed in metal Schottky contacts on n-type GaN (Zhou et al., 2007) or PEDOT:PSS Schottky contacts on ZnO (Nakano et al., 2008). The superior photovoltages of the TCP/epi.-GaN heterojunctions are attributed to the following advantages conveyed by our process and substance properties: the ambient temperature fabrication resulted in less process damage and GaN exhibits less electron affinity (3.3 eV) than ZnO (4.4 eV) (Wu et al., 1999).

However, the rather small shunt resistivity and large series resistance that are observed, especially in the PANI/epi.-GaN heterojunction solar cell, are clearly due to the deterioration of V_{OC} and FF. The optimization of the deposition process of TCP and introduction of a metal comb-shaped electrode on the TCP layer will improve V_{OC} and FF. Figure 5(b) depicts external quantum efficiency of the PANI/epi.-GaN heterojunction solar cell. In order to visualize the capabilities of the photovoltaic device, the transmittance of PANI and the solar light intensity are also plotted as a function of wavelength.

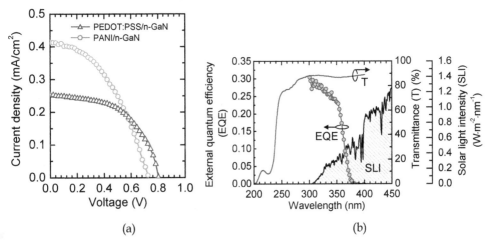

(a) (b)

Fig. 5. (a) Photovoltaic characteristics of PANI/epi.-GaN and PEDOT:PSS/epi.-GaN heterojunction solar cells. (b) External quantum efficiency of PANI/epi.-GaN heterojunction solar cell, transmittance of PANI (T), and solar light intensity (SLI) as a function of wavelength.

	Polymer thickness (nm)	Schottky contact area (mm²)	Photovoltaic characteristics				Resistivity	
			V_{OC} (V)	J_{SC} (mA/cm²)	FF	P_{max} (mW/cm²)	R_{sh} (kΩ/cm²)	R_s (Ω/cm²)
PANI	170	7.1	0.73	0.41	0.42	0.13	21.2	310.3
PEDOT:PSS	420	3.0	0.80	0.25	0.54	0.11	36.8	17.4

Table 3. Photovoltaic characteristics of PEDOT:PSS/epi.-GaN and PANI/epi.-GaN.

5.4 Frequency-dependent capacitance and its application to deep-level optical spectroscopy (DLOS)

In this study, we found that the capacitance of the TCP/epi.-GaN heterojunction exhibits significant dependence on the frequency of measurement. Figure 6 shows the capacitance-frequency (C-f) characteristics of the samples. The characteristics were measured under zero-bias conditions. As seen in the graph, the capacitance is constant at a lower frequency; however, it starts to drop at a specific frequency and then rapidly decreases towards the higher frequencies (cut-off). The frequencies at which the capacitance begins to drop are located at ~20 Hz and ~6 kHz for the PEDOT:PSS/epi.-GaN and PANI/epi.-GaN samples, respectively. It is obvious that the difference in the specific frequencies between the two samples is due to differences in the intrinsic properties of the TCPs. Conductivity in TCPs is generated by a polaron in the π-conjugated bond; this polarized state causes a Debye-type dielectric dispersion response against an applied alternating electric field (Cole et al., 1941).

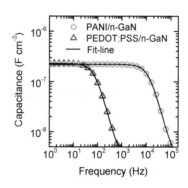

Fig. 6. C-f characteristics of TCP/epi.-GaN heterojunction solar cells.

Referring to a previous study on the frequency-dependent capacitance of PANI film (Mathai et al., 2002), the characteristics can be analyzed by assuming an equivalent circuit consisting of a frequency-independent capacitive element, C_0, in parallel with a resistive element, R, both in series with a constant low-value resistance. Based on this model, the frequency-dependent capacitance of TCP, C_p, is given by the following equation:

$$C_p = C_0 + \frac{1}{(2\pi fR)^2 C_0} \tag{6}$$

where f is the applied bias frequency.

Furthermore, considering that the capacitance of the depletion layer, C_d, is in series with C_P, then the measured total capacitance of the sample, C_{total}, can be expressed by

$$C_{total} = \frac{C_p \cdot C_d}{C_p + C_d}. \tag{7}$$

The solid lines shown in Figure 5 represent the results of the least-square fit of the analytical curve produced based on Equations (6) and (7). The excellent fitting results indicate that the assumed model is adequate. The values of R and C_0, which were derived from the fitting, were 5.3×10^2 Ω and 2.1×10^{-9} F·cm^{-2}, respectively, for PANI/epi.-GaN and 8.4×10^4Ω and 2.3×10^{-9} F·cm^{-2}, respectively, for PEDOT:PSS/epi.-GaN. The large difference in the R values between the two samples is reasonable if we take into account the large difference in the conductivity between PEDOT:PSS (5.7×10^{-1} S/cm) and PANI (3.4×10^2 S/cm).

We describe below that the transparent Schottky contact fabricated by TCP is applicable not only to the photovoltaic device but also to defect density investigation. Nakano et al. applied deep-level optical spectroscopy (DLOS) to the PANI/epi.-GaN samples (Nakano et al., 2010, 2011a, 2011b). DLOS allows the deep-level density in semiconductors to be estimated by detecting the change in capacitance, which is caused by discharging the deep-levels by exciting electrons with monochromatic light. The measurement process was as follows. The residual electrons in the deep levels were excluded by applying a reverse bias (-2 V) and extending the depletion layer. Then, the bias was removed for 1 second to fill the deep levels with electrons in the dark. After that, the same reverse bias was again applied to form the depletion layer followed by monochromatic light illumination that excites electrons in the deep levels up to the conduction band. The difference in the capacitance between the filled state and post-excited states (discharged) was detected as ΔC. The density of the deep-levels is estimated by $2N_D \Delta C/C_i$, where N_D is the donor concentration and C_i is the initial capacitance that is obtained in the filled state in the dark. Figure 7 shows the resulting DLOS spectra. Interestingly, both the spectra acquired at 1 and 10 kHz bias frequency show no characteristic peaks; however, when the bias frequency was increased to 100 kHz, several peaks appeared in the spectrum. This specific frequency, 100 kHz, corresponds to the point where the total capacitance dropped down to a negligible level compared to the capacitance at 1 and 10 kHz. This means that the C_i became smaller comparable to ΔC, thus, $2N_D \Delta C/C_i$ is effectively enhanced enough to be detectable.

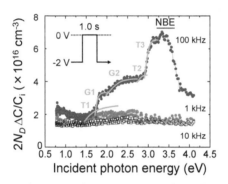

Fig. 7. DLOS spectra of PANI/epi.-GaN heterojunction solar cell.

In Figure 7, five photoemission states are clearly revealed with onsets at ~1.40, ~1.70, ~2.08, ~2.64, and 2.90 eV below the conduction band, which are denoted as T1, G1, G2, T2, and T3, in addition to the near-band-edge (NBE) emissions of GaN at 3.3–3.5 eV. For all the deep levels, electron emission to the conduction band is a dominant process due to their positive photocapacitance transients. The T1, T2, and T3 levels are identical to the deep-level defects that have been commonly reported for GaN, whereas the G1 and G2 levels look like the specific deep levels characteristic of AlGaN/GaN heterointerfaces that were reported recently (Nakano et al., 2008). Using the TCP Schottky contact, we successfully revealed the deep-level states in the near-surface region of the n-GaN layer. These experimental results and further detailed investigations can provide important information on the electronic properties that is needed to improve the performance of the device in optical and electronic fields.

5.5 Future perspective of TCP/nitride semiconductor heterojunction solar cells

In order to increase the output power of TCP/nitride semiconductor heterojunction solar cells, the nitride portion is required to be substituted from GaN to $In_xGa_{1-x}N$. The presumed difficulty in developing the TCP/n-$In_xGa_{1-x}N$ heterojunction is the lowering of the barrier height since the electron affinity significantly increases with an increase of the In content.

One of the most plausible solutions for this issue is to insert a several-tens-nanometer-thick GaN or AlN layer between TCP and n-In$_x$Ga$_{1-x}$ N. With this device structure, it is expected that the barrier height at the TCP/nitride semiconductor interface will be maintained at a high value and an internal electric field should be formed.

The cost of the sapphire substrate will become a high barrier for reducing the production cost of III-nitride based solar cells. Matsuki et al. have shown that high quality GaN can be grown on mica plates (Matsuki et al., 2005), which are inexpensive and flexible. Applying such a novel alternative to sapphire for the epitaxial growth substrate will be effective for developing large area TCP/nitride semiconductor heterojunction solar cells.

TCPs have a high transparency from 250 nm to the visible wavelength region, as described in section 5.1. Thus TCP/nitride semiconductor heterojunction photovoltaic devices also have a high potential for applications in ultraviolet sensors.

6. Conclusion

We have fabricated TCP/nitride semiconductor heterojunction solar cell structures by the spin-coating method using PEDOT:PSS or PANI as the TCP layer and Si-doped GaN as the semiconductor layer. The devices exhibited high quality rectifying properties and have an approximately 1 eV barrier height. Both the PANI/epi.-GaN and PEDOT:PSS/epi.-GaN heterojunction solar cells exhibited ultraviolet-sensitive photovoltaic action. The observed open-circuit voltage was superior to previously reported values for metal/GaN Schottky photo-detectors. A characteristic frequency-dependent behaviour of the interface capacitance was found for the TCP/epi.-GaN solar cells. The C-f characteristics were analyzed based on the dielectric dispersion theory and the intrinsic capacitance and resistance were obtained. The considerable reduction of the interface capacitance in the high frequency region allowed for highly-sensitive detection of deep levels in GaN by DLOS measurements.

7. Acknowledgments

The authors wish to acknowledge collaborations and discussions with Professor Michio Kondo, Dr. Takuya Matsui, Dr. Kenji Itaka, Professor Shunro Fuke, and Professor Hideomi Koinuma. This study was partially supported by the New Energy and Industrial Technology Development Organization (NEDO) Project, Research and Development on Innovative Solar Cells.

8. References

Bolto, B. A., McNeill, R., & Weiss, D. E. (1963). Electronic conduction in polymers III. Electronic properties of polypyrrole, *Australian Journal of Chemistry*, Vol. 16, No. 5, (June 1963), pp. 1090-1103, ISSN 0004-9425

Brown, T. M., Kim, J. S., Friend, R. H., Cacialli, F., Daik R., & Feast, W. J. (1999). Built-in field electroabsorption spectroscopy of polymer light-emitting diodes incorporating a doped poly(3,4-ethylene dioxythiophene) hole injection layer, *Applied Physics Letters*, Vol. 75, No. 12, (July 1999), pp. 1679-1681, ISSN 0003-6951

Chang, C. S., Chang, S. J., Su Y. K., Lin, Y. C., Hsu Y. P., Shei, S. C., Chen, S. C., Liu C. H., & Liaw U. H. (2003). InGaN/GaN light-emitting diodes with ITO p-contact layers prepared by RF sputtering, *Semiconductor Science and Technology*, Vol. 18, No. 4, (February 2003), pp. L21-L23, ISSN 1361-6641

Chen, X., Matthews, K. D., Hao, D., Schaff, W. J., & Eastman, L. F. (2008). Growth, fabrication, and characterization of InGaN solar cells, *Physica Status Solidi (a)*, Vol. 205, No. 5, (July 2003), pp. 1103-1105, ISSN 1862-6300

Crowell, C. R. (1965). The Richardson constant for thermionic emission in Schottky barrier diodes, *Solid-State Electronics*, Vol. 8, No. 4, (April 1965), pp. 395-399, ISSN 0038-1101

Cole, K. S. & Cole, R. H. (1941). Dispersion and absorption in dielectrics I. Alternating current characteristics, *J. Chem. Phys.*, Vol. 9, No. 4, (April 1941), pp. 341-351, ISSN 0021-9606

Dahal, R., Pantha, B., Li. J., Lin, J. Y., & Jiang, H. X. (2009). InGaN/GaN multiple quantum well solar cells with long operating wavelengths, *Applied Physics Letters*, Vol. 94, No. 6, (February 2009), pp. 063505-1-063505-3-1693, ISSN 0003-6951

Grabowski, S. P., Schneider, M., Nienhaus, H., Mönch, W., Dimitrov, R., Ambacher, O., & Stutzmann, M. (2001). Electron affinity of $Al_xGa_{a1-x}N(0001)$ surfaces, *Applied Physics Letters*, Vol. 78, No. 17, (April 2001), pp. 2503-2505, ISSN 0003-6951

Green, M. A., Emery, K., Hishikawa, Y., & Warta, W. (2010). Solar cell efficiency tables (version 37). *Progress in Photovoltaics: Research and Applications*, Vol. 19, No. 1, (December 2010), pp. 84-92, ISSN 1099-159X

Green, M. A. (2001). Third generation photovoltaics: Ultra-high conversion efficiency at low cost. *Progress in Photovoltaics: Research and Applications*, Vol. 9, No. 2, (December 2010), pp. 123-125, ISSN 1099-159X

Ha, Y. H., Nikolov, N., Pollack, S. K., Mastrangelo, J., Martin, B. D., & Shashidhar, R. (2004). Towards a transparent, highly conductive poly(3, 4-ethylenedioxythiophene), *Advanced Functional Materials*, Vol. 14, No. 6, (June 2004), pp. 615-622, ISSN 1616-301X

Heeger J. A. (2001). Nobel Lecture: Semiconducting and metallic polymers: The fourth generation of polymeric materials, *Review of Modern Physics*, Vol. 73, No. 3, (July 2001), pp. 681-700, ISSN 0034-6861

Jang J., Ha, J., & Kim, K. (2008). Organic light-emitting diode with polyaniline-poly(styrene sulfonate) as a hole injection layer, *Thin Solid Films*, Vol. 516, No. 10, (August 2007), pp. 3152-3156, ISSN 1616-301X

Kampen, T. U. (2006). Electronic structure of organic interfaces – a case study on perylene derivatives, *Applied Physics A*, Vol. 82, No. 3, (September 2005), pp. 457-470, ISSN 0947-8396

Kane, E. O. (1962). Theory of Photoelectric Emission from Semiconductors, *Physical Review*, Vol. 127, No. 1, (July 1962), pp. 131-141, ISSN 1943-2879

Kim, J. K., Jang, H. W., Jeon, C. M., & Lee, J.-L. (2002). GaN metal–semiconductor–metal ultraviolet photodetector with IrO_2 Schottky contact, *Applied Physics Letters*, Vol. 81, No. 24, (December 2002), pp. 4655-4657, ISSN 0003-6951

Kuwahara, Y., Takahiro F.; Yasuharu, F.; Sugiyama, T.; Iwaya, M.; Takeuchi, T.; Kamiyama, S.; Akasaki, I.; & Amano, H. (2010). Realization of nitride-based solar cell on freestanding GaN substrate, *Applied Physics Express*, Vol. 3, No. 11, (October 2010), pp. 111001-1-111001-3, ISSN 1882-0778

Kuwahara, Y., Takahiro, F., Sugiyama, T., Iida, D., Isobe, Y., Fujiyama, Y., Morita, Y., Iwaya, M., Takeuchi, T., Kamiyama, S., Akasaki, I., & Amano, H. (2011). GaInN-based solar cells using strained-layer GaInN/GaInN superlattice Active Layer on a Freestanding GaN Substrate, *Applied Physics Express*, Vol. 4, No. 2, (January 2011), pp. 021001-1-021001-3, ISSN 1882-0778

Lee, K., Cho, S., Park, S.-H., Heeger, A. J., Lee, C.-W., & Lee, S.-H. (2006). Metallic transport in polyaniline, *Nature*, Vol. 441, (4 May 2006), pp. 65-68, ISSN 0028-0836

Mathai, C. J., Saravanan, S., Anantharaman, M. R., Venkitachalam, S., & Jayalekshmi, S. (2002). Characterization of low dielectric constant polyaniline thin film synthesized by ac plasma polymerization technique, *J. Phys. D: Applied Physics*, Vol. 35, No. 3, (January 2002), pp. 240-245, ISSN 0022-3727

Matsuki, N., Kim, T.-W., Ohta, J., & Fujioka, H. (2005). Heteroepitaxial growth of gallium nitride on muscovite mica plates by pulsed laser deposition, *Solid State Communications*, Vol. 136, No. 6, (August 2005), pp. 338-341, ISSN 0038-1098

Matsuki, N., Irokawa, Y., Matsui, T., Kondo, M., & Sumiya, M. (2009). Photovoltaic Action in Polyaniline/n-GaN Schottky Diodes, *Applied Physics Express*, Vol. 2, No. 9, (August 2009), pp. 092201-1-092201-3, ISSN 1882-0778

Matsuki, N., Irokawa Y., Nakano Y., & Sumiya M. (2010). π-Conjugated polymer/GaN Schottky solar cells, *Solar Energy Materials & Solar Cells*, Vol. 95, No. 9, (May 2010), pp. 284-287, ISSN 0927-0248

Matsuki, N., Irokawa Y., Nakano Y., & Sumiya M. (2011). Heterointerface properties of novel hybrid solar cells consisting of transparent conductive polymers and III-nitride semiconductors, *Journal of Nonlinear Optical Physics & Materials*, Vol. 19, No. 4, (December 2010), pp. 703-711, ISSN 0218-8635

Matsuoka, T., Okamoto, H., Nakano, M., Harima, H., & Kurimoto, E. (2002). Optical bandgap energy of wurtzite InN, *Applied Physics Letters*, Vol. 81, No. 7, (June 2002), pp. 1246-1248, ISSN 0003-6951

Mott, N. F. (1939). The theory of crystal rectifiers, *Proceedings of the Royal Society*, Vol. 171, No. 944, (May 1939), pp. 27-38, ISSN 364-5021

Myers, D. R., Kurtz, S. R., Whitaker, C., & Townsend, T. (2000). Preliminary investigations of outdoor meteorological broadband and spectral conditions for evaluating photovoltaic modules and systems. *Program and Proceedings: NCPV Program Review Meeting 2000*, 16-19 April 2000, Denver, Colorado. BK-520-28064. Golden, CO: National Renewable Energy Laboratory; pp. 69-70; NREL Report No. CP-560-28187

Nakano, M., Makino, T., Tsukazaki, A., Kazunori U., Ohtomo, A., Fukumura, T., Yuji, H., Nishimoto, Y., Akasaka, S., Takamizu, D., Nakahara, K., Tanabe, T., Kamisawa, A., & Kawasaki M. (2007). Mg$_x$Zn$_x$O-based Schottky photodiode for highly color-selective ultraviolet light detection, *Applied Physics Letters*, Vol. 91, No. 14, (October 2007), pp. 142113-1-142113-3, ISSN 0003-6951

Nakano, M., Makino, T., Tsukazaki, A., Ueno K., Ohtomo, A., Fukumura, T., Yuji H., Akasaka, S., Tamura, K., Nakahara, K., Tanabe, T., Kamisawa, A., & Kawasaki, M. (2008). Transparent polymer Schottky contact for a high performance visible-blind ultraviolet photodiode based on ZnO, *Applied Physics Letters*, Vol. 93, No. 12, (December 2008), pp. 123309-1-123309-3, ISSN 0003-6951

Nakano, Y., Irokawa, Y., & Takeguchi M. (2008). Deep-level optical spectroscopy investigation of band gap states in AlGaN/GaN hetero-interfaces, *Applied Physics Express*, Vol. 1, No. 9, (August 2008), pp. 091101-1-091101-3, ISSN 1882-0778

Nakano, Y., Matsuki, N., Irokawa, Y., & Sumiya, M. (2010). Electrical characterization of n-GaN epilayers using transparent polyaniline Schottky contacts, *Physica Status Solidi C*, Vol. 7, No. 7-8, (April 2010), pp. 2007-2009, ISSN 1610-1634

Nakano, Y., Matsuki, N., Irokawa, Y., & Sumiya, M. (2011a). Deep-level characterization of n-GaN epitaxial layers using transparent conductive polyaniline Schottky contacts, *Japanese Journal of Applied Physics*, Vol. 50, No. 1, (January 2011), pp. 01AD02-1-01AD02-4, ISSN 0021-4922

Nakano, Y., Lozac'h, M., Matsuki, N., Sakoda, K., & Sumiya, M. (2011b). Photocapacitance spectroscopy study of deep-level defects in freestanding n-GaN substrates using transparent conductive polymer Schottky contacts, *Journal of Vacuum Science and Technology B*, Vol. 29, No. 2, (January 2011), pp. 023001-1-023001-4, ISSN 1071-1023

Satoh, K., Kakehi, Y., Okamoto, A., Murakami, S., Moriwaki, K., & Yotsuya, T. (2007). Electrical and optical properties of Al-doped ZnO–SnO$_2$ thin films deposited by RF magnetron sputtering, *Thin Solid Films*, Vol. 516, No. 17, (October 2007), pp. 5814-5817, ISSN 1616-301X

Schottky, W. (1939). Zur Halbleitertheorie der Sperrschicht- und Spitzengleichrichter (The semiconductor theory of the barrier layer rectifiers and tip rectifiers), *Zeitschrift für Physik*, Vol. 113, No. 5-6, (May 1939), pp. 367-414, ISSN 0939-7922

Shirakawa, H., Edwin L., MacDiamid, A. G., Chiang, C., & Heeger, A. (1977). Synthesis of electrically conducting organic polymers: halogen derivatives of polyacetylene, (CH), *Journal of the Chemical Society, Chemical Communications*, No. 16, (May 1977), pp. 578-580, ISSN 0022-4936

Shim, K.-H., Paek, M.-C., Lee, B. T., Kim C., & Kang J. Y. (2001). Preferential regrowth of indium–tin oxide (ITO) films deposited on GaN(0001) by rf-magnetron sputter, *Applied Physics A*, Vol. 72, No. 1, (February 2001), pp. 471-474, ISSN 0947-8396

da Silva, W. J., Hümmelgen, I. A., & Mello, R. M. Q. (2009). Sulfonated polyaniline/n-type silicon junctions, *Journal of Material Science: Materials in Electronics*, Vol. 20, No. 2, (February 2009), pp. 123-126, ISSN 0022-2461

Solomon, S., Qin, D., Manning, M., Chen, Z., Marquis, M., Averyt, K., Tignor M. B. M., & Miller Jr., H. L. (Ed.). (2007). *Climate Change 2007, The Physical Science Basis*, Contribution of Working Group I to the Fourth Assessment Report of the IPCC, ISBN 978 0521 88009-1

Song, J. O., Ha, J.-S., & Seong, T.-Y. (2010). Ohmic-contact technology for GaN-based light-emitting diodes: Role of p-type contact, *IEEE Transactions on Electron Devices*, Vol. 57, No. 1, (January 2010), pp. 42-59, ISSN 0018-9383

Sze, S. M. (1981). *Physics of Semiconductor Devices*, Wiley Interscience Publication, ISBN 0-471-05661-8, New York, USA

Takeuchi, T., Wetzel, C., Yamaguchi, S., Sakai, H., Amano, H., & Akasaki I. (1998). Determination of piezoelectric fields in strained GaInN quantum wells using the quantum-confined Stark effect. *Applied Physics Letters*, Vol. 73, No. 12, (July 1998), pp. 1691-1693, ISSN 0003-6951

Tracy, K. M., Hartlieb P. J., Einfeldt, S., Davis, R. F., Hurt, E. H., & Nemanich, R. J. (2003). Electrical and chemical characterization of the Schottky barrier formed between clean n-GaN (0001) surfaces and Pt, Au, and Ag, *Journal of Applied Physics*, Vol. 94, No. 6, (September 2003), pp. 3939-3948, ISSN 0021-8979

Tung, R. T. (2001). Recent advances in Schottky barrier concepts, *Material Science and Engineering B*, Vol. 35, No. 1-3, (November 2001), pp. 1-138, ISSN 0921-5107

Wang, W. & Schiff, E. A. (2007). Polyaniline on crystalline silicon heterojunction solar cells, *Applied Physics Letters*, Vol. 91, No. 13, (September 2007), pp. 133504-1-133504-3, ISSN 0003-6951

Wu, C.I. & Kahn, A. (1999). Electronic states and effective negative electron affinity at cesiated p-GaN surfaces, *Journal of Applied Physics*, Vol. 86, No. 6, (September 1999), pp. 3209-3212, ISSN 0021-8979

Wu, J., Walukiewicz, W., Shan, W., Yu, K. M., Ager, J. W., Li, S. X., Haller, E. E., Lu, H., & Schaff, W. J. (2003). Temperature dependence of the fundamental band gap of InN, *Journal of Applied Physics*, Vol. 94, No. 7, (July 2003), pp. 4457-1260, ISSN 1089-7550

Wu, J., Walukiewicz. W, Li., S. X., Armitage, R., Ho, J. C., Weber, E., R., Haller, E. E., Lu, H. Schaff, W. J., Barcz, A. & Jakieka R. (2004). Effects of electron concentration on the optical absorption edge of InN, *Applied Physics Letters*, Vol. 84, No. 15, (April 2004), pp. 2805-2807, ISSN 0003-6951

Wu, J. (2009). When group-III nitrides go infrared: New properties and perspectives, *Journal of Applied Physics*, Vol. 106, No. 1, (July 2009), pp. 011101-1-011101-28, ISSN 0021-8979

Yamaguchi, M. (2003). III–V compound multi-junction solar cells: present and future, *Solar Energy Materials & Solar Cells*, Vol. 75, No. 1-2, (April 2003), pp. 261-269, ISSN 0927-0248

Yamaura, J., Muraoka, Y., Yamauchi, T., Muramatsu, T., & Hiroi, Z. (2003). Ultraviolet light selective photodiode based on an organic-inorganic heterostructure, *Applied Physics Letters*, Vol. 83, No. 11, (July 2003), pp. 2097-2099, ISSN 0003-6951

Zheng X., Horng, R.-H., Wuu, D.-S., Chu, M.-T., Liao, W.-Y., Wu, M.-H., Lin, R.-M., & Lu, Y.-C. (2008). *Applied Physics Letters*, Vol. 93, No. 26, (December 1998), pp. 261108-1-261108-3, ISSN 0003-6951

Zhou, Y., Ahyi, C., Tin, C.-C., Williams, J., Park, M., Kim, D.-J., Cheng, A.-J., Wang D., Hanser, A., Edward, A. P., Williams, N. M., & Evans K. (2007). Fabrication and device characteristics of Schottky-type bulk GaN-based "visible-blind" ultraviolet photodetectors, *Applied Physics Letters*, Vol. 90, No. 12, (March 2007), pp. 121118-1-121118-3, ISSN 0003-6951

5

High Efficiency Solar Cells via Tuned Superlattice Structures: Beyond 42.2%

AC Varonides

Physics & Electrical Engineering Dept, University of Scranton, Scranton, PA, USA

1. Introduction

Modern PV devices are a direct outcome of solid state devices theory and applications of the last forty years. They are devices made of crystalline structures and basically, when illuminated with solar light, they convert solar photons into electric current. In the following a quick explanation of how this happens is presented. What is a solar cell? What is the basic function behind a cell's operation? Typically, in an illuminated p-n junction, photons are absorbed and electron-hole pairs are generated. These carriers diffuse in opposite directions (separated by the existing electrostatic field at the junction), and within their respective diffusion lengths. Electrons at the p-side diffuse through the junction potential and holes (similarly) get to the opposite directions. Under open-circuit conditions, the voltage across the cell is given by the following formula:

$$V_{oc} = kT \ln(1 + \frac{I_L}{I_o}) \tag{1}$$

Where k is Boltzmann's constant, T (in Kelvin) is the cell temperature, I_L is the light-generated current, and I_o is the p-n junction's reverse saturation current (see below). Cell theory and p-n junctions under a bias are briefly discussed in the next section.

2. Background theory: The p-n junction

Photonic device (solar cells included) operation is based on a p-n junction: two regions of the semiconducting material doped p and n type respectively and brought together in contact form a p-n junction. At thermal equilibrium, the p-n dope bulk semiconducting crystal, in order to keep its equilibrium, develops an internal field and develops its own built-in potential; the latter is total due to p- and n-type carrier migration across the junction.

Donor and acceptor atoms embedded in the lattice of the host material provide electrons and holes (as potential current carriers) that are free to wander in the crystal. In principle these carriers move randomly in the lattice, however, guiding these carriers accordingly could lead to non-zero currents coming off such semiconductors, and therefore to current producing devices. A semiconductor sample doped with donors and acceptors becomes a p-n junction and therefore a device with two regions tending to overlap at their boundary.

If the interface is at (say) x = 0 position, free electrons and free holes diffuse through the interface and inevitably form space charge regions as shown figure 1 below:

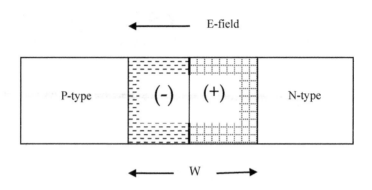

Fig. 1. pn-junction (e.g. of a Si sample) with the depletion W region shown: both sides of the interface are shown, with their space charge distributions respectively.

A static electric field develops at the interface (figure above) emanating from the (+) region and prohibiting respective carriers to further access the PN regions. From basic pn-junction theory, we can solve for the electric field and the potential developed by means of Poisson's equation. If the limits of the depletion region are $-x_p$ and x_n ($W = x_p + x_n$) respectively, we can derive expressions for both field and potential developed at the junction:

$$E(x) = \frac{qN_d}{\varepsilon}(x - x_n) \qquad 0 < x < x_n \tag{2}$$

$$E(x) = -\frac{qN_a}{\varepsilon}(x + x_p) \qquad -x < x < 0 \tag{3}$$

Where maximum field value is $E_{max} = E$(at x = 0) = $- (q\, N_d/\varepsilon)\, x_n$; q is the electronic charge, $N_{d,a}$ stands for donor and acceptor atom concentrations (per volume) respectively, ε is the total sample's dielectric constant (or the product of the relative times the free space dielectric constants, e.g. ε_r = 11.7 for Si). Based on expressions (2, 3) and on the fact that potential generated at the junction is the negative integral of the electric field across the depletion region, we can in principle derive the potential V(x) across the junction: it can be shown that V(x) is as follows:

$$V(x) = \frac{qN_a}{2\varepsilon}(x + x_p)^2 \; ; \text{In the p-region, and} \tag{4}$$

$$V(x) = -\frac{qN_d}{2\varepsilon}[(x - x_n)^2 - x_n^2] + \frac{qN_a}{2\varepsilon}x_p^2 \; ; \text{In the n-region and} \tag{5}$$

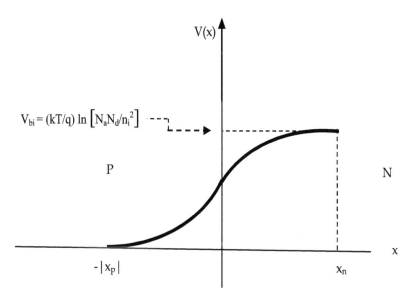

Fig. 2. Potential V as a function of x across the depletion region. Note (a) the two branches of V across both sides of the junction's boundary (x = 0) in accordance with (4) and (5) (b) the built-in voltage V_{bi} at the right edge of the junction [1, 2, 3]. Note also that built-in voltage is normally computed as shown in the inset.

It is a straightforward matter to produce explicit results about widths in the junction area (w, x_n, -x_p) in terms of device doping levels and built-in voltage values. The built-in voltage is determined from (4) at x = x_n:

$$V_{bi} = V(x = x_n) = \frac{q}{2\varepsilon}(N_a x_n^2 + N_d x_p^2) \tag{6}$$

3. Diode currents

The fundamental current equation for p-n junctions is derived based on considering that the built-in voltage is reduced down to $V_{bi} - V_a$, by the forward bias voltage V_a, helping majority carriers to escape and diffuse in the neighboring regions while, once electrons and holes reach the edges of the depletion region to the p and n regions, they diffuse accordingly according to a decaying exponential law of the type exp(x/$L_{n,p}$); the latter includes distance x and the diffusion length for electrons and/or holes respectively. Excess minority carriers *diffuse* in both regions according to the following expressions:

$$\delta p(x) = p_{no}(e^{x/Lp} - 1)e^{(x+x_n)/Lp} \tag{7}$$

$$\delta n(x) = n_{po}(e^{x/Lp} - 1) \tag{8}$$

Where p_{no} is holes in the n-region, L_p is the diffusion length of holes in the n-region, and where δp represents excess holes in the n-region. Diffusion currents can be calculated by means of the diffusion equation along with suitable boundary conditions:

$$J_p = -qD_p \frac{d\delta p(x)}{dx}\bigg| \quad (x = x_n) \tag{9}$$

$$Jn = qD_n \frac{d\delta p(x)}{dx}\bigg| \quad (x = -x_p) \tag{10}$$

Based on the above expressions, current density of the p-n junction due to a forward bias V_a is found to be as follows (see also (1)):

$$J = J_o(e^{V_a/V_t} - 1) \tag{11}$$

(Where V_t is the thermal voltage (kT/q))

4. p-n junctions as solar cells

Fundamentally, solar cell modeling correlates incident solar photon flux Φ_{ph} (# of photons cm^{-2} s^1) with generation and recombination carrier rates in the interior of the device. Photo-generated concentrations of diffusing carriers are typically modeled through the diffusion equation (under appropriate boundary conditions):

$$\frac{d^2\delta p_n}{dx^2} - \frac{\delta p_n}{L_p} + \alpha(1-R)\Phi_{ph}e^{-\alpha(x+d)} = 0 \tag{12}$$

Photon-collection efficiency is usually defined as the ratio of total current over solar photo-flux (cm^{-2} s^{-1}):

$$\eta_{col} = \frac{J_p + J_n - J_{rec}}{q\Phi_{ph}} \tag{13}$$

The numerator in (13) is total photo-induced current in the p and n-regions minus recombination current. Boundary conditions include continuity of carrier concentrations at the junction x (j), and the dependence of the first derivative of carrier concentration on recombination velocity s_p, at the edge of the window layer as shown in the figure below:

$$\left(\frac{d\delta p}{dx}\right)_{x=-d} = \frac{s_p}{D_p}(p(-d) - p_{no}) \cong \frac{s_p}{D_p}(p(-d)) \tag{14}$$

Figure-3 shows a generally accepted modeling geometry of a p-n junction solar cell. These two regions are separated by the depletion region (of thickness w): majority electrons from the n-region migrate to the p-region, and majority holes reciprocate from the latter region.

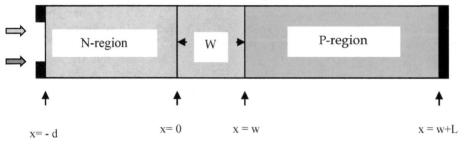

Fig. 3. Typical modeling geometry of a solar cell: w is the depletion width, J is the exact interface, L is the width of the p-region and d is the n-region (window layer). Note that the n-region is the window for solar photons.

Minority holes generated in the window layer (x from –d to 0) are:

$$\delta p_n(x) = C \exp(-x/L_p) + \frac{\alpha \Phi_{ph}(1-R)(L_p^2/D_p)}{1-(\alpha L_p)^2} e^{-\alpha(x+d)} \tag{15}$$

Note that at x = 0: $\delta p_n(0) = \dfrac{\alpha \Phi_{ph}(1-R)(L_p^2/D_p)}{1-(\alpha L_p)^2} e^{-\alpha(x+d)}$

Maximum hole- current density generated in the n-region is:

$$J_p(x=0) = -qD_p \times \begin{bmatrix} \dfrac{1}{L_p}\left(\dfrac{\alpha F_{ph}(1-R)(L_p^2/D_p)}{(\alpha L_p)^2 - 1}\right) \times \\[2ex] \left[\tanh(\dfrac{d}{L_p}) + \dfrac{\alpha L_p}{\cosh(d/L_p)} + \dfrac{s_p L_p}{D_p} \times \dfrac{p(-d)}{\cosh(d/L_p)}\right] + \alpha e^{-\alpha d} \end{bmatrix} \tag{16}$$

The surface recombination velocity s_n at the edge of the p-region is

$$-D_n(\frac{dn}{dx})_{x=L+W} = s_n(n_p - n_{po}) \tag{17}$$

The diffusion equation reads as follows:

$$\frac{d^2 n_p}{dx^2} - \frac{n_p - n_{po}}{L_n} + \alpha(1-R)\Phi_{ph}e^{-\alpha(x+d)} = 0 \tag{18}$$

Solution of (18) is of similar kind with (12) along with boundary conditions (17):

$$\delta n(x) = A\cosh(x/L_n) + B\sinh(x/L_n) + \frac{\alpha \Phi_{ph}(1-R)(L_n^2/D_n)}{1-(\alpha L_n)^2} e^{\alpha(x+d)} \tag{19}$$

The total current out of the cell is the sum of all currents minus recombination components from each region, especially recombination at the w region. Excess carriers in solar cells (as in any photonic device) are minority electrons and holes in the p and n regions respectively. When a cell is illuminated, solar photons excite electron hole pairs in all regions: the p-, n- and depletion regions. The latter may be become of great significance for the following reason: excited electrons and holes do split away from each other due to the existing electrostatic field. This means that these excess carriers will reach the edges of the depletion region in a very short time. Note also that typically, mean diffusion lengths of these carriers are much longer than t he actual width of the depletion area (even in pin devices). This makes the depletion region especially attractive for illumination: electrons and holes will separate from each other quickly, and they will diffuse in the bulk parts of the cell very fast assisted by the electrostatic field. In addition, space availability in the mid-region provides a chance for excess layer s that can be tuned to desired solar photons for subsequent absorption, thus enhancing device performance. This is why multi-layers are used in the intrinsic region (long depletion region in p-n junctions). If tuned quantum wells are grown somewhere in the middle, incident solar illumination will push electrons in the quantum wells and to tunneling or thermionic escape. The notion of additional band gaps integrated in the intrinsic region has been adopted successfully recently. For instance, successful cells with more than one band gaps have been designed and realized, where two or three cells are connected in series forming tandem cells with the advantage of voltage increase. This is possible due to the series connection of the tandem cells. Tandems provide excess voltage but they lack in current, in other words, due to the differences of the layers involved, current matching will be enforced due to the series connection. If these structures can ensure relatively high current outputs, then, along with increased voltage one should expect efficiency improvements. In the next we outline the behavior of a cell in tandem: top cell of AlAs/GaAs and bottom cell of a pin GaAs/Ge/Alloy for long wavelengths.

5. Heterojunction cells

Improved cell design has to include more than one band-gap for larger number of absorbed photons. P-i-n (from now on pin diode) diode designs offer wide intrinsic regions between the p- and n- regions of a p-n junction, where photo-carriers have a great chance to be generated and quickly swept away to the ends of the two-lead diode. This is possible due to the electrostatic field that develops at the depletion region. Illumination of the structure at the intrinsic region or a pin increases the chances of more photo-excited carriers. On the other hand, for a pin diode exposed to solar light and with a thin p-layer, minority electrons from the p-region may cross very fast ($\tau_n \sim$ fraction of μs) the junction at the p-i interface and be swept away to the load by the electric field in the mid-region of the cell. More than one band gaps in the mid region may lead to quantum wells where quantum size effects may take over as long as thickness values are in the order of 5 to 8 nm. Superlattice-like structures may be grown in the intrinsic region in order to accommodate both short and long solar wavelengths. It is commonly accepted that thin bulk window layers grown on top of a wide mid-region with quantum wells may offer a two-fold advantage (a) short wavelengths absorbed at the top and longer wavelengths absorbed in the mid region where a superlattice structure is essentially tuned at specific wavelengths. Thus, by growing a superlattice in the middle of a pin region (rather by changing the mid-region into a multi-quantum well (mqw) sequence) one may reach the main objective: to capture more solar photons with energies higher than the band gap of the host material. The figure below depicts an intrinsic multi-quantum well area, where discrete energy levels cause a widening of the host material's gap (commonly GaAs with gap at 1.42 eV).

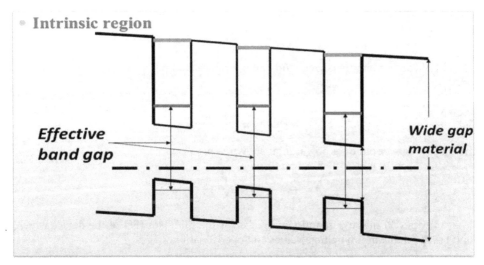

Fig. 4. Detail from a superlattice structure (typically GaAs/alloy and GaAs/Ge (as proposed in this study). The dashed line represents the Fermi level at thermal equilibrium. The optical gap can be tuned to desired energy values.

Figure 5 shows a superlattice covering the mid region of a pin cell:

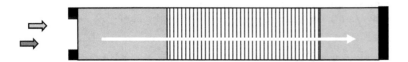

Fig. 5. A p-i-n GaAs/alloy superlattice developed in the mid-region of a pin cell: the middle section depicts: the top layer (blue) is the wide gap alloy (e.g. AlAs) and the bottom layer is GaAs (host material); GaAs is also grown in the superlattice as the low gap medium.

Such a cell design (shown above) is an expanded p-n junction with a wide superlattice mid-region occupying the intrinsic or low doped region between p and n. To reduce cost such a structure can be compromised by inserting a short period tuned superlattice as a small percentage of the device as a total.

6. Top cell (AlAs/GaAs)

Modeling of the top region may be performed in two ways, by considering the equivalent circuit of the device and/or by solving for excess carriers and subsequent electric currents and current densities in the solid state. In this brief outline we are considering the first approach by starting from the basic illuminated diode equation and adopting standard results regarding maximum power, short circuit current and open-circuit voltage values. Starting from the fundamental solar cell equation, we can derive maximum power conditions:

$$I_m = \frac{\beta V_m}{1 + \beta V_m} I_L \tag{20}$$

Where I_m, V_m are maximum current and voltage values, and where $\beta = q\,(kT)^{-1}$.
And

$$V_m = V_{oc} - \frac{1}{\beta} \ln(1 + \beta V_m) \tag{21}$$

Efficiency (as power out over power in) is shown to be:

$$\eta = \frac{P_o}{P_{in}} = \frac{I_L V_{oc}}{P_{in}} \left[1 - (\frac{V_m}{V_{oc}}) \frac{\ln(\beta V_m)}{\beta V_m} \right] \tag{22}$$

It is clear from (22) that the quantity in brackets is the fill factor (FF) of the device which is found based on maximum voltage values and open circuit voltage:

$$FF = 1 - \frac{\ln(\frac{V_m}{kT})}{(\frac{V_{oc}}{kT})} = 1 - \frac{\ln(\beta V_m)}{\beta V_{oc}} \tag{23}$$

Highly efficient solar cells have been found to have open-circuit voltages within a range from 1 to 1.08V. The table below indicates how open circuit voltage controls maximum voltage (voltage at maximum power point). Assuming short circuit current at 30mA/cm², under one-sun (100mW/cm²), the efficiency is depicted below by Table 1:

V_{oc}(V)	V_m(V)	FF (%)	η (%)
1.02	0.926	0.907	27.75
1.03	0.938	0.906	27.99
1.04	0.948	0.905	28.23
1.05	0.958	0.904	28.50

Table 1. Open circuit and maximum voltages, Fill Factor (FF) and collection efficiency (300 Kelvin). The cell is the top AlAs/GaAs that serves as a window to the solar flux.

It is of advantage to suggest an undoped GaAs-Ge multi-quantum well (MQW) in a standard pin-design, namely, p-intrinsic (MQW)-n geometry that includes lattice-matched GaAs and Ge layers in the intrinsic region of the PV device. This formation could offer the advantage of 1eV absorption (at the appropriate quantum well width), without compromises in device transport properties, such as mobility or conductivity. GaAs-based layers provide (a) high mobility and absorption values and (b) a chance for fine-tuning of the optical gap with specific solar photon wavelength. Recently, high efficiency cell designs have been proposed where two pn cells are grown in tandem (series connection), where the top cell is a bulk (e.g. GaAs/AlAs cell) and the bottom is a superlattice-based pin cell optimized at long wavelengths. Such devices offer efficiency increase by acting simultaneously: top unit near 20% and bottom unit near 30% (when they operate on their own) lead to structures (quantum cells) with overall efficiencies in excess of 35% (under one sun and with recombination effects and scattering included). As seen in Figure 5 below, the superlattice approach offers a tool for capturing solar photons at desired wavelengths with the appropriate quantum mechanical tuning. In other words, ground eigen-states in quantum wells match specific wavelengths (corresponding to photons with the same energy);

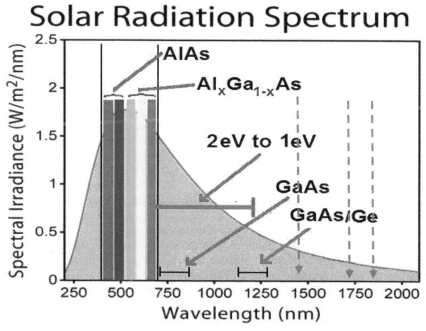

Fig. 6. Regions of the solar spectrum covered by the superlattice cell and the top cell (visible). The superlattice can be tuned at ~1 eV. Dashed arrows indicate region of feasible absorption from the superlattice region

As seen from the figure above, almost full spectrum absorption can be achieved with materials that absorb at desired photon energies. Specifically, visible photons may be absorbed by means of a GaAs/AlAs bulk cell, and IR radiation absorption can be achieved via GaAs (1.42 eV) and Ge (0.67 eV) respectively with superlattice or superlattice sections tuned at desired wavelengths. The n-region of the pin cell can be selected to be Ge in the bulk, ensuring absorption at the tail of the solar spectrum (for Ge: wavelength absorbed at $\lambda = 1.24/0.67 = 1.85$ µm, see last arrow in the figure above). How is the current formed in the superlattice layer? The answer hides in the quantum nature of this region: quantum wells quantized the energy of the captured electrons (and light and heavy holes in the valence band); photo-excited electrons escape thermionically from the wells and form excess current in the conduction band. On the other hand, incident IR photons are expected to be absorbed in the MQW area. Projected excess carrier population (electrons with recombination included) is of the order of 10^{12} to 10^{13} cm^{-2} per eigen-state. Thermionic current density values have been found to be near order of 30mA/cm^2 and open-circuit voltage values above 1V, at one sun. Overall (for a composite cell see figure 3) collection efficiency values are initially projected well in excess of 35%, which is a key for immediate improvement to even higher collection efficiency. Total current density is dominated by the lowest of the two sub-cell currents, and open-circuit voltage values are the sum of the two sub-cell V_{oc} values. Total current from the bottom cell is the sum of thermionic and nearest neighbor hopping currents. Preliminary results reach estimates of efficiencies from each of the two (lattice-matched) sub-cells in excess of 21% per cell (predicted synergy of the two sub-cells in excess of 40%). Loss mechanisms at interfaces and quantum wells and their role in overall efficiency determination will also be included. Advantages of the design are:

i.   Solar spectrum matching in both visible and IR ranges through layer band gap-matching selection.
ii. Lattice-matching
iii. Increased carrier transport due to GaAs. It is conceivable that even the 40%-plus target of conversion efficiency can be reached with such designs

Heterostructure and (most recently) multijunction solar devices exhibit better performance in transport properties, when compared to bulk solar cells: especially in quantum well devices, photo-excitation causes carrier accumulation in discrete energy levels, with subsequent escape to the conduction band (minus recombination losses) via standard mechanisms such as tunneling, thermal escape or nearest neighbor hopping conduction. Full spectrum absorption and triple junction solar cells have become key factors for high efficiency collection in PV structures of various geometries. Most recently, successful photovoltaic device (PV) designs have shown high efficiency values well above 30%, and efficiency levels in excess of 40% have been reached by means of triple junction metamorphic solar cells and under high sun concentration (good candidate for concentrated PV or CPV). Multijunction solar cells offer a great advantage over their bulk counterparts: by incorporating lattice-matched alloys, one may succeed in designing a device with more than one energy gaps thus increasing the number of absorbed solar photons. During the last decade, various groups have modeled and developed *multijunction* solar cells in order to increase overall collection efficiencies. Emphasis has been given in two types of PV devices (a) lattice-matched solar cells and (b) metamorphic (lattice-mismatched) solar cells. In particular, III-V multijunction solar cells have shown the greatest progress in overall efficiency. The broader impact of this project is a new design proposal for high efficiency solar cells. The target is to exceed 45% collection efficiency for very efficient photovoltaic devices. It is more than clear that once such a cell is realized, the field of concentration photovoltaics (CPV) will benefit greatly: solar cells with (a) record high efficiency values (b) under several hundred suns (Fresnel optics at 500+ suns) and (c) small in size (low area hence less material) is already attracting interest for mass production in many places in the world. In recent years, it has been proposed by us a new design for a high efficiency and lattice-matched solar cell (HESC), where both visible and infrared portions of the solar spectrum are absorbed according to the structure's geometric material arrangement: simultaneous absorption of both short and long wavelengths. In this on-going research enterprise, the synergy between a highly efficient triple junction cell and a highly efficient superlattice or a multi-quantum well region, is presented as a new and innovative way for further efficiency increase. It is well established by now, that triple junction solar cells are exceeding the upper threshold of collection efficiency to ever higher levels, namely above 38% with latest threshold at 41.1% (Fraunhofer Institute, Germany). Currently, a cell that will operate above the 40% threshold is in target, with ultimate target the efficiency at or near 50%. The cell design is based on a p-i-n bulk device model with three distinct areas, two of which are complete PV-heterostructures on their own; in other words, these two regions could *stand alone* as *two independent solar cell structures with quite acceptable performance* (of the order of 21% and more as it has been demonstrated by our group recently). The power output of the PV composite device is a function of the individual power outputs from each sub-cell in the PV unit. On the other hand, triple junction solar cells seem to lead the way to high efficiency photovoltaics especially in the area of concentrated photovoltaics (CPV), where small cell area and therefore less material (hence lower material costs) may lead to high PV performance. The latter are triple junctions of lattice-matched and non-lattice matched III-V heterostructures with two tunnel junctions between the layers.

7. Suggestions for modeling

Fully develop a theoretical model of PV composite PV devices by first principle calculations and computations based on realistic device parameters; propose a composite PV structure with two major cells: a triple junction and multi-layer tuned cell, with the prospect of high efficiency near 50%. Modeling tools include several established math software packages. Seek for a composite photovoltaic device that combines properties of direct-gap crystalline semiconductors and absorption in the entire spectrum, mainly in the visible and in the infrared (NIR/IR) wavelength ranges, and which is configured as a two-part solar cell: a top triple junction and a multi-layer p-i-n bottom unit tailored to IR infrared wavelengths. The solar spectrum (a 6,000 °K, see in Figure 5) offers the option of finding suitable band gaps for highest absorption. Material selection shows a blue shift in the absorption via wide gap materials as shown (AlAs). Low gap materials offer wavelength matching in the IR range (note the dashed arrows indicating optical gaps corresponding to various wavelengths. It is of advantage to exploit quantum wells grown on n-type or low-doped substrates.

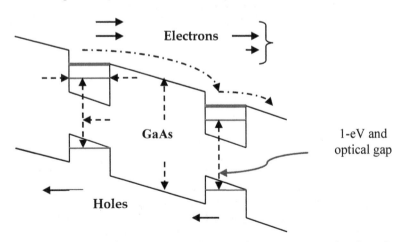

Fig. 7. Tuned quantum wells at 1eV solar photons: shown are energy levels and optical gap increase

Superlattice structures in both cases mentioned above are at the designer's disposal, in the sense that appropriate quantum well geometries may lead to desired solar photons absorption. Enhancement of cell performance can be achieved by replacing the intrinsic region with *tuned* multi-quantum well (MQW) layers, designed for specific wavelengths. Thermionic emission, hopping conduction and tunneling are dominant mechanisms of photo-carrier transport in heterostructures (against losses due to recombination processes). Photo-excited layers thermally escape from quantum wells (minus recombination losses): incident solar photons typically generate 10^{12} to 10^{13} net photo-excited carriers per unit area (cm^2), after recombination effects have been taken into account. This population is expected to migrate to the conduction band assisted by the escape mechanisms named above and the built-in electrostatic field in the p-i-n region.

By selecting suitable geometry of the quantum wells that leads to one or two energy levels in the quantum wells, ground state of electron-hole pairs at 1eV may be formed, and a second state at the very edge of the GaAs layer conduction band (see Fig 2): this event has been shown to act in favor of nearest neighboring hopping electrons from site to site (QW). Thus a three-fold advantage of the superlattice/MQW region is that (1) excess (in addition to carriers from the bulk part of the device) carriers are trapped and thermally escape to the conduction band and (2) nearest neighbor hopping conduction (NNH) becomes a second conduction mechanism and (3) band gaps of other materials may be represented via energy levels in quantum wells. The total current from the intrinsic region will be the sum of the thermionic and the NNH current components (minus recombination losses). Subsequent well width selection may lead to further refinement of solar photon absorption. Near infrared and infrared portions of the solar spectrum can be covered by suitable width selections, with equal amount of modeling effort (from the point of view of computations, it is a mere change of parameters for slightly different optical gaps). It is also interesting to note at this point that quantum well width could be modeled as a random variable, leading to a random distribution of optical gap values (as function of well width) and hence a smeared distribution of optical gap values and absorbed photon wavelengths, for the benefit of the photovoltaic device. Thus, IR photon absorption in the neighborhood of 1eV is feasible. In addition, the superiority of transport properties of the proposed quantum-PV device should be noted compared to its III-N-V "high" efficiency counterpart: our proposed superlattice cell is mainly a GaAs device perturbed by thin Ge layers, and therefore this region exhibits much higher electron mobility. In the absence of tunneling (thick potential barriers) total currents are in essence the sum of (a) bulk currents from the mainly bulk pin device (b) thermionic and (c) hopping current components, due to free electrons in the GaAs conduction band, assisted by the overall electrostatic field in the intrinsic region. Recent modeling and simulation have shown that the top cell retains visible absorption (AlAs/ (Al) GaAs/GaAs at ~21%) or to include a highly efficient triple junction cell (in this proposal, our own choice (InP/GaAs/GaAs at 30% efficiency). The bottom multi-well cell operates at longer wavelengths (1eV or 1,240nm), and therefore the whole of the unit absorbs in both regimes visible and IR respectively. Since germanium and gallium arsenide layers are lattice-matched, it is conceivable that a superlattice would fit in between the p- and n-regions of the device. Advantages of such designs are summarized below:

1.   less material to grow
2.   small area (exposed to sun-light cell aperture)

3. less complexity in the structure overall
4. reduced scattering of drifting and diffusing carriers
5. reduced carrier trapping and recombination (carriers in MQW region separate from their corresponding holes as being away from the quantum wells)
6. faster growth conditions attainable
7. lower fabrication costs

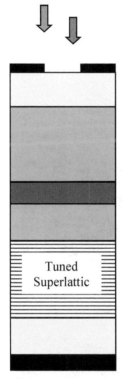

Fig. 8. Proposed cell structure: top cell p-n junction, tunnel junction (TJ) (purple) and p-i-n bottom cell with superlattice in the middle; P region (green), N region (yellow). Top cell is the window facing the sun (anti-reflected coating and surface texturing not shown).

As seen from the figure above, there are several options for further design and optimization (a) top region offers the possibility of another superlattice tuned at selected wavelengths (b) layers and alloys other than GaAs can be used (in the lattice-matched fashion) (c) tuned superlattice (bottom cell) can be split is more than one narrow units tuned at desired solar spectrum peaks (d) cell can be of small area (less material used) or of large area for higher exposure.

8. Some thoughts on concentrated photovoltaics (CPV)

Concentrated light on small solar cells can become of great advantage: a small size cell (~2 mm² area) may be placed at the focal point of a Fresnel optical system. Concentrated light causes higher carrier absorption from the bulk of the device and therefore higher current generation. Currently III-V multijunction cells have shown to have the highest collection efficiency. Efficiency $\eta(\%)$ increases logarithmically with solar power up to about 500 suns (one sun = 100mW/cm^2). Currently, it seems that CPV cells show the highest efficiency (consistently above 38%) with latest record efficiencies at 41.1% (Fraunhofer Institute). As it can be seen from equation (11) and the efficiency expression: $\eta(\%) = (V_{oc} J_{sc} FF)/P_{in}$, the efficiency of a solar cell increases logarithmically with J_{sc}. Such a behavior has been observed, in fact, $\eta(\%)$ increases with increasing current generation (maximum value in the neighborhood of 550 suns).

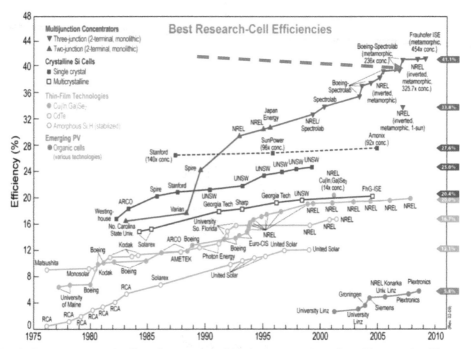

Fig. 9. Current status of cell performance and improvement since the mid seventies. Note that MJ cells have taken the lead in the high efficiency race. Latest (2010) results: 41.1% collection efficiency (Fraunhofer Institute at 454 suns) [© 2009 Spectrolab, Inc. All rights reserved].

As seen from the figure above, multijunction cells, with more than one band gaps, take the lead in current and voltage production (recall that efficiency varies with open-circuit voltage and short-circuit current). CPV systems have given a boost of solar power production globally because they combine (a) highly efficient cells with small exposure area and (b) less costly optical system and components. As of 2009, CPV systems operate at 28 – 30% total efficiency (cell plus optics) and seem to be coming dynamically in the global PV market.

9. Conclusions: The immediate future

Photovoltaics is the child of progress in condensed matter physics, and has matured to the point that solar energy has been competing with fossil fuel energy sources. Small in size highly efficient solar cells are the answer for our future energy needs. The 40% threshold has already been reached and current research shows that 50% photovoltaics will soon be a reality. It seems that such high conversion will be succeeded by means of small size highly efficient solar cells. By small size we mean from the mm level down to nano-sized PV particles mounted to Fresnel-type optical systems with high solar concentration. Global energy production based on high efficiency PV will solve the energy needs of all nations and will slow down planet pollution. No nuclear waste and zero chance for accidents will guide common sense in immediate future. The concept of tuned superlattices was outlined and its advantages have been presented. Well-understood and lattice-matched materials, such as GaAs/Alloy and Ge, along with improved growth techniques pave the way to high efficiency photovoltaic devices. Integrated circuit techniques are also available for cells of minute size (e.g. 5 mm²), which is a dramatic reduction of material and hence of cost. Reduced size photovoltaic cells, under high solar concentration (currently from 450 to 500 suns), have opened the avenue for a competitive PV industry in the near future. Concentrated Photovoltaics (CPV-farms) will eventually dominate the world energy production. PV system price range has been steadily reducing from $0.40/KWh (mid-1990's) to mere $0.20/KWh in 2008. Market penetration of the PV industry increases steadily (under 1GW in the US to 6GW by the year 2015). It is expected that the average KWh will be ~10 cents by or before 2015, with a steady GW plant installation. High efficiency solar cells (~50%) and parallel optical system advancement (total system at 30%), will lead to a very strong PV industry, for the benefit of all.

Current modeling of the top structure has indicated top efficiency values in excess of 21% (power out vs. power in) while for the bottom cell preliminary calculations indicate collection efficiency in excess of 25%. The bottom cell is a GaAs-superlattice-Ge structure, where quantum size effects occur. Photo-excited carriers in the middle region are electrons trapped in quantum wells (thin germanium layers sandwiched by gallium arsenide layers). Thin Ge layers (20 nm) are tuned at 1eV. They act as quantum traps and confine electrons in a discrete set of energy levels (one or two at the most). From these traps photo-electrons escape to the conduction band (minus the lost ones). Some advantages of our design over other high-efficiency full-spectrum solar cells are: (a) No excess tunnel junctions are needed to connect the cells (b) The superlattice region includes germanium layers tuned to absorb photons near 1eV (or more, depending on the quantum well thickness) (c) High mobility of carriers in both cells (top, bottom); the latter is a direct advantage over existing III-N-V *high efficiency* competing (nitrogen based) solar cell structures (d) Perfect lattice matching among the layers (e) Parallel carrier transport via (i) tunneling (ii) hopping and (iii) thermionic carrier escape. In the case at hand, tunneling is not a part of the action; instead thermionic emission currents are of importance. Maximum efficiency over 40% is expected via the synergistic action of the two cells.

10. References

[1] SM Sze, High-Speed Semiconductor Devices, John Wiley and Sons, 1990
[2] M Yamaguchi, Solar Energy Materials & Solar Cells 90 (2006) 3068–3077
[3] M Yamaguchi et al, Solar Energy 82, 173 (2008)
[4] R. King, et al, 20th European PVSEC, 2005
[5] K Nishioka, Solar Energy Materials & Solar Cells 90 (2006) 1308–1321, RR King, Nature Photonics, May 2008

[6] Es Yang, Microelectronic Devices, McGraw-Hill, 1988

[7] JF Geisz, S Kurz, MW Wanlass, JS Ward, A Duda, DJ Friedman, JM Olson, WE McMahon, TE Moriarty, and JT Kiehl, Applied Phys. Letters 91, 023502 (2007)

[8] T Kirchartz, Uwe Rau, et al, Appl. Phys Lett. 92, 123502 (2008)

[9] T Mei, Journal of Appl Phys 102, 053708 (2007)

[10] AC Varonides and RA Spalletta, Physica Stat. Sol. 5, No. 2 441 (2008)

[11] GFX Strobl at al, Proc. 7th European Space Power Conference, 9-13 May, Italy, 2005

[12] H. L. Cotal, D. R. Lillington, J. H. Ermer, R. R. King, S.R. Kurtz, D. J. Friedman, J. M. Olson, et al, 28th IEEE PVSC, 2000, p. 955

[13] T Kieliba, S Riepe, W Warta, Journal of Appl. Phys. 100, 093708 (2006)

[14] JF Geisz and DJ Friedman, Semicond. Sci. Technol. 17, 789 (2002)

[15] W Hant, IEEE Trans Electron Devices, VOL ED-26, NO 10, 1573 (1979)

[16] AC Varonides, Physics E 14, 142 (2002)

[17] E Istrate, EH Sargent, Rev Mod Phys, 78, 455 (2006)

[18] H E Runda et al, Nanoscale Res Lett (2006) 1:99

[19] W Li, BE Kardynal, et al, Appl. Phys Lett 93, 153503 (2008)

[20] AC Varonides, Thin Solid Films, Vol. 511-512, July 2006, pp 89-92

[21] BL Stein and ET Yu, Appl. Phys. Lett. 70 (25), 23 June, 1997

[22] AC Varonides, RA Spalletta, WA Berger, WREC-X and Exhibition, 19-25 July 2008, Glasgow, Scotland, UK.

[22] R. King, *Multijunction Cells*, Industry Perspective, Technology Focus, Nature photonics | VOL 2 | MAY 2008 | www.nature.com/naturephotonics

[23] T Kirchartz, BE Pieters, K Taretto, U Rau, Journal of Appl. Physics 104, 094513 (2008)

[24] AC Varonides and RA Spalletta, Thin Solid Films 516, 6729-6733 (2008)

[25] K Jandieri, S D Baranovskii, W Stolz, F Gebhard, W Guter,

[33] M Hermle and A W Bett, J. Phys. D: Appl. Phys. 42 (2009) 155101

[34] R Jones, CPV Summit, Spain, 2009

6

Photons as Working Body of Solar Engines

V.I. Laptev[1] and H. Khlyap[2]
[1]Russian New University,
[2]Kaiserslautern University,
[1]Russian Federation
[2]Germany

1. Introduction

Models of solar cells are constructed using the concepts of band theory and thermodynamic principles. The former have been most extensively used in calculations of the efficiency of solar cells (Luque & Marti, 2003; Badesku et al., 2001; De Vos et al., 1993, 1985; Landsberg & Tonge, 1989, 1980; Leff, 1987). Thermodynamic description is performed by two methods. In one of these, balance equations for energy and entropy fluxes are used, whereas the second (the method of cycles) comes to solutions of balance equations (Landsberg & Leff, 1989; Novikov, 1958; Rubin, 1979; De Vos, 1992).Conditions are sought under which energy exchange between radiation and substance produces as much work as possible. Work is maximum when the process is quasi-static. No equilibrium between substance and radiation is, however, attained in solar cells. We therefore believe that the search for continuous sequences of equilibrium states in solar energy conversion, which is not quasi-static on the whole, and an analysis of these states as separate processes aimed at improving the efficiency of solar cells is a problem of current interest. Examples of such use of the maximum work principle have not been found in the literature on radiant energy conversion (Luque & Marti, 2003; Badesku et al., 2001; De Vos et al., 1993, 1992, 1985; Landsberg & Tonge, 1989, 1980; Leff, 1987; Novikov, 1958; Rubin, 1979).

2. Theory of radiant energy conversion into work

2.1 Using model for converting radiant energy into work

We use the model of solar energy conversion (De Vos, 1985) shown in Fig. 1. The absorber of thermal radiation is blackbody 1 with temperature T_A. The blackbody is situated in the center of spherical cavity 2 with mirror walls and lens 3 used to achieve the highest radiation concentration on the black surface by optical methods. Heat absorber 4 with temperature $T_0 < T_A$ is in contact with the blackbody.

The filling of cavity 2 with solar radiation is controlled by moving mirror 5. If the mirror is in the position shown in Fig. 1, the cavity contains two radiations with temperatures T_A and T_S. If the mirror prevents access by solar radiation, the cavity contains radiation from blackbody 1 only. Radiations in excess of these two are not considered. In this model, solar energy conversion occurs at $T_0 = 300$ and $T_S = 5800$ K. The temperature of the blackbody is $T_A = 320$ K.

2.2 Energy exchange between radiation and matter
2.2.1 Energy conversion without work production

It is known that the solar radiation in cavity 2 with volume V has energy $U_S = \sigma V T_S^4$ and entropy $S_s = 4\sigma V T_S^3/3$, where σ is the Stefan-Boltzmann constant (Bazarov, 1964). The black body absorbs the radiation and emits radiation with energy $U_A = \sigma V T_A^4$ in cavity 2. If $T_A = 320$ K, these energies stand in a ratio of $U_S/U_A = (T_S/T_A)^4 \approx 10^6$, while $S_S/S_A = (T_S/T_A)^3 \approx 6 \times 10^3$. As the volumes of radiations are equal, the amount of evolved heat ΔQ is proportional to the difference $T_A^4 - T_S^4$ and is equal to the area under the isochore st on the entropy diagram drawn on the plane formed by the temperature (T) and entropy (S) axes in Fig. 2. The solar energy U_S entering the cavity and heat ΔQ are in ratio:

$$\eta_U = \Delta Q/U_S = (U_S - U_A)/U_S = 1 - (T_A/T_S)^4. \tag{1}$$

Our model considers the value η_U as an efficiency of the photon reemission for a black body if the radiation and matter do not perform work in this process.

One should note that the efficiency of the photon absorption can be defined as (Wuerfel, 2005)

$$\eta_{abs} = 1 - (\Omega_{emit}/\Omega_{abs})(T_A/T_S)^4,$$

where Ω is a solid angle for the incident or emitted radiation. In our case, the ratio $\Omega_{emit}/\Omega_{abs}$ can be ignored because value of η_U is close to one, for $(T_A/T_S)^4 = (320/5800)^4 \approx 10^{-5}$. Thereafter we have to assume that $\eta_U = \eta_{abs}$. The consequence is that solar energy can be almost completely transmitted to the absorber as heat if no work is done. Then a part of evolved heat ΔQ can be tranformed into work.

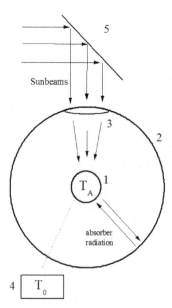

Fig. 1. Model of solar energy conversion from (Landsberg, 1978). Designations: 1. black body, 2. spherical cavity, 3. lens, 4. heat receiver, 5. movable mirror added by the author.

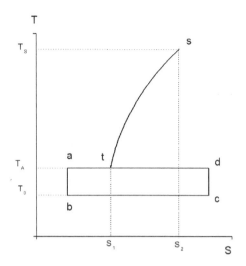

Fig. 2. Entropy diagram showing isochoric cooling of radiation (line st) in the cavity 2. The amount of evolved radiant heat is proportional to the area sts_1s_s. The amount of radiant heat converted into work is proportional to the area abcd. The work is performed by matter in a heat engine during Carnot cycle abcd.

2.2.2 Work production during the Carnot cycles

The absorbed radiant heat is converted into work by Carnot cycles involving matter as a working body. One such cycle is the rectangle abcd in Fig. 2. Work is performed during this cycle with an efficiency of

$$\eta_0 = 1 - T_0/T_A = 0.0625 \qquad (2)$$

between the limit temperatures T_0=300 K and T_A=320 K.

It is common to say that the matter is the working body in this cycle. But radiation can be involved in the isothermic process ad in Fig. 2, because the efficiency of a Carnot cycle does not depend on kind and state of the working body. We will not discuss the properties of a matter-radiant working body. Let us simply note that a matter-radiant working body is possible. In this case upper limit of temperature T_A can reach 5800 K. The curve AB on Fig. 3 is the efficiency of this Carnot cycle where the matter cools down and heats up between temperatures T_A, T_0 and radiation has temperature T_A. In this cyclic process the matter and radiation are in equilibrium.

Let us show the absorption of radiation on an entropy diagram (Fig. 4) as an isothermal transfer of radiation from the volume V_2 of the cavity (state s) to the volume V_1 of the black body (state p). One can even reduce the radiation to state p* in Fig. 4. We will not discuss the properties of points p and p* here. Let us simply note that radiation reaches heat equilibrium with the black body (state e) from these points either through the adiabatic process p*e or through the isochoric process pe.

Let us represent the emission of radiation as its transfer from the volume of the black body (state e) to the volume V_2 of the cavity along the isotherm T_A (state t). As the radiation fills the cavity, it performs a work equal to the difference between the evolved and absorbed heat. The radiation performs a considerable work if it reaches state t* on Fig. 3. Our calculations show that work is performed along the path sp*et* with an efficiency of

$$\eta_C = 1 - T_A/T_S = 0.945 \tag{3}$$

when $T_A = 320$ K. It is important to note that, when radiation returns to its initial state s along the adiabat t*s, it constitutes a Carnot cycle with the same efficiency η_C.

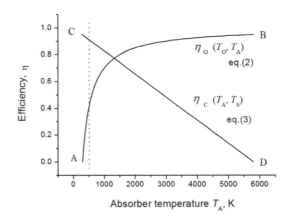

Fig. 3. Efficiencies of Carnot cycles in which the radiation takes place. The efficiency η_0 of work of radiation and matter in a cycle with temperatures limited at T_0 and T_A is shown as curve AB. Line CD shows the efficiency of a cyclic process where work is performed by radiation only.

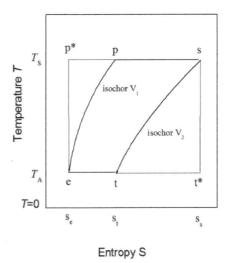

Fig. 4. Entropy diagram showing some thermodynamic cycles for conversion of solar heat into work in cavity 2 with the participation of a black body. Isotherms represent the absorption and emission of radiant energy. Lines pe, p*e correspond to the cooling of radiation in the black body. Line st indicates the temperature and entropy of radiation in cavity.

Fig. 3 compares work efficiencies η_0 and η_C during Carnot cycles described above. Radiation performs work during the Carnot cycle with a greater efficiency than η_0. η_0 and η_C values are calculated from Eqs. (2),(3) as a function of temperature T_A. We see that the efficiency η_C of conversion of heat into work in process with radiation only decreases with increasing temperature of the absorber, but the work efficiency η_0 of matter and radiation increases. Efficiencies are equal to 0.77 at T_A = 1330K.

Fig. 3 is divided in two parts by an isotherm at 500 K. On the left side is the region with temperatures where solar cells are used. The efficiency there of conversion of heat into work can reach value of 0.39 for a Carnot cycle with matter and be above 0.91 during a Carnot cycle with radiation. It is important to note, that other reversible and irrevesible cycles between these limit temperatures have efficiency smaller than efficiencies η_C or η_0.

The efficiency of parallel work done by radiation in the Carnot cycle sp*et*s (Fig. 4) and the matter in the Carnot cycle abcda (Fig. 3) is $\eta_0\eta_C$. It follows from (2) and (3) that

$$\eta_0 \eta_C = (1-T_0/T_A)(1-T_A/T_S) = 0.0591.$$

After mathematical operations, it takes the form

$$\eta_0 \eta_C = \eta_0+\eta_C -(1-T_0/T_S)= \eta_0+\eta_C-\eta_{OS},$$

where

$$\eta_{OS} = (1-T_0/T_S) \tag{4}$$

is the efficiency of the Carnot cycle in which the isotherm T_S corresponds to radiation and isotherm T_0, to the matter. Efficiency η_{OS} is independed from an absorber temperature T_A which divides adiabates in two parts. Upper parts of adiabates correspond to the change of radiation temperature, bottom parts to that of matter. It is important that such a Carnot cycle allows us to treat radiant heat absorption and emission as an isothermal and adiabatic processes performed by the matter. Efficiency η_{OS} is limiting for solar-heat engine. It is equal to 0.948 for limit temperatures T_0=300 K and T_S=5800 K. This Carnot cycle is not described in literature.

2.2.3 Work production during unlike Carnot cycles

Solar energy is converted as a result of a combination of different processes. Their mechanisms are mostly unknown. For this reason, one tries to establish the temperature dependence of the limit efficiency of a reversible combined process with the help of balance equations for energy and entropy flows. For solar engine, it takes the form (Landsberg, 1980; 1978)

$$\eta_{AS} = 1 - 4T_A/3T_S + T_A^4/3T_S^4. \tag{5}$$

For example, η_{AS} = 0.926 when T_A=320 K. Fig. 5 compares work efficiencies η_{AS} and η_C during the cycles with radiant working body at the same limit temperatures. η_{AS} and η_C values are calculated from Eqs. (3),(5) as a function of absorber temperature T_A. We see that $\eta_{AS} < \eta_C$, that is not presenting controversy to the Carnot theorem. The efficiencies η_{AS} and η_C of conversion of radiant heat into work decreases with increasing absorber temperature. The maximum difference $\eta_C-\eta_{AS}$ is approx. 18% when T_A=3500 K (Landsberg, 1980; 1978).

The maximal value of the efficiency if for a black body at a temperature $T_A < T_S$ were possible to absorb the radiation from the sun without creating entropy is shown in (Wuerfel,

2005). It follows from a balance of absorbed and emitted energy and entropy flows under the condition of reversibility. The efficiency of a reversible process in which radiation and matter perform work is equal to

$$\eta_L = 1-(T_A/T_S)^4 - 4T_0 [1 - (T_A/T_S)^3]/3T_S. \tag{6}$$

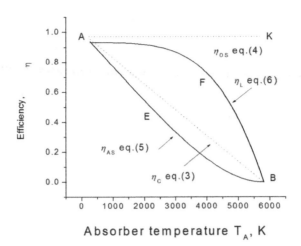

Absorber temperature T_A, K

Fig. 5. Consideration of Carnot efficiencies and efficiencies of reversible processes other than the Carnot cycle. Dot lines AB,AK denote Carnot efficiencies η_C and η_{OS} at lower limit temperatures 320 and 300 K respectively. Solid curves AFB and AEB are efficiencies η_L and η_{AS} of non-Carnot engines at the same limit temperatures. Line AK and curve AFB discribe cycles with a radiant-matter working bodies. In the same time line AB and curve AEB describe cycles with radiant working body only.

For example, η_L = 0.931 when T_A=320 K and T_0=300 K. Condition $T_0=T_A$ excludes temperature T_0 from expression (6) which in this case is described by the Eq. (5). It means that the work can be obtained during a cycle with a radiant and matter working bodies. Dependencies (5), (6) are shown in the Fig. 5 by curves AFB and AEB, respectively. Line AB presenting η_C from Eq. (3) and line AK presenting η_{OS} from Eq. (4) are also shown.

3. Elementary and matter-radiant working bodies

Two types of working body are considered:
• Elementary working body – matter or radiation in one cycle;
• Matter-radiant working body – matter and radiation in one cycle.

3.1 Energy conversion without irrevocable losses
According to Carnot theorem, an efficiency of a Carnot engine does not depend on a chemical nature, physical and aggregate states of a working body. The work presents a peculiarity of applying this theorem for solar cells. The statement is that the maximal efficiency of solar cells can be achieved with help of a combined working body only. Let's consider it in detail.

For example, the maximal efficiencies of the solar energy conversion are equal 94.8% at the limit temperatures 300 K and 5800 K (η_{OS} in the Table 1). Under these temperatures the efficiencies of the solar energy conversion can be equal 5.91% ($\eta_0\eta_C$ in Table 1). The one belongs to a Carnot cycle, in which a matter and radiation are found as a combined working body, i.e. matter and radiation as a whole system. The other belongs to 2 cycles running parallel. A matter performs the work with a low efficiency 6.25% (η_0 in Table 1), but the radiation performs the work with a high efficiency 94.5% (η_C in Table 1). In these cases matter and radiation are elementary working bodies. The efficiencies of these parallel processes is

$$\eta_0\eta_C = 0.948 * 0.0625 = 0.0591 = 5.91\%.$$

Table 1 shows that a matter performs the work with a low efficiency in solar cells. However, the efficiency of the radiant work at the same absorber temperature is considerably higher. For example, a radiation performs the work with efficiency 92.6% during a non-Carnot cycle (η_{AS} in Table 1), but a matter produces work only with an efficiency 6.25% (η_0 in Table 1) at the absorber temperature 320 K. This difference is caused by various limit temperatures of the cycles (Table 1). The efficiencies of these processes running parallelis smaller than that of $\eta_0\eta_C$:

$$\eta_0\eta_{AS} = 0.926 * 0.0625 = 0.0579 = 5.79\%.$$

However, at the same temperatures the efficiency of solar energy conversion achieves 94.8% (η_{OS} in Table 1), if a work is performed during a cycle with the matter-radiant working body.

Classification of engines	Efficiency at T_A = 320 K and other parameters of cycles					
	Cycle	Working body	Limit tempe-ratures,K	Symbols	Limit, %	Calcilated equation
	Carnot engines					
heat	Carnot	Elementary / *matter or radiation*	300-320	η_0	6.25	2
solar	Carnot	elementary	320-5800	η_C	94.5	3
ideal solar-heat	Carnot	matter-radiation / *matter and radiation in one cycle*	300-5800	η_{OS}	94.8	4
	non-Carnot engines					
solar	non-Carnot	elementary	320-5800	η_{AS}	92.6	5
solar- heat	non-Carnot	matter-radiation	300-5800	η_L	93.1	6
	combined engines					
combined	Carnot, Carnot	elementary	300-320 320-5800	$\eta_0\eta_C$	5.91	2,3
combined	Carnot Carnot	elementary	300-320 300-5800	$\eta_0\eta_{OS}$	5.93	2,4
combined	Carnot, non-Carnot	elementary	300-320 320-5800	$\eta_0\eta_{AS}$	5.79	2,5
combined	Carnot, non-Carnot	elementary	300-320 300-5800	$\eta_0\eta_L$	5.82	2,6

Table 1. Classification and efficiencies of the engines with the elementary and matter-radiant working bodies

A Carnot cycle with the efficiency η_{OS} and the matter-radiant working body has been considered by the author earlier in the chapter, its efficiency is given by eq.4. Further we will call an engine with the matter-radiant working body an ideal solar-heat one. The elementary working bodies perform the work by solar or heat engines. Their properties are listed in Table 1. The advantage of the cyclic processes in comparison with the matter-radiant working body is obvious.

So, the elementary working bodies perform the work with the efficiencies η_0, η_C, η_L and η_{AS}. The matter-radiant working bodies perform the work with the efficiency η_{OS}. According to the Table. 1, one can confirm:

- a Carnot cycle with the matter-radiant working body has a maximally possible efficiency of solar energy conversion. It is equal 94.8% and does not depend on absorber temperature T_A. Engine where work is done during such cycle we will call an ideal solar-heat engine.
- electrical energy can be obtained under operating an ideal heat-solar engine with a very high efficiency and without additional function of low efficiency heat engine.

Therefore, high efficiency solar cells should be designed as solar-heat engine only.

3.2 Energy conversion with irrevocable losses

The absorption of radiation precedes the conversion of solar heat into work. In our model, the black body absorbs solar radiation and generates another radiation with a smaller temperature. Heat is evolved in this process; it is either converted into work or irrevocable lost. In this work the photon absorption in solar cells is divided into processes with and without work production. For the sake of simplicity, let us assume that heat evolved during solar energy reemission is lost with an efficiency of η_U from Eq. (1). The work of the cyclic processes is performed with the efficiencies of η_0, η_C, η_{AS}, η_{OS} and η_L (Tabl. 1). Then the conversion of solar heat with and without work production is performed with the efficiencies, for example, $\eta_C\eta_U$ or $\eta_0\eta_C\eta_U$. These and other combinations of efficiencies are compared in (Laptev, 2008).

It is important to note that the irrevocable energy losses of absorber at temperature 320 K do not cause the researchers' interest in thermodynamic analysis of conversion of solar heat into work. Actually, efficiency of the solar energy reemission as the irrevocable energy losses μ_U is close to 1 for $(T_A/T_S)^4 = (320/5800)^4 \approx 10^{-5}$. So efficiency of the solar energy reemission at 320 K has a small effect on efficiency of solar cell. The Tables 1,2 list efficiencies of the solar cells with and without irrevocable losses calculated in this work. The difference between these values does not exceed 0.01%. Values of $\mu_0\mu_C$ and $\mu_0\mu_C\mu_U$ may serve as examples. It might be seen that irrevocable energy losses are not to be taken into account in the thermodynamic analysis of conversion of solar heat into work. However, the detailed analysis of efficiencies of conversion of solar heat into work enabled us to reveal a correlation between the reversibility of solar energy reemission and efficiency of solar cell. The following parts of the chapter are devoted to this important aspect of conversion of solar heat into work.

3.3 Combinations of reversible and irreversible energy conversion processes

The temperature dependencies of μ_L from Eq. (5) and $\mu_0\mu_U$ from Eqs. (1),(2) are shown by lines LB, CB in Fig. 6. Let us also make use of the fact that every point of the line LB is (by definition) a graphical illustration of the sequence of reversible transitions from one energy state of the system to another, because each reversible process consists of the sequence of reversible transitions only.

Cycle parameters			Efficiency at T_A=320 K			
working body	cycle	Limit temperatures,K	Symbols	Limit, %	eq-tion	
		non-working conversion				
-	reemission	320-5800	η_U	99.99	1	
		heat, solar and heat-solar endoreversible engines				
elementary*	Carnot	300-320	$\mu_0\mu_U$	6.25	1, 2	
elementary	Carnot	320-5800	$\mu_C\mu_U$	94.5	1, 3	
elementary	non-Carnot	320-5800	$\mu_{AS}\mu_U$	92.6	1, 5	
matter-radiation**	non-Carnot	300-5800	$\mu_L\mu_U$	93.1	6	
matter-radiation	Carnot	300-5800	$\mu_{OS}\mu_U$	94.8	1, 4	
		Combined endoreversible engines				
elementary	Carnot	300-320/320-5800	$\mu_0\mu_C\mu_U$	5.91	1, 2, 3	
elementary	Carnot/non-Carnot	300-320/320-5800	$\mu_0\mu_{AS}\mu_U$	5.79	1, 2, 5	
elementary	Carnot	300-320/300-5800	$\mu_0\mu_{OS}\mu_U$	5.93	1, 2, 4	
elementary	Carnot	300-320/320-5800/300-5800	$\mu_0\mu_C\mu_{OS}\mu_U$	5.60	1, 2, 3, 4	
elementary and matter-radiation	Carnot/non-Carnot	300-320/300-5800/320-5800	$\mu_0\mu_{OS}\mu_{AS}\mu_U$	5.49	1, 2, 4, 5	

* elementary working body – matter or radiation;
** matter-radiant working body – matter and radiation in one cycle.

Table 2. Classification and efficiencies of the engines with solar energy conversion as a irrevocably losses

Based on this statement one can say that every point of the line CB is (by definition) a graphical illustration of the sequence of reversible and irreversible energy transitions. First represent a Carnot cycle abcd in the Fig.2, second represent a process of cooling of radiation running according to the line st. This engine performs the work with the efficiency $\mu_0\mu_U$. The combination of reversible and irreversible processes allows us to call this engine an endoreversible. Engines with the efficiencies μ_L and μ_0 are reversible.

Curves CB и LB in Fig.6 have some dicrepancy because of the entropy production without a work production in endoreversible engine. Indeed reversible energy conversion with the efficiency μ_L or μ_0 occurs without entropy production. The energy conversion with efficiency $\mu_0\mu_U$ is accompanied by the entropy production during the solar energy reemissions. The latter ones do not take place in the work production and cause irrevocably losses.

So, the entropy is not performed during a reversible engines. Endoreversible engines perform the entropy. Thus, the author supposes that difference between efficiencies ($\mu_L - \mu_0\mu_U$) of these engines is proportional to the quality (or number) of irreversible transitions. In most cases the increase of number of irreversible transitions in conversion of radiant heat into irrevocably losses calls a reduce of the efficiency of engine from μ_L down to $\mu_0\mu_U$.

Then, with help of the Fig. 6 one can find a ratio of reversible and irreversible transitions in a solar cell performing the work with the efficiency $\mu_0 \mu_U$. For example, let point a on Fig. 6 denote the conversion of solar energy with an efficiency of 30%. Let us draw an isotherm running through the point a; the intersections of this line with the line LB and the y-axis give us the values $\mu_L = 0.93$ and $T_A = 430$ K. Then, by the lever principle, the fraction of irreversible transitions q_{ir} in point a is

$$q_{ir} = (\mu_L - 30)100/ \mu_L = 68\%.$$

The fraction of irreversible transitions q_{rev} in point a will be 100–68=32%.

In other words, when in solar-heat engine (with the efficiency μ_L) 68% energy transitions can be made irreversible then the radiant work production will be ceased. If the reversible transitions (their number is now 32% of the total transitions' number) may form a Carnot cycle with the efficiency μ_0, then one can say that the reduce of number of reversible transitions leads to a cease of radiant work production, makes it possible to continue matter work production and reduces efficiency of engine from 93% down to 30%. The radiant heat evolved under these conditions partially takes a form of irrevocably losses as irreversible non-working transitions.

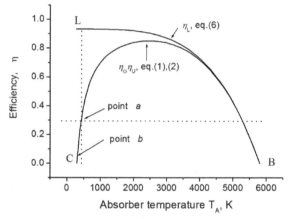

Fig. 6. Efficiency μ_L of solar-heat reversible engine (line LB) as a function of the absorber temperature T_A compared with the efficiency $\mu_0\mu_U$ of an heat endoreversible engine (line CB). Every point of the line LB correspond to 100% reversible transitions in energy conversion without entropy production. Every point of the line CB correspond to 100% irreversible transitions in energy conversion without work production. Every point of the region LBC between lines LB and CB correspond the reversible and irreversible transitions. Their parts can be found according to the lever's law (see text).

Let's consider a case of parallel work of matter and radiation. One could assume the work of radiation should increase an efficiency of solar-heat engine. However, it is not correct. The efficiencies of endoreversible processes whose elements are Carnot cycles with the efficiencies μ_C and μ_0 are shown in Fig. 7 by the CEB and CFB lines. They are situated below the LB line, whose every point is the efficiency μ_L of a reversible process by definition. We can therefore calculate the contributions of the Carnot cycle to the efficiencies $\mu_0 \mu_U$ and μ_0

μ_C μ_U if the efficiency μ_L (μ-coordinates of line LB points) is taken as one. For example, the efficiency μ_0 μ_U at 320 K (point b in Fig.6) is 6.25% (Tabl.2). Let us draw an isotherm running through the point b; the intersections of this line with the line LB and the y-axis give us the value $\mu_L = 0.931$. Then, by the lever principle, point b corresponds to a fraction

$$q_{ir} = (\mu_L - 0.0625)100/\ \mu_L = 93.3\%$$

of the irreversible transitions and the fraction $q_{rev} = 6.7\%$ of the reversible ones in conversion solar energy with the efficiency μ_0 $\mu_U = 6.25\%$. If matter and radiation produce work with the efficiency μ_0 μ_C $\mu_U = 5.91\%$ at 320 K (Tabl.2), the fraction of irreversible transitions as irrevocably losses is

$$q_{ir} = (\mu_L - 5.91)100/\ \mu_L = 93.6\%.$$

The fraction of irreversible transitions q_{rev} of two Carnot cycles will be equal 100-93.6=6.4%. Thus, when matter and radiation produce work simultaneously in Carnot cycles, only less then 7% of energy transitions in solar energy conversion are working cycles. Remaining 97% of energy transitions in solar energy conversion are non-working processes. So, the efficiency of solar-heat engine sufficiently depends on the reversability of non-working processes, rather than on the efficiencies of the work cycles. Let's explain in more details.

3.4 Reversible and irreversible non-working processes
As it was mentioned above, the reason of descrepancy between curves μ_L, $\mu_0\mu_U$ and $\mu_0\mu_C\mu_U$ at $T_A < T_S$ is entropy production during the processes without work production. Then the μ-coordinates of the points belonging to the plane between line μ_L and the lines $\mu_0\mu_U$, $\mu_0\mu_C\mu_U$ on Fig. 7 are proportional to the shares of the reversible and irreversible transitions without work production. We propose the following method for calculating relative contributions of reversible and irreversible processes without work production.

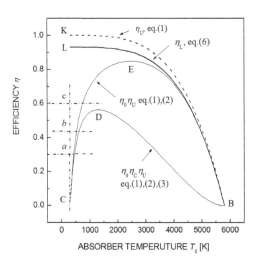

Fig. 7. The comparison of the well-known thermodynamic efficiency limitations of the solar energy conversion with and without entropy production.

The temperature dependencies of μ_L, $\mu_0\mu_U$, $\mu_0\mu_C\mu_U$ derived from Eqs. (1),(2),(3) and(6) are shown by lines LB, CEB and CDB in Fig. 7. Remember that the μ-coordinates of the points belonging to the lines $\mu_0\mu_U$, $\mu_0\mu_C\mu_U$ are proportional to the shares of reversible transitions which are one or two Carnot cycles with the efficiencies μ_0 and μ_C. At the same time remaining 100% irreversible transitions are (form) irreversible processes without work production as irrevocably losses.

The lines CEB and CDB on Fig. 7 are not only the illustration of possible ways of converting solar energy. For example, it is known that the limiting efficiencies of solar cells on the basis of a p,n-transition are 30% (Shockley & Queisser, 1961) and 43% (Werner et al., 1994). at T_A=320 K. They are shown by points a and b on Fig. 7 and lie above the lines $\mu_0\mu_U$, $\mu_0\mu_C\mu_U$, but below the line μ_L. One can say that processes without work production may be the combination of reversible and irreversible processes running parallel. Then the μ-coordinates of the points above the lines CEB, CDB are proportional to the shares q of reversible and irreversible transitions, which are appearing as processes without work production. According to the lever's principle, fraction of irreversible transitions q_{ir} in conversion of solar energy is

$$q_{ir} = (\mu_L - \mu_0\mu_C\mu_U)100/\mu_L.$$

For example, according to Tabl. 1,2, μ_L=93.1% and $\mu_0\mu_C\mu_U$=5.91% at T_A=320 K. Then q_{ir} = (93.1-5.91)100/93.1=93.7%. The fraction of reversible transitions q_{rev} is 100-93.7=6.3%.

Point a in Fig.7 denotes the conversion of solar energy with an efficiency μ=30% at T_A=320 K. It lies above the line $\mu_0\mu_C\mu_U$. Then the value

$$q_{ir} = (\mu_L - \mu)100/(\mu_L - \mu_0\mu_C\mu_U)$$

is a part of irreversible transitions in the conversion of solar energy without work production. In our case it is (93.1-30)100/(93.1-5.91)=70.0%. The fraction of reversible transitions will be equal to 100-70.0=30.0%. Note that there are no reversible transitions of Carnot cycles.

The band theory proposes mechanisms of converting solar energy into work with an efficiency of 43% (Landsberg & Leff, 1989). Let us denote last value by the point b on the isotherm in Fig. 7. Then the fraction of irreversible transitions q_{ir} without work production is (93.1-43)100/(93.1-5.91)= 57.5%, and the fraction of reversible transitions q_{rev} without work production will be equal 100-57.5=42.5%. So that, to increase efficiency of endorevesible solar-heat engine from 30 up to 43% at T_A=320 K, it's necessary to reduce a fraction of irreversible transitions without work production from 72.4% down to 57.5%.

As an example of reverse calculations let's find out efficiency of solar-heat engine with 50% irreversible transitions without work production. Denote the value is to be found as x and write down the equation

$$q_{ir} = (93.1-x)100/(93.1-5.91)= 50\%.$$

Solving it reveal that x=49.5%. Note that for q_{ir} = 100% efficiency $\mu_0\mu_C\mu_U$=50% may be achieved only under T_A=700 K (see Fig.7). This temperature is much higher than the temperatures of solar cells' exploitation (the temperatures of solar cells' working conditions) Thus, the author supposes the key for further increase of solar cells' efficiency is in the study and perfecting processes without work production.

3.5 Ideal solar-heat engines

According to the Eqs. (4),(6), there are two efficiency limits of solar energy conversion in reversible processes for a pair of limiting temperatures T_0 and T_S, namely, μ_{OS}, μ_L. In the Fig. 8 are shown their dependencies on absorber temperature T_A. Obviously, under all values of T_A the condition $\mu_{OS} > \mu_L$ is fulfilled. The suggestion can be made that processes can occur whose efficiency is between these limits. By virtue of the second law of thermodynamics, they cannot be reversible.

For instance, radiant energy conversion with the efficiency $\mu_{OS}\mu_U$ is irreversible, because radiation performs an irreversible process with the efficiency μ_U along the line st shown in Fig. 2. We see from Fig. 8 that the $\mu_{OS}\,\mu_U$ values (line AB) at temperatures $T_A \ll T_S$ are indeed larger than μ_L (line LB) and smaller than μ_{OS} (line AD).

In each point of the line AB the irreversible processes appear as a sequence of transitions without work production. Because the efficiency $\mu_{OS}\,\mu_U$ of the processes with their participation larger then μ_L, one can suppose that the fraction of irreversible transitions along the line LB is equal to zero, and along the line AB is equal to unity. In the plane between lines AB и LB the fraction of irreversible transitions q_{ir} may be calculated according to the lever principle:

$$q_{ir} = (\mu_{OS}\mu_U - \mu)100/(\mu_{OS}\mu_U - \mu_L).$$

In each point of the line AB the fraction of irreversible transitions q^0_i of total number of irrevesible transitions is

$$q^0_{ir} = (\mu_{OS}\mu_U - \mu_L)100/(\mu_S\mu_U).$$

According to the Tables 1, 2, $q^0_{ir}=1.8\%$ at $T_A=320$ K. The processes consisting of these transitions, do not produce work.

So, along the line AD (Fig. 8) conversion of solar energy is perfomed during a Carnot cycle, along the line AB – during a Carnot cycle and the processes without work production. In the plane between these lines only work is perfomed. The working process is a non-Carnot cycle. Between the lines AB and LB non-working processes are reversible and irreversible. There the fraction of irrevesible transitions decreases with closing the line LB. Along the line LB all energy transitions (with and without work production) are reversible, i.e. $q^0_{ir}=0$.

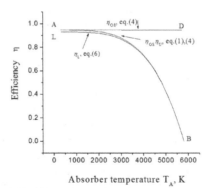

Fig. 8. Comparison of the efficiency $\mu_{OS}\mu_U$ of combinations of reversible working and irreversible non-working processes (the AB line) with the efficiency μ_{OS} of a reversible Carnot working process (the AD line) and the efficiency μ_L of reversible non-Carnot working processes (the LB line). Lines AD и AB coincide at $T_A=0$ K.

Irreversible processes without work production appear under the line LB. Their fraction increase as we move off the line LB. The previous chapter gives examples of calculating their contributions to solar energy conversion.

The value μ_L is a keyword in separation of solar energy conversion in processes with and without work production. It is referred as Landsberg efficiency (Wuerfel, 2005). According to this tradition, solar-heat engine with the efficiency μ_L are called Landsberg engine. It has a minimal efficiency of all solar-heat engines.

Solar-heat engine with the maximal efficiency μ_{OS} we call ideal solar-heat engine. Then solar-heat engine with the efficiency between μ_{OS} и $\mu_{OS}\mu_U$ may be noted as ideal non-Carnot solar-heat engine, and a solar-heat engine with the efficiency between $\mu_{OS}\mu_U$ and μ_L as an endoreversible solar-heat engine.

Solar-heat engine have cycles with the matter-radiant working body. We will not discuss the properties of the matter-radiant working body here. Let us simply note that all energy transitions in solar-heat engines are performed by the matter-radiant working body (Table 3). In the endoreversible solar-heat engine the matter-radiant working body makes working transitions, their fraction of total number of energy transitions is 100-1.8=98.2% because the processes without work production have only 1.8% energy transitions. Engine with the matter-radiant working body is not produced yet.

Engine	Efficiency at 320 K	contributions of energy transitions, %		matter and radiation state
		working	non-working	
Idel Carnot solar-heat	94.80	100	-	equilibrium
Ideal non-Carnot solar-heat	94.79	100	-	equilibrium
Endoreversible solar-heat	93.10-94.79	98.2	1.8	Non-equilibrium
Landsberg reversible	93.10	< 7	>93	equilibrium
Combined endoreversible	<6	< 7	>93	Non-equilibrium

Table 3. Comparison between the engine efficiencies and the contributions of working and non-working transitions

Existent solar cells are endoreversible combined engines with the non-Carnot cycles and elementary working bodies (matter or radiation). The working bodies perform working energy transitions, their fraction (as it is shown in the Table 3) can achieve only 7 % of transitions total number. We cannot overcome this 7% barrier but it doesn't mean that there is no possibility to increase the efficiency of such a machine. According to the Table 3, the non-working transitions are more than 93% from the total number of energy transitions. They can be either irreversible or revesible. If the first are made as the last, the engine efficiency can be increased from 6 up to 93.1%. So the aim of perfection in endoreversible solar-heat engine processes without work production is arisen. In ideal case it is necessary to obtain Landsberg engine.

So, efficiencies of the combined endoreversible engines with the non-Carnot cycles and elementary working bodies (matter or radiation) mostly depend on reversibility of processes without work production, running in the radiation absorber rather than on the work production processes. It follows that the contribution of reversible processes is smaller than the contribution of irreversible processes in devices that are currently used. According to the maximum work principle, their efficiency can be increased by increasing the fraction of reversible processes in solar energy absorbers. The following approach to solution of this problem can be suggested.

4. Antenna states and processes

Antenna states of atomic particles or their groups are the states between them resulting from the reemission of photons without work production and heat dissipation. Below is shown that a comparison of the efficiencies of the reversible and irreversible photon reemissions occuring in parallel made it clear that it is impossible to attain very high efficiencies in the conversion of solar heat into work without the reversible photon reemissions as antenna process.

4.1 Photon reemissions as non-working processes

Matter and solar radiation are never in equilibrium in solar cells and quasi-static conversion of the solar energy is not achieved. For this reason, the irreversible thermodynamic engines are described using the method of endoreversible thermodynamics of solar energy conversion (Novikov, 1958; Rubin, 1979; De Vos, 1992). Remember that endoreversible engines are irreversible engines where all irreversibilities are restricted to the coupling of engine to the external world. It is assumed that the inner reversible part of an endoreversible engine is a Carnot cycle. We have also considered the non-Carnot cycles and found out that photon absorption in solar cells can be considered as the external reversible part of an endoreversible engine.

The photon absorption in solar cells is separated into processes with and without work production. These processes are sequencies of transitions from one energy state of the system to another. The energy transitions between particle states are called "photon reemission" if they do not take part in performing work and heat dissipation.

The photon reemission is divided into reversible and irreversible processes. These non-working processes are regarded as a continuous series of equilibrium states outside the irreversible or endoreversible engine, are isolate into separate processes, and are used to obtain a higher efficiency of the generally non-quasistatic solar energy conversion. We can consider those equilibrium processes as a base of the „exoreversible" additional device for the irreversible or endoreversible engine.

So, we have called the processes without work production photon reemission. Let us now shown that they play a special role in the conversion of solar heat into work.

4.2 Antenna states of the absorber particles

Let us consider these particle states in a radiant energy absorber, transitions between them result from the absorption of photons. The states of atomic particles or their groups, as well as energy transitions between them, are called "working" if they take part in performing work. It is known that solar energy conversions are not always working processes. The states of atomic particles and the energy transitions between them are said to be "antenna" ones if they take part in the absorption and emission of radiant energy without work production, i.e. in the photon reemission. Carnot cycles are examples of working processes, while cycles described below involving the photon reemission are examples of antenna processes.

It is clear that antenna and working states are equilibrium ones if cycles of the radiant energy conversion are not accompanied by entropy production, i.e. if it takes place along line LB on Fig. 9 with an efficiency μ_L from Eq. (5). Let us take their total amount to be 100%. Now assume that the conversion of radiant energy consists of reversible working processes with an efficiency μ_O from Eq. (2) for matter, with an efficiency μ_{AS} from Eq. (5) for radiation and irreversible with an efficiency μ_U from Eq. (1) for irreversible antenna processes, i.e. that the conversion of radiant energy corresponds to line CEB. In this case, the μ-coordinates of all points belonging to the line CEB divided by μ_L are equal to the fraction of working states, while $1-\mu/\mu_L$ is equal to the fraction of antenna non-equilibrium states.

Fig. 9 shows that the number of antenna non-equilibrium states in solar cells decreases as the efficiency μ grows along lines CE and CF. The isothermal growth of the efficiency μ implies that the total amount of antenna states remains constant, but certain antenna states have become equilibrium states and that reversible transitions that do not generate work have appeared. If the temperature dependence of μ is determined by experiment, Fig. 9 becomes in our opinion an effective tool for interpreting and modeling the paths along which work is performed by solar cells. We can consider Fig. 9 as the reversibility diagram of the conversion of solar heat into work.

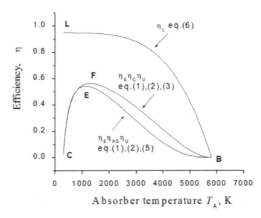

Fig. 9. Comparison of the efficiency μ_L of Landsberg engine (the LB line) with the efficiencies of combined engines (the CEB, CFB lines). A difference between the efficiencies $\eta_O\eta_{AS}\eta_U$ and $\mu_O\mu_C\mu_U$ is that in the first case radiation working cycle is not a Carnot cycle. Lines CEB and CFB are practically coincide at temperatures used for operation of solar cells.

Obviously, knowledge of physical and chemical nature of antenna states of particles is important for revealing optimal schemes for radiant energy conversion. The question about technical realization of antenna states is not subject of this chapter. This is a material science problem. One only should note the model of absorber with antenna states does not contradict to thermodynamic laws and represents a way of solution of following aim: how to separate any given cycle into infinite number of infinitely small arbitrary cycles.

4.3 Antenna processes

The absorption and generation of photons in the abcd (Fig. 2) and spet (Fig. 4) processes is accompanied by the production of work. No work is done in the st process (Fig. 2). This difference can be used to divide all processes with the absorption and generation of photons into two types. The first type includes processes in which the matter and radiation do work, that is, participate in the working process. The second type includes processes in which the substance absorbs and emits radiant energy without doing work, that is, participates in an antenna process. The spontaneous evening out of the temperatures of the matter and radiation is an example of antenna processes, and radiant energy conversion in a solar cell is a combination of processes of the first and second type.

The highest efficiency of a combination of work and antenna processes is described by the line AB in Fig. 8. If the work is performed during a non-Carnot cycle then efficiency of the work and antenna processes is described by the LB line. The difference between $\eta_{OS}\eta_U$ and η_L is 94.8-93.1=1.7%. For analysis of antenna processes one should choose the temperature dependence of the value η_L because in this case a conversion process is reversible.

The CEB line in Fig.9 shows the efficiency of a combination of work done by matter (in a Carnot cycle with the efficiency η_0), by radiation (in a non-Carnot cycle with the efficiency η_{AS}) and a background antenna process with the efficiency η_U. The η coordinates of the CEB line divided by η_L describe the largest contribution of working processes to the efficiency of solar energy conversion if η_L is taken as one. The contribution of the antenna process is then $1 - \eta_0\eta_{AS}\eta_U/\eta_L$.

It follows that the efficiency of conversion higher than $\eta_0\eta_{AS}\eta_U$ (the region between the LB and CEB lines, Fig. 9) can be obtained by perfecting the antenna process only. The efficiency of solar energy conversion is maximum in this case if all antenna processes are reversible. The efficiency is minimum if all antenna processes are irreversible.

The η coordinates of points between the LB and CEB lines are then proportional to the fraction of reversible antenna processes. An irreversible antenna process corresponds to the $1 - \eta/\eta_L$ value. The character of this dependence will not be discussed here. Only note that Fig. 9 can be used to calculate the fractions of work and antenna processes, as well as contributions of reversible and irreversible antenna processes in solar energy conversion if experimental data on the temperature dependence of the solar cell efficiency are available.

So, the antenna processes in Landsberg engine are reversible. When antenna processes are made irreversible, then the efficiency of transformation of radiant energy decreases from η_L=94.8% down to $\eta_0\eta_{AS}\eta_U$=5.79% at T_A=320 K. We have to select the opposite way.

4.4 Photon reemissions as reversible antenna processes

The notion of the reversibility of a process is applicable to any thermodynamic system, although we have not found it applied to quantum systems in physical literature. Let us recall that the transition of a system from equilibrium state 1 to equilibrium state 2 is called reversible (i.e., bi-directional) if one can return from state 2 to state 1 without making any changes in the surrounding environment, i.e. without compensations. The transition of a system from state 1 to state 2 is said to be irreversible if it is impossible to return from state 2 to state 1 without compensations.

Let us consider these particle states in an absorber, transitions between them result from the absorption of photons. The state of radiation will not change if the photon reemissions (121) or (12321) can take place (Fig. 10). These reemission processes will be reversible ones if the equilibrium state of matter is not violated by a photon absorption. According to these definitions, the reemission (123) or (1231) will alter the frequency distribution of photons (Fig. 11). For this reason, such reemission processes are irreversible.

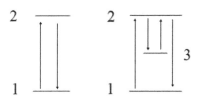

Fig. 10. The reversible reemissions in the systems with 2 and 3 energy levels.

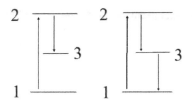

Fig. 11. The irreversible reemissions in the systems with 3 energy levels.

4.5 Photon reemissions as quasi-static process

Every quasi-static process is reversible and infinitely slow. Are the continuous series of the photon reemissions $(121)_n$ and $(12321)_n$ quasi-static processes? For the sake of simplicity, let us approximate the spatial arrangement of particles by a periodic chain of elements that have a length of $2\mu m$ (absorber thickness) and that are separated by 2 Å (atomic distance). Use the identity of electronic states 1, 2, 3 in particles A and present the sequence of electronic excitations in the particles' chain with help of reemission cycles in the Figs. 10, 11, for example, as follows:

$$A_1 + h\upsilon_{12} \rightarrow$$
$$\rightarrow A_{1\rightarrow2\rightarrow1} + h\upsilon_{21} \rightarrow$$
$$\rightarrow A_{1\rightarrow2\rightarrow3\rightarrow1} + h\upsilon_{23} + h\upsilon_{31} \rightarrow$$
$$\rightarrow A_{1\rightarrow3\rightarrow1} + h\upsilon_{31} \rightarrow$$
$$\rightarrow A_{1\rightarrow3\rightarrow2\rightarrow1} + h\upsilon_{21} \rightarrow ....$$
$$\rightarrow A_{1\rightarrow2\rightarrow1} + h\upsilon_{21} \qquad (7)$$

In this expression the first particle is excited by the photon with energy $h\upsilon_{21}$ and emits the same photon. This emitted photon excites the second particle, which emits two photons with energies $h\upsilon_{23}$ and $h\upsilon_{31}$. They, in turn, excite the third particle and so on.

The time period during which solar radiation will excite only the first particle in the chain, while subsequent particles are excited only by reemitted photons, is equal to 10^{-4} s if the lifetime of the excited state is $\sim 10^{-8}$ s. The time needed for the wave to travel down the chain and excite only the last particle is $\sim 10^{-14}$ s. These time intervals are related to each other as 320 years to one second. Therefore, the energy exchange due to reemission in a chain of particles is an infinitely slow process in comparison with the diffusion of electromagnetic radiation in the chain. Even the multiple repetition of reemission (121) by one particle during $\sim 10^{-6}$ s is an infinitely slow process, the time taken by such reemission exceeds the lifetime of the excited states by a factor of 100.

The radiant energy conversion may be characterized by the variable ratio of numbers of reversible and irreversible processes. In the chain (7) this ratio depends on (as shown in Figs. 10, 11) the order of states 1, 2, 3 alternation in cycle of each particle A. The index of A points it out. In general case we do not know either the reemission number, or states number in the cycle of reemission, as well as their alternation. But formally we can make a continuum consisting of all reversible cycles of the chain (7) and photons of reemission. Separate the reversible cycles and photons by brackets and right down the following expression:

$$(A_{1\to2\to1} + h\upsilon_{21} + A_{1\to3\to1} + h\upsilon_{31} + A_{2\to3\to2} + h\upsilon_{32} + ...) +$$

$$A_{1\to2\to3\to1} + h\upsilon_{23} + h\upsilon_{31} + A_{1\to3\to2\to1} + h\upsilon_{32} + h\upsilon_{21} \qquad (8)$$

The particles with irreversible reemission cycles are out of brackets.

The sequence of particles states in brackets may be thought as infinite one. (Note that if only one particle were in the brackets, then manifold repetition of its cycle might be considered as an infinite sequence of equilibrium states). In the solid absorber one can select a big number (~10^{19}) of chains consisting of these particles. While the quasistatic process is by definition an infinite and continous sequence of equlibrium states, then there is a fundamental possibility to consider local quasistatic processes in the whole irreversible process of radiant energy transformation.

Thus the sequences of photon reemissions of absorber particles in solar cells can take the form of quasi-static processes. Their isolation out from the general process of the conversion of solar energy into work (which is a non-equilibrium process on the whole) does not contradict the laws of thermodynamics.

4.6 Antenna processes and temperature of radiation

Let us call the reemission of solar energy during an antenna process *retranslation*, if the temperature of the radiation remains constant, and *transformation*, if the temperature of the radiation changes. The retranslation of radiation by a black body is, by definition, a reversible process. Transformation can take place both in a reversible and irreversible way. Therefore, the efficiency of the work performed by a solar cell can be improved by increasing the fraction of retranslating and transforming reversible antenna processes as well as working equilibrium states in an absorber.

4.7 Antenna processes and photon cutting

A photon cutting process (Wegh et al., 1999) as photon reemission is an example for an antenna process. It includes the emission of two visible photons for each vacuum ultraviolet photon absorbed. For us important, that materials with introduced luminescent activators ensure the separation of the photon antenna reemissions and the photon work production processes. In our opinion, ultra-high efficiencies can be reached if: firstly, we solve the problem of separating the photon antenna reemissions and the photon work production processes. Secondly, one should know a way of transformation of irreversible photon antenna reemissions into reversible. Let us consider the activator states in an absorber as state 3. A photon cutting process is irreversible in case of the reemissions $(1231)_n$. They can be luminescent. A photon cutting process can be reversible if an activator performs the reemissions $(1231321)_n$. They cannot be luminescent. An absorber with such activators can resemble a black body. Therefore, photon cutting processes can also play a role in transformation of irreversible reemission into reversible. This way is one of many others.

4.8 Antenna processes in plants

Let us leave the discussion of technological matters relating to the manipulation of antenna processes aside for the time being. We will devote a subsequent publication to this subject. Let us only remark here that the conversion of solar energy involving the participation of antenna molecules figures in the description of photosynthesis in biology. Every chlorophyll molecule in plant cells, which is a direct convertor of solar energy, is surrounded by a complex of 250-400 pigment molecules (Raven et al., 1999). The thermodynamic aspects of photosynthesis in plants were studied in (Wuerfel, 2005; Landsberg, 1977), yet the idea of antenna for solar cells was not proposed. We hope that the notions of the antenna and working states of an absorber particles will make it possible to attain very high efficiencies of the radiant energy convertors, especially in those cases when solar radiation is not powerful enough to make solar cells work efficiently yet suffices to drive photosynthesis in plants.

4.9 Conclusion

This leads us to conclude that reemission of radiant energy by absorbent particles can be considered a quasi-static process. We can therefore hope that the concept of an antenna process, which is photon absorption and generation, can be used to find methods for attaining the efficiency of solar energy conversion close to the limiting efficiency without invoking band theory concepts.

5. Thermodynamic scale of the efficiency of chemical action of solar radiation

Radiant energy conversion has a limit efficiency in natural processes. This efficiency is lower in solar, biological and chemical reactors. With the thermodynamic scale of efficiency of chemical action of solar radiation we will be able to compare the efficiency of natural processes and different reactors and estimate their commercial advantages. Such a scale is absent in the well_known thermodynamic descriptions of the solar energy conversion, its storage and transportation to other energy generators (Steinfeld & Palumbo, 2001). Here the thermodynamic scale of the efficiency of chemical action of solar radiation is based on the Carnot theorem.

Chemical changes are linked to chemical potentials. In this work it is shown for the first time that the chemical action of solar radiation **S** on the reactant **R**

$$R + S \leftrightarrow P + M \tag{9}$$

is so special that the difference of chemical potentials of substances R and P

$$\mu_R - \mu_P = f(T) \tag{10}$$

becomes a function of their temperature even in the idealized reverse process (9), if the chemical potential of solar radiation is accepted to be non-zero. Actually, there are no obstacles to use the function $f(T)$ in thermodynamic calculations of solar chemical reactions, because in (Kondepudi & Prigogin, 1998) it is shown that the non-zero chemical potential of heat radiation does not contradict with the fundamental equation of the thermodynamics. The solar radiation is a black body radiation.

Let us consider a volume with a black body R, transparent walls and a thermostat T as an idealized solar chemical reactor. The chemical action of solar radiation S on the reactant R will be defined by a boundary condition

$$\mu_R - \mu_P = \mu_m - \mu_S = f(T), \tag{11}$$

where μ_m is a chemical potential of heat radiation emitted by product P. Then the calculation of the function $f(T)$ is simply reduced to the definition of a difference $(\mu_m - \mu_S)$, because chemical potential of heat radiation does not depend on chemical composition of the radiator, and the numerical procedure for μ_m and μ_S is known and simple (Laptev, 2008). The chemical potential as an intensive parameter of the fundamental equation of thermodynamics is defined by differentiation of characteristic functions on number of particles N (Laptev, 2010). The internal energy U as a characteristic function of the photon number

$$U(V,N) = (2.703Nk)^{4/3}/(\sigma V)^{1/3}$$

is calculated by the author in (Laptev, 2008, 2010) by a joint solution of two equations: the known characteristic function

$$U(S,V) = \sigma V(3S/4 \; \sigma \; V)^{4/3}$$

(Bazarov, 1964) and the expression (Couture & Zitoun, 2000; Mazenko, 2000)

$$N = 0.370 \; \sigma T^3 V/k = S/3.602 \; k, \tag{12}$$

where T, S, V are temperature, entropy and volume of heat radiation, σ is the Stephan-Boltzmann constant, k is the Boltzmnn constant. In total differential of $U(V, N)$ the partial derivative

$$(\partial U/\partial N)_V \equiv \mu \text{ heat radiation} = 3.602kT \tag{13}$$

introduces a temperature dependence of chemical potential of heat radiation (Laptev, 2008, 2010). The function $U(S, V, N)$ is an exception and is not a characteristic one because of the relationship (12).

The Sun is a total radiator with the temperature $T_S = 5800$ K. According to (13), the chemical potential of solar radiation is $3.602kT = 173.7$ kJ/mol. Then the difference $f(T) = \mu_m - \mu_S$ is a function of the matter temperature T_m. For example, $f(T) = -165.0$ kJ/mol when $T_m = 298.15$ K, and it is zero when $T_m = T_S$. According to (13), the function $f(T)$ can be presented as proportional to the dimensionless factor:

$$f(T) = \mu_m - \mu_S = -\mu_S (1 - T_m/T_S).$$

According to the Carnot theorem, this factor coincides with the efficiency of the Carnot engine $\eta_C(T_m, T_S)$. Then the function

$$f(T)/\mu_S = -\eta_c(T_m, T_S) \tag{14}$$

can characterize an efficiency of the idealized Carnot engine-reactor in known limit temperatures T_m and T_S.

In the heat engine there is no process converting heat into work without other changes, i.e. without compensation. The energy accepted by the heat receiver has the function of compensation. If the working body in the heat engine is a heat radiation with the limit temperatures T_m and T_S, then the compensation is presented by the radiation with temperature T_m which is irradiated by the product P at the moment of its formation. We will call this radiation a compensation one in order to make a difference between this radiation and heat radiation of matter.

The efficiency of heat engines with working body consisting of matter and radiation is considered for the first time in (Laptev, 2008, 2010). During the cycle of such an engine–reactor the radiation is cooled from the temperature T_S down to T_m, causing chemical changes in the working body. The working bodies with stored energy or the compensational radiation are exported from the engine at the temperature T_m. The efficiency of this heat engine is the base of the thermodynamic scale of solar radiation chemical action on the working body.

Assume that the reactant R at 298.15 K and solar radiation S with temperature 5800 K are imported in the idealized engine–reactor. The product P, which is saving and transporting stored radiant energy, is exported from the engine at 298.15 K. Limit working temperatures of such an engine are 298.15 K and 5800 K. Then, according to relationships (10), (11), (14), the equation

$$(\mu_m - \mu_S) / \mu_S = - \eta_c (T_m, T_S) \tag{15}$$

defines conditions of maintaining the chemical reaction at steady process at temperature Tm in the idealized Carnot engine–reactor.

According to the Carnot theorem, the way working body receives energy, as well as the nature of the working body do not influence the efficiency of the heat engine. The efficiency remains the same under contact heat exchange between the same limit temperatures. The efficiency of such an idealized engine is

$$\eta_0 (T_m, T_S) = (1 - T_m/T_S). \tag{16}$$

Then the ratio of the values η_C and η_0 from (15), (16)

$$\zeta = \eta_c / \eta_0 = (\mu_P - \mu_R) / [\mu_S (1 - T_m/T_S)]. \tag{17}$$

is a thermodynamic efficiency ζ of chemical action of solar radiation on the working body in the idealized engine–reactor.

We compare efficiencies ζ of the action of solar radiation on water in the working cycle of the idealized engine-reactor if the water at 298.15 K undergoes the following changes:

$$H_2O_{water} + S_{solar\ rad.} = H_2O_{vapor} + M_{heat\ rad.,.vapor} ; \tag{18}$$

$$H_2O_{water} + S_{solar\ rad.} = H_{2gas} + \frac{1}{2}O_{2gas} + M_{heat\ rad.,\ H_2} + \frac{1}{2}M_{heat\ rad.,}O_2; \tag{19}$$

$$H_2O_{water} + S_{solar\ rad.} = H^+_{gas} + OH^-_{gas} + M_{heat\ rad.,\ H^+} + M_{heat\ rad.,}OH^-. \tag{20}$$

Chemical potentials of pure substances are equal to the Gibbs energies (Yungman & Glushko, 1999). In accordance with (17),

$$\zeta_{(18)} = [-228.61 - (-237.25)] / 173.7 / 0.95 = 0.052$$

$$\zeta_{(19)} = \zeta_{(18)} [0 + \frac{1}{2} \ 0 - (-228.61)] / 173.7 / 0.95 = 0.052 \times 1.39 = 0.072$$

$$\zeta_{(20)} = \zeta_{(18)} [1517.0 - 129.39 - (-228.61)] / 173.7 / 0.95 = 0.052 \times 9.79 = 0.51$$

So, in the engine–reactor the reaction (20) may serve as the most effective mechanism of conversion of solar energy.

In the real solar chemical reactors the equilibrium between a matter and radiation is not achieved. In this case the driving force of the chemical process in the reactor at the temperature T will be smaller than the difference of Gibbs energies

$$\Delta G_T = (\mu_P - \mu_R) + (\mu_m - \mu_S). \tag{21}$$

For example, water evaporation at 298.15 K under solar irradiation is caused by the difference of the Gibbs energies

$$\Delta G_{298.15} = -228.61 - (-237.25) + (-165.0) = -156.4 \text{ kJ/mol}.$$

At the standard state (without solar irradiation)

$$\Delta G^0_{298.15} = (\mu_P - \mu_R) = -228.61 - (-237.25) = 8.64 \text{ kJ/mol}.$$

The changes of the Gibbs energies calculated above have various signs: $\Delta G_{298.15} < 0$ and $\Delta G^0_{298.15} > 0$. It means that water evaporation at 298.15 K is possible only with participation of solar radiation. The efficiency ζ of the solar vapor engine will not exceed $\zeta_{(18)} = 5.2\%$. There is no commercial advantage because the efficiency of the conventional vapor engines is higher. However, the efficiency of the solar engine may be higher than that of the vapor one if the condition $\Delta G_{298.15} < 0$ and $\Delta G^0_{298.15} > 0$ is fulfilled. The plant cell where photosynthesis takes place is an illustrative example.

If the condition is $\Delta G_{298.15} < 0$ and $\Delta G^0_{298.15} < 0$ the radiant heat exchange replaces the chemical action of solar radiation. If $\Delta G_{298.15} > 0$ and $\Delta G^0_{298.15} > 0$, then neither radiant heat exchange, nor chemical conversion of solar energy cause any chemical changes in the system at this temperature. The processes (19) and (20) are demonstrative. Nevertheless, at the temperatures when ΔG becomes negative, the chemical changes will occur in the reaction mixture. So, in solar engines–reactors there is a lower limit of the temperature T_m. For example, in (Steinfeld & Palumbo, 2001) it is reported that chemical reactors with solar radiation concentrators have the minimum optimal temperature 1150 K.

The functions $\Delta G(T)$ and $\zeta(T)$ describe various features of the chemical conversion of solar energy. As an illustration we consider the case when the phases R and P are in thermodynamic equilibrium. For example, the chemical potentials of the boiling water and the saturated vapor are equal. Then both their difference $(\mu_P - \mu_R)$ and the efficiency ζ of the chemical action of solar radiation are zero, althought it follows from Eq. (21) that $\Delta G(T) < 0$. Without solar irradiation the equation $\Delta G(T) = 0$ determines the condition of the thermodynamic equilibrium, and the function $\zeta(T)$ loses its sense.

The thermodynamic scale of efficiency $\zeta(T)$ of the chemical action of solar radiation presented in this paper is a necessary tool for choice of optimal design of the solar engines_reactors. It is simple for application while its values are calculated from the experimentally obtained data of chemical potentials and temperature. Varying the values of chemical potentials and temperature makes it possible to model (with help of expressions (17), (21)) the properties of the working body, its thermodynamic state and optimal conditions for chemical changes in solar engines and reactors in order to bring commercial advantages of alternative energy sources.

6. Thermodynamic efficiency of the photosynthesis in plant cell

It is known that solar energy for glucose synthesis is transmitted as work (Berg et al., 2010; Lehninger et al., 2008; Voet et al., 2008; Raven et al., 1999). Here it is shown for the first time that there are pigments which reemit solar photons whithout energy conversion in form of heat dissipation and work production. We found that this antenna pigments make 77% of all pigment molecula in a photosystem. Their existance and participation in energy transfer allow chloroplasts to overcome the efficiency threshold for working pigments as classic heat engine and reach 71% efficiency for light and dark photosynthesis reactions. Formula for efficiency calculation take into account differences of photosynthesis in specific cells. We are also able to find the efficiency of glycolysis, Calvin and Krebs cycles in different organisms.

The Sun supplies plants with energy. Only 0.001 of the solar energy reaching the Earth surface is used for photosynthesis (Nelson & Cox, 2008; Pechurkin, 1988) producing about 1014 kg of green plant mass per year (Odum, 1983). Photosynthesis is thought to be a low-effective process (Ivanov, 2008). The limiting efficiency of green plant is defined to be 5% as a ratio of the absorbed solar energy and energy of photosynthesis products (Odum,1983; Ivanov, 2008). Here is shown that the photosynthesis efficiency is significantly higher (71% instead of 5%) and it is calculated as the Carnot efficiency of the solar engine_reactor with radiation and matter as a single working body.

The photosynthesis takes place in the chloroplasts containing enclosed stroma, a concentrated solution of enzymes. Here occure the dark reactions of the photosynthesis of glucose and other substances from water and carbon dioxide. The chlorophyll traps the solar photon in photosynthesis membranes. The single membrane forms a disklike sac, or a thylakoid. It encloses the lumen, the fluid where the light reactions take place. The thylakoids are forming granum (Voet et al., 2008; Berg et al., 2010). Stacks of grana are immersed into the stroma.

When solar radiation with the temperature T_S is cooled in the thylakoid down to the temperature T_A, the amount of evolved radiant heat is a fraction

$$\eta_U = 1 - (T_A/T_S)^4$$

of the energy of incident solar radiation (Wuerfel, 2005). The value η_U is considered here as an efficiency of radiant heat exchange between the black body and solar radiation (Laptev, 2006).

Tylakoids and grana as objects of intensive radiant heat exchange have a higher temperature than the stroma. Assume the lumen in the tylakoid has the temperature T_A = 300 K and the stroma, inner and outer membranes of the chloroplast have the temperature T_0 = 298 K. The solar radiation temperature T_S equals to 5800 K.

The limiting temperatures T_0, T_A in the chloroplast and temperature T_S of solar radiation allow to imagine a heat engine performing work of synthesis, transport and accumulation of substances. In idealized Carnot case solar radiation performs work in tylakoid with efficiency

$$\eta_C = 1 - T_A/T_S = 0.948,$$

and the matter in the stroma performs work with an efficiency

$$\eta_0 = 1 - T_0/T_A = 0.0067.$$

The efficiency $\eta_0\eta_C$ of these imagined engines is 0.00635.

The product $\eta_0\eta_C$ equals to the sum $\eta_0 + \eta_C - \eta_{0S}$ (Laptev, 2006). Value η_{0S} is the efficiency of Carnot cycle where the isotherm T_S relates to the radiation, and the isotherm T_0 relates to the matter. The values η_{0S} and η_C are practically the same for chosen temperatures and $\eta_{0S}/\eta_0\eta_C$ = 150. It means that the engine where matter and radiation performing work are a single working body has 150 times higher efficiency than the chain of two engines where matter and radiation perform work separately.

It is known (Laptev, 2009) that in the idealized Carnot solar engine–reactor solar radiation S produces at the temperature T_A a chemical action on the reagent R

$$R_{reagent} + S_{solar\ radiation} \leftrightarrow P_{product} + M_{thermal\ radiation\ of\ product}$$

with efficiency

$$\zeta = (\mu_P - \mu_R)/[\mu_S/(1 - T_A/T_S)], \tag{22}$$

where μ_P, μ_R are chemical potentials of the substances, μ_S is the chemical potential of solar radiation equal to $3.602kT_S = 173.7$ kJ/mol. The efficiency of use of water for alternative fuel synthesis is calculated in (Laptev, 2009).

Water is a participant of metabolism. It is produced during the synthesis of adenosine triphosphate (ATP) from the adenosine diphosphate (ADP) and the orthophosphate (P_i)

$$ADP + P_i = ATP + H_2O. \tag{23}$$

Water is consumed during the synthesis of the reduced form of the nicotinamide adenine dinucleotide phosphate (NADPH) from its oxidized form ($NADP^+$)

$$2NADP^+ + 2H_2O = 2NADPH + O_2 + 2H^+_{thylakoid} \tag{24}$$

and during the glucose synthesis

$$6CO2 + 6H_2O = C_6H_{12}O_6 + 6O_2. \tag{25}$$

Changes of the Gibbs energies or chemical potentials of substances in the reactions (23)–(25) are 30.5, 438 and 2850 kJ/mol, respectively (Voet et al., 2008).

The photosynthesis is an example of joint chemical action of matter and radiation in the cycle of the idealized engine–reactor, when the water molecule undergoes the changes according to the reactions (23)–(25). According to (22), the photosynthesis efficiency ζ_{Ph} in this model is

$$\zeta_{(5)} \times 1/2\zeta_{(6)} \times 1/6\zeta_{(7)} = 71\%.$$

The efficiency ζ_{Ph} is smaller than the Landsberg limiting efficiency

$$\eta_L = \eta_U - 4T_0/3T_S + 4T_0 T_A^3/3\ T_S^4, \tag{26}$$

known in the solar cell theory (Wuerfel, 2005) as the efficiency of the joint chemical action of the radiation and matter per cycle. ζ_{Ph} and the temperature dependence η_L are shown in Fig. 12 by the point F and the curve LB respectively. They are compared with the temperature dependence of efficiencies $\eta_0\eta_C\eta_U$ (curve CB) and $\eta_{0S}\eta_U$ (curve KB). Value η_U is close to unity because $(T_A/T_S)^4 \sim 10^{-5}$.

We draw in Fig. 12 an isotherm t–t' of η values for $T_A = 300$ K. It is found that $\eta_{0S} = 94.8\%$ at the interception point K, $\eta_L = 93.2\%$ at the point L and $\eta_0\eta_C\eta_U = 0.635\%$ at the point C. The following question arises: which processes give the chloroplasts energy for overcoming the point C and achieving an efficiency $\zeta_{Ph} = 71\%$ at the point F?

First of all one should note that the conversion of solar energy into heat in grana has an efficiency η_g smaller than η_U of the radiant heat exchange for black bodies. From (26) follows that the efficiency ζ_{Ph} cannot reach the value η_L due to necessary condition $\eta_g < \eta_U$.

Besides in the thylakoid membrane the photon reemissions take place without heat dissipation (Voet et al., 2008; Berg et al., 2010). The efficiency area between the curves LB and CB relates to photon reemissions or antenna processes. They can be reversible and irreversible. The efficiencies of reversible and irreversible processes are different. Then the point F in the isotherm t-t' is the efficiency of engine with the reversible and irreversible antenna cycles.

The antenna process performs the solar photon energy transfer into reaction centre of the photosystem. Their illustration is given in (Voet et al., 2008; Berg et al., 2010). Every photosystem fixes from 250 to 400 pigments around the reaction center (Raven et al., 1999). In our opinion a single pigment performs reversible or irreversible antenna cycles. The antenna cycles form antenna process. How many pigments make the reversible process in the photosynthetic antenna complex?

One can calculate the fraction of pigments performing the reversible antenna process if the line LC in Fig. 12 is supposed to have the value equal unity. In this case the point F corresponds to a value $x = \zeta / (\eta_L - \eta_0 \eta_C \eta_U) = 0.167$. This means that 76.7% of pigments make the revesible antenna process. 23.3% of remaining pigments make an irreversible energy transfer between the pigments to the reaction centres. The radiant excitation of electron in photosystem occurs as follows:

$$\text{chlorophyll } a + \text{photon} \leftrightarrow \text{chlorophyll } a^+ + e^-. \tag{27}$$

The analogous photon absorption takes place also in the chlorophylls b, c, d, various carotenes and xanthophylls contained in different photosystems (Voet et al., 2008; Berg et al., 2010). The excitation of an electron in the photosystems P680 and P700 are used here as illustrations of the reversible and irreversible antenna processes.

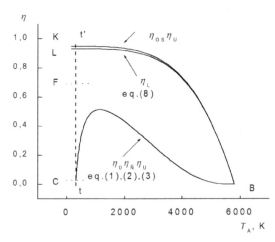

Fig. 12. The curve CB is the efficiency of the two Carnot engines (Laptev, 2005). The curve LB is the efficiency of the reversible heat engine in which solar radiation performs work in combination with a substance (Wuerfel, 2005). The curve KB is the efficiency of the Carnot solar engine_reactor (Laptev, 2006), multiplied by the efficiency η_U of the heat exchange between black bodies. The isotherm t-t' corresponds to the temperature 300 K. The calculated photosynthesis efficiency is presented by the point F in the isotherm.

Schemes of working and antenna cycles are shown in Fig. 13. Working pigment (a) is excited by the photon in the transition $1 \rightarrow 3$. Transition $3 \rightarrow 2$ corresponds to the heat compensation in the chloroplast as engine–reactor. The evolved energy during the transition $2 \rightarrow 1$ is converted into the work of electron transfer or ATP and NADPH synthesis.

When the antenna process passes beside the reaction centre, the photosystems make the reversible reemissions. Fig. 13 presents an interpretation of absorption and emission of photons in antenna cycles. The reemission $2 \rightarrow 3 \rightarrow 2$ shows a radiant heat exchange. The reemissions $1 \rightarrow 2 \rightarrow 1$ and $1 \rightarrow 3 \rightarrow 1$ take place according to (27). Examples are the pigments in chromoplasts.

According to the thermodynamic postulate, the efficiency of reversible process is limited. In our opinion, just the antenna processes in the pigment molecules of the tylakoid membrane allow the photosystems to overcome the forbidden line (for a heat engine efficiency) CB in Fig. 12 and to achieve the efficiency $\zeta_{Ph} = 71\%$ in the light and dark photosynthesis reactions.

There are no difficulties in taking into account in (22) the features of the photosynthesis in different cells. The efficiency of glycolyse, Calvin and Krebs cycles in various living structures may be calculated by the substitution of solar radiation chemical potential in the expression (22) by the change of chemical potentials of substances in the chemical reaction.

The cell is considered in biology as a biochemical engine. Chemistry and physics know attempts to present the plant photosynthesis as a working cycle of a solar heat engine (Landsberg, 1977). The physical action of solar radiation on the matter of nonliving systems during antenna and working cycles of the heat engine is described in (Laptev, 2005, 2008). In this article the Carnot theorem has been used for calculation of the thermodynamic efficiency of the photosynthesis in plants; it is found that the efficiency is 71%.

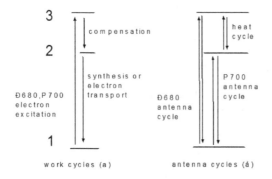

Fig. 13. The interpretation of energy transitions in the work (a) and antenna (b) cycles. Level *1* shows the ground states, levels *2, 3* present excited states of pigment molecules.

One can hope that the thermodynamic comparison of antenna and working states of pigments in the chloroplast made in this work will open new ways for improving technologies of solar cells and synthesis of alternative energy sources from the plant material.

7. Condensate of thermal radiation

Thermal radiation is a unique thermodynamic system while the expression $dU = TdS - pdV$ for internal energy U, entropy S, and volume V holds the properties of the fundamental equation of thermodynamics regardless the variation of the photon number (Kondepudi & Prigogin, 1998. Bazarov, 1964). Differential expression $dp/dT=S/V$ for pressure p and temperature T is valid for one-component system under phase equilibrium if the pressure does not depend on volume V (Muenster, 1970). Thermal radiation satisfies these conditions but shows no phase equilibrium.

The determinant of the stability of equilibrium radiation is zero (Semenchenko, 1966). While the „zero" determinants are related to the limit of stability, there are no thermodynamic restrictions for phase equilibrium of radiation (Muenster, 1970). However, successful attempts of finding thermal radiation condensate in any form are unknown. This work aims to support enthusiasm of experimental physicists and reports for the first time the phenomenological study of the thermodynamic medium consisting of radiation and condensate.

It is known (Kondepudi & Prigogin, 1998; Bazarov, 1964), that evolution of radiation is impossible without participating matter and it realizes with absorption, emission and scattering of the beams as well as with the gravitational interaction. Transfer of radiation and electron plasma to the equilibrium state is described by the kinetic equation. Some of its solutions are treated as effect of accumulation in low-frequency spectrum of radiation, as Bose-condensation or non-degenerated state of radiation (Kompaneets, 1957; Dreicer, 1964; Weymann, 1965; Zel'dovich & Syunyaev, 1972; Dubinov A.E. 2009). A known hypothesis about Bose-condensation of relic radiation and condensate evaporation has a condition: the rest mass of photon is thought to be non-zero (Kuz'min & Shaposhnikov, 1978). Nevertheless, experiments show that photons have no rest mass (Spavieri & Rodrigues, 2007).

Radiation, matter and condensate may form a total thermal equilibrium. According to the transitivity principle of thermodynamic equilibrium (Kondepudi & Prigogin, 1998; Bazarov, 1964), participating condensate does not destroy the equilibrium between radiation and matter. Suppose that matter is a thermostat for the medium consisting of radiation and condensate. A general condition of thermodynamic equilibrium is an equality to zero of virtual entropy changes δS or virtual changes of the internal energy δU for media (Bazarov, 1964; Muenster, 1970; Semenchenko, 1966). Using indices for describing its properties, we write $S=S_{rad}+S_{cond}$, $U=U_{rad}+U_{cond}$. The equilibrium conditions $\delta S_{rad}+\delta S_{cond}=0$, $\delta U_{rad}+\delta U_{cond}=0$ will be completed by the expression $T\delta S=\delta U+p\delta V$, and then we get an equation

$$(1/T_{cond}-1/T_{rad})\delta U_{cond} +(p_{cond}/T_{cond})\delta V_{cond}+(p_{rad}/T_{rad})\delta V_{rad} = 0.$$

If $V_{rad}+V_{cond}=V=const$ and $\delta V_{rad}= -\delta V_{cond}$, then for any values of variations δU_{cond} and δV_{cond} we find the equilibrium conditions: $T_{rad}=T_{cond}=T$ and $p_{rad}=p_{cond}=p$. When condensate is absolutely transparent for radiation, it is integrated in condensate, so that $V_{rad}=V_{cond}=V$ and $\delta V_{rad}=\delta V_{cond}$. Thus, conditions

$$T_{rad} = T_{cond}, \qquad p_{rad} = - p_{cond} \qquad\qquad (28)$$

are satisfied for any values of variations δU_{cond} and δV_{cond}.

The negative pressure arises in cases, when $U-TS+pV=0$ and $U>TS$. We ascribe these expressions to the condensate and assume the existence of the primary medium, for which the expression $S_0=S_{cond}+S_{rad}$ is valid in the same volume. Now we try to answer the question about the medium composition to form the condensate and radiation from indefinitely small local perturbations of entropy S_0 of the medium. Two cases have to be examined.

Suppose that the primary medium is radiation and for this medium $U_{00,rad}-TS_0+p_{rad}V=0$. Then the condition $U_{00,rad}<U_{rad}+U_{cond}$ is satisfied for values of p and T necessary for equilibrium. According to this inequality and the Gibbs stability criterion (Muenster, 1970), the medium consisting of the condensate and radiation is stable relatively to primary radiation, i.e. the condensation of primary radiation is a forced process.

In contrary, the equilibrium state of condensate and radiation arises spontaneously from the primary condensate, because $U_{00,cond}>U_{rad}+U_{cond}$, if $U_{00,cond}-TS_0+p_{cond}V=0$. However, the condensate has to lower its energy before the moment of the equilibrium appearance to prevent self-evaporation of medium into radiation. Such a process is possible under any infinitely small local perturbations of the entropy S_0. Really, the state of any equilibrium system is defined by the temperature T and external parameters (Kondepudi & Prigogin, 1998; Bazarov, 1964; Muenster, 1970).

While the state of investigated medium is defined by the temperature only, the supposed absence of external forces allows the primary condensate to perform spontaneous adiabatic extension with lowering energy by factor $\Delta U_{cond}=U_{0,cond}-U_{00,cond}=p_{cond}\Delta V$. When the energy rest will fulfill the condition $U_{0,cond}=U_{rad}+U_{cond}$, the required condition $U_{0,cond}-TS_0+p_{cond}V=0$ for arising equilibrium between condensate and radiation will be achieved.

Let's consider the evolution of the condensate being in equilibrium with radiation. Once the medium is appeared, this medium consisting of the equilibrium condensate and radiation can continue the inertial adiabatic extension due to the assumed absence of external forces. When $V_{rad}\equiv V_{cond}$, the second law of thermodynamics can be written as $u=Ts-p$, where u and s are densities of energy and entropy, respectively. Fig. 14 plots a curve of radiation extension as a cubic parabola $s_{rad}=4\sigma T^3/3$, where σ is the Stefan-Boltzmann constant. Despite the fact that the density of entropy of the condensate is unknown, we can show it in Fig.1 as a set of positive numbers $\lambda=Ts$, if each λ_i is ascribed an equilateral hyperbola $s_{cond}=\lambda_i/T$. Fig.14 illustrates both curves.

We include the cross-section point c_i of the hyperbola cd and the cubic parabola ab in Fig. 14 in the interval $[c_0, c_k]$. Assume the generation of entropy along the line cd outside this interval and the limits of the interval are fixing the boundary of the medium stability. Absence of the entropy generation inside the interval $[c_0, c_k]$ means that the product $2s_i(T_k-T_0)$ is $\int_{T_0}^{T_k}dT(s_{cond}+s_{rad})$. By substituting s we can see that these equalities are valid only at $T_0=T_k=T_i$. So, if the condensate and radiation are in equilibrium, the equality $s_{cond}=s_{rad}$ is also valid.

Thus, when the equilibrium state is achieved the medium extension is realized along the cross-section line of the parabola and hyperbolas. Equalities $s_{cond}=s_{rad}=4\sigma T^3/3$ are of fundamental character; all other thermodynamic values for the condensate can be derived from these values. For example, we find that $\lambda_i=4\sigma T_i^4/3$. For the condensate $u-Ts+p=0$ is valid. Then, according to (28),

$$u_{cond}=Ts-p_{cond}=5\sigma T^4/3. \tag{29}$$

For the equilibrium medium consisting of the condensate and radiation $u_{cond}=5u_{rad}/3$ and $u_0=u_{rad}+u_{cond}=8\sigma T^4/3=2\lambda$. For the primary condensate before its extension $u_{00}=u_{rad}+u_{cond}-p=3\sigma T^4$. For thermal radiation $u_{rad}=3p$ and the pressure is always positive (Kondepudi & Prigogin, 1998; Bazarov, 1964).

The extension of the medium is an inerial process, so that the positive pressure of radiation p_{rad} lowers, and the negative pressure of the condensate p_{cond} increases according to the condition (1). Matter is extended with the medium. As it is known in cosmological theory (Kondepudi & Prigogin, 1998; Bazarov, 1964), the plasma inertial extension had led to formation of atoms and distortion of the radiation-matter equilibrium. Further local unhomogeneities of matter were appeared as origins of additional radiation and, consequently, matter created a radiation excess in the medium after the equilibrium radiation-matter was disturbed This work supposes that radiation excess may cause equilibrium displacement for the medium, thus radiation and condensate will continue extending inertially in a non-equilibrium process.

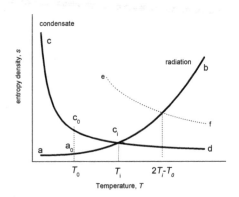

Fig. 14. Schematically plots the density of entropy for radiation (curve ab) and for a condensate (curves cd and ef). The positive pressure of the radiation and negative pressure of condensate are equal by absolute value at the points c_i and e_k at the interceptions of these curves.

We assume that the distortion of the equilibrium radiation-condensate had been occurred at the temperature T_i of the medium at the point c_i in Fig. 14. The radiation will be extended adiabatically along the line $c_i a$ of the cubic parabola without entropy generation. While the condition $V_{rad}=V_{cond}$ is satisfied if the equilibrium is disturbed, the equality $s_{cond}=s_{rad}$ points out directions of the condensate extension without entropy generation. As it is shown in Fig.14, the unchangeable adiabatic isolation is possible if the condensate extends along the isotherm T_i without heat exchange with radiation. Differentiation of the expression $U_{cond}-T_iS_{cond}+p_{cond}V=0$ with T=const and S=const gives that p_{cond} is also constant.

The medium as a whole extends in such a manner that the positive pressure p_{rad} of radiation decreases, and the negative pressure p^*_{cond} remains constant. As radiation cools down, the ratio p^*/p_{rad} lowers, the dominant p^* of the negative pressure arises, and the medium begins to extend with positive acceleration.

The thermodynamics defines energy with precision of additive constant. If we assume this constant to be equal to TS_{cond}, then the equality $u^*_{cond}=U^*_{cond}/V = -p^*_{cond}$, is valid; this equality points out the fixed energy density of the condensate under its expansion after distortion of the medium equilibrium.

The space is transparent for relic radiation which is cooling down continuously under adiabatic extension of the Universe. Assuming existence of the condensate of relic radiation we derive an expression for a fixed energy density of the condensate u^* with the beginning of accelerated extension of the Universe. The adiabatic medium with negative pressure and a fixed energy density 4 GeV/m³ is supposed to be the origin of the cosmological acceleration. The nature of this phenomenon is unknown (Chernin, 2008; Lukash & Rubakov, 2008; Green, 2004) . What part of this energy can have a relic condensate accounting the identical equation of state $u = -p$ for both media?

The relic condensate according to (29) has the energy density 4 GeV/m³ when the temperature of relic radiation is about 27 K. If the accelerated extension of the cosmological medium arises at $T^* \leq 27$ K, the part of energy of the relic condensate in the total energy of the cosmological medium is $(T^*/27)^4$. According to the Fridman model T^* corresponds to the red shift ≈0.7 (Chernin, 2008) and temperature 4.6 K. Then the relic condensate can have a 0.1% part of the cosmological medium.

As a conclusion one should note that the negative pressure of the condensate of thermal radiation is Pascal-like and isotropic, it is constant from the moment as the equilibrium with radiation was disturbed by the condensate and is equal (by absolute value) to the energy density with precision of additive constant. The condensate of thermal radiation is a physical medium which interacts only with the radiation and this physical medium penetrates the space as a whole. This physical medium cannot be obtained under laboratory conditions because there are always external forces for a thermodynamic system in laboratory. While this paper was finalized the information (Klaers et al., 2009) showed the photon Bose-condensate can be obtained. This condensate has no negative pressure while it is localized in space. It seems very interesting to find in the nature a condensate of thermal radiation with negative pressure. Possible forms of physical medium with negative pressure and their appearance at cosmological observations are widely discussed. The radiation can consist of other particles, then the photon, among them may be also unknown particles. We hope that modelling the medium from the condensate and radiation will be useful for checking the hypotheses and will allow explaining the nature of the substance responsible for accelerated extension of the Universe. The medium from thermal radiation and condensate is the first indication of the existence of physical vacuum as one of the subjects in classical thermodynamics and the complicated structure of the dark energy.

8. Electrical properties of copper clusters in porous silver of silicon solar cells

Technologies for producing electric contacts on the illuminated side of solar cells are based on chemical processes. Silver technologies are widely used for manufacturing crystalline silicon solar cells. The role of small particles in solar cells was described previously (Hitz, 2007; Pillai, 2007; Han, 2007; Johnson, 2007). The introduction of nanoparticles into pores of photon absorbers increases their efficiency. In our experiments copper microclusters were chemically introduced into pores of a silver contact. They changed the electrical properties of the contact: dark current, which is unknown for metals, was detected.

In the experiments, we used 125 x×125-mm commercial crystalline silicon wafers Si<P>/SiN$_x$ (70 nm)/Si<B> with a silver contact on the illuminated side. The silver contact was porous silver strips 10–20 μm thick and 120–130 μm wide on the silicon surface. The diameter of pores in a contact strip reached 1μm. The initial material of the contact was a silver paste (Dupont), which was applied to the silicon surface through a tungsten screen mask. After drying, organic components of the paste were burned out in an inert atmosphere at 820–960° C. Simultaneously, silver was burned in into silicon through a 70-nm-thick silicon nitride layer. After cooling in air, the wafer was immersed in a copper salt solution under the action of an external potential difference; then, the wafer was washed with distilled water and dried with compressed nitrogen until visible removal of water from the surface of the solar cell (Laptev & Khlyap, 2008).

The crystal structure of the metal phases was studied by grazing incidence X-ray diffraction. A 1-μm-thick copper layer on the silver surface has a face-centered cubic lattice with space group *Fm3m*. The morphology of the surface of the solar cell and the contact strips before and after copper deposition was investigated with a KEYENCE-5000 3D optical microscope. Fig. 15 presents the result of computer processing of images of layer-by-layer optical scanning of the surface after copper deposition.

Fig. 15. Contact strip morphology. Scanning area 430 x 580 μm²; magnification 5000x.

The copper deposition onto the silver strips did not change the shape and profile of the contact, which was a regular sequence of bulges and compressions of the contact strip. The differences in height and width reached 5 μm. In some cases, thin copper layers caused slight compression of the contact in height. The profiles of the contacts were studied using computer programs of the optical microscope. It was found that copper layers to 1 μm in thickness on the silver contact could cause a decrease in the strip height by up to 10%.

The chemical composition of the contact and the depth distribution of copper were investigated by energy dispersive X-ray analysis, secondary ion mass spectrometry, and X-ray photoelectron spectroscopy. The amount of copper in silver pores was found to decrease with depth in the contact. Copper was found at the silicon–silver interface. No copper diffusion into silicon was detected.

The resistivity of the contacts was measured at room temperature with a Keithley 236 source-measure unit. Two measuring probes were placed on the contact strips at a distance of 8 mm from each other. A probe was a tungsten needle with a tip diameter of 120 μm. The measurements were made on two samples in a box with black walls and a sunlight simulator. Fig. 16 presents the results of the experiments.

Line 1 is the current–voltage diagram for the initial silver contact strip on the silicon wafer surface. The other lines are the current–voltage diagrams for the contacts after copper deposition. All the lines confirm the metallic conductance of the contact strips. The current–voltage diagrams for the contacts with copper clusters differ by the fact that they do not pass through the origin of coordinates for both forward and reverse currents. A current through a metal in the absence of an external electric field is has not been observed. In our experiment, the light currents were 450 µA in the contact where copper clusters were only in silver pores and 900 µA in the contact where copper clusters were both in silver pores and on the silver surface.

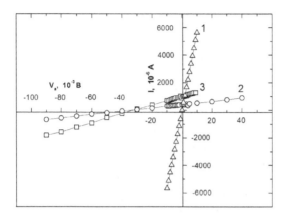

Fig. 16. Electrical properties of (1) a silver contact strip, (2) a contact strip with copper clusters in silver pores, and (3) a strip with a copper layer on the surface and copper clusters in silver pores.

It is worth noting that the electric current in the absence of an external electric field continued to flow through these samples after the sunlight simulator was switched off. The light and dark currents in the contact strips are presented in Fig. 17. It is seen that the generation of charge carriers in the dark at zero applied bias is constant throughout the experiment time. The dark current in the silver contact is caused by the charge carrier generation in the contact itself. The source of dark-current charge carriers are copper clusters in silver pores and on the silver surface.

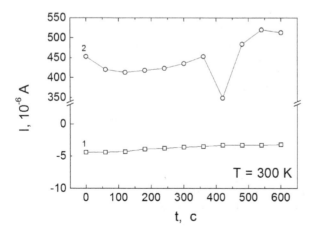

Fig. 17. Time dependence of the (1) dark and (2) light currents at zero applied bias in contact strips with copper clusters in silver pores.

The current in the silver contact with copper clusters while illuminating the solar cell is caused by the generation of charge carriers in the semiconductor part of the silicon wafer. The number of charge carriers generated in the $p–n$ junction is two orders of magnitude larger than the number of charge carriers in copper clusters since the light current is so larger than the dark current (Fig. 17).

The copper deposition onto silver does not lead to the formation of a silver–copper solid solution. The contact of the crystal structures gives rise to an electric potential difference. This is insufficient for generation of current carriers.

However, the contact of the copper and silver crystal structures causes compression of the metal strip and can decrease the metal work function of copper clusters.

We consider that the charge carrier generation in the dark by copper clusters in the contact strip as a component of the solar cell is caused by the deformation of the strip. It is known (Albert & Chudnovsky, 2008), that deformation of metal cluster structures can induce high-temperature superconductivity. Therefore, it is necessary to investigate the behavior of the studied samples in a magnetic field.

Solar energy conversion is widely used in electric power generation. Its efficiency in domestic and industrial plants depends on the quality of components (Slaoui A & Collins, 2007). Discovered in this work, the dark current in the silver contact on the illuminated side of a silicon solar cell generates electricity in amount of up to 5% of the rated value in the absence of sunlight. Therefore, the efficiency of solar energy conversion plants with copper–silver contacts is higher even at the same efficiency of the semiconductor part of the solar cell.

9. Metallic nanocluster contacts for high-effective photovoltaic devices

High efficiency of solar energy conversion is a main challenge of many fields in novel nanotechnologies. Various nanostructures have been proposed early (Pillai et al., 2007; Hun et al., 2007; Johnson et al., 2007; Slaoui & Collins, 2007). However, every active element cannot function without electrodes. Thus, the problem of performing effective contacts is of particular interest.

The unique room-temperature electrical characteristics of the porous metallic nanocluster-based structures deposited by the wet chemical technology on conventional silicon-based solar cells were described in (Laptev & Khlyap, 2008). We have analyzed the current-voltage characteristics of Cu-Ag-metallic nanocluster contact stripes and we have registered for the first time dark currents in metallic structures. Morphological investigations (Laptev & Khlyap, Kozar et al., 2010) demonstrated that copper particles are smaller than 0.1 μm and smaller than the pore diameter in silver.

Electrical measurements were carried out for the nanoclustered Ag/Co-contact stripe (Fig.18, inset) and a metal-insulator-semiconductor (MIS) structure formed by the silicon substrate, SiN_x cove layer, and the nanocluster stripe. Fig. 18 plots experimental room-temperature current-voltage characteristics (IVC) for both cases.

As is seen, the nanocluster metallic contact stripe (function 3 in Fig. 18) demonstrates a current-voltage dependence typical for metals. The MIS-structure (functions 1 and 2 in Fig. 18) shows the IVC with a weak asymmetry at a very low applied voltage; as the external electric field increases, the observed current-voltage dependence transforms in a typical "metallic" IVC. More detailed numerical analysis was carried out under re-building the experimental IVCs in a double-log scale.

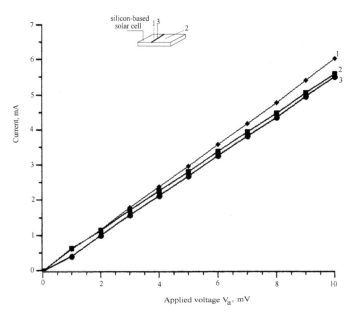

Fig. 18. Room-temperature current-voltage characteristics of the investigated structures <8see text above): functions 1 and 2 are "forward" and "reverse" currents of the MIS-structure (contacts 1-2), and the function 3 is a IVC for the contacts 1-3.

Fig. 19 illustrates a double-log IVCs for the investigated structure. The numerical analysis has shown that both "forward" and 'reverse" currents can be described by the function

$$I = f(V_a)^m,$$

where I is the experimental current (registered under the forward or reverse direction of the applied electric field), and V_a is an applied voltage. The exponential factor m changes from 1.7 for the "forward" current at V_a up to 50 mV and then decreases down to ~1.0 as the applied bias increases up to 400 mV; for the "reverse" current the factor m is almost constant (~1.0) in the all range of the external electric field.

Thus, these experimental current-voltage characteristics (we have to remember that the investigated structure is a metallic cluster-based quasi-nanowire!) can be described according to the theory (Sze & Ng, 2007) as follows: the first section of forward current

$$I = T_{tun}A_{el}(4\varepsilon/9L^2)(2e/m^*)^{1/2}(V_a)^{3/2}$$

(ballistic mode) and the second one as

$$I = T_{tun}A_{el}(2\varepsilon v_s/L^2)V_a,$$

and the reverse current is

$$I = T_{tun}A_{el}(2\varepsilon v_s/L^2)V_a$$

(velocity saturation mode). Here T_{tun} is a tunneling transparency coefficient of the potential barrier formed by the ultrathin native oxide films, A_{el} and L are the electrical area and the length of the investigated structure, respectively, ε is the electrical permittivity of the structure, m^* is the effective mass of the charge carriers in the metallic Cu-Ag-nanoclucter structure, and v_s is the carrier velocity (Kozar et al., 2010). These experimental data lead to the conclusion that the charge carriers can be ejected from the pores of the Cu-Ag-nanocluster wire in the potential barrier and drift under applied electric field (Sze & Ng, 2007; Peleshchak & Yatsyshyn, 1996; Datta, 2006; Ferry & Goodnick, 2005; Rhoderick, 1978).

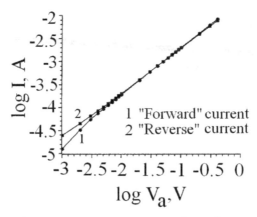

Fig. 19. Experimental room-temperature current-voltage characteristic of the examined structure in double-logarithmic scale.

10. References

Albert J. & Chudnovsky E.M. (2008). Voltage from mechanical stress in type-II superconductors: Depinning of the magnetic flux by moving dislocations, *Appl. Phys. Lett.*, Vol. 93, Issue 4, pp. 042503-1-3, doi:10.1063/1.2960337, 0003-6951(print), 1077-3118 (online).

Badesku V., Landsberg P. T. & De Vos A., Desoete B. (2001). Statistical thermodynamic foundation for photovoltaic and photothermal conversion. IV. Solar cells with larger-than-unity quantum efficiency revisited, *Journal of Applied Physics*, Vol. 89, No.4, (February 2001), pp.2482-2490, ISSN 0021-8979.

Bazarov I.P. (1964).*Thermodynamics*, Pergamon Press, ISBN 978-0080100050, Oxford, Great Britain.

Berg J.M., Tymoczko J.L. & Stryer, L. (2010). *Biochemistry*, Freeman W. H. & Company, ISBN: 1429229365, ISBN-13: 9781429229364, New York, USA.

Chernin A.D. (2008). Dark energy and universal antigravitation. *Phys. Usp.* Vol. 51, No. 3, pp. 253–282, ISSN: 1063-7869(Print), 1468-4780(Online).

Couture L. & Zitoun R. (2000). *Statistical Thermodynamics and Properties of Matter*, Gordon and Breach, ISBN 9789056991951, Amsterdam, Holland.

Curzon F. & Ahlborn B. (1975). Efficency of a Carnot engine at maximum power output, *American Journal of Physics*, Vol. 43, Issue 1, January 1975, pp. 22-24, ISSN 0002-9505.

Datta S. (2006). *Quantum transport: Atom to Transistor,* Cambridge Univ. Press, ISBN 0-521-63145-9, Cambridge, Great Britain.

De Vos A.et al. (1993). Entropy fluxes, endoreversibility, and solar energy conversion, *Journal of Applied Physics,* Vol. 74, No. 6, (June 1993), pp.3631-3637, ISSN 0021-8979.

De Vos A. (1992). *Endoreversible thermodynamics of solar energy conversion,* Oxford Univ. Press, ISBN 978-0198513926, Oxford, Great Britain.

De Vos A. (1985). Efficiency of some heat engine at maximum-power conditions, *Am. J. Phys.,* Vol. 53, Issue 6, pp. 570-573, ISSN 0002.9505.

Dreicer H. (1964). Kinetic Theory of an Electron-Photon Gas, *Phys. Fluids,* Vol. 7, No. 5, pp. 732-754, Print: ISSN 1070-6631 Online: ISSN 1089-7666.

Dubinov A.E. (2009). Exact stationary solution of the Kompaneets kinetic equation. Точное стационарное решение кинетического уравнения Компанейца, Technical Physics Letters, Vol.35, No.3, pp.260-262. *Письма в ЖТФ,* том 35, вып.6, С.25-30, ISSN: 0320 - 0116.

Ferry D. & Goodnick S. (2005). *Transport in Nanostructures,* Cambridge Univ. Press, ISBN 0-521-66365-2, Cambridge, Great Britain

Green B. (2004). *The fabric of the cosmos: space, time, and the texture of realiti.* Alfred A. Knoff, ISBN 978-5-397-00001-7, New York, USA.

Han H., Bach U. & Cheng Y., Caruso R.A. 2007. Increased nanopore filling: Effect on monolithic all-solid-state dye-sensitized solar cell. *Applied Physics Letters,* Vol. 90, No.21, (May 2007), pp.213510-1-3, ISSN 0003-6951.

Hitz B. (2007). Mid-IR Fiber Laser Achieves ~10 W, *Photonics Spectra,* Vol. 41, No. 9, pp. 21-23. ISSN: n.d.

Ivanov K.P. (2008). Energy and Life. Энергия и жизнь, *Успехи современной биологии (Usp. Sovrem. Biol),* Vol. 128, No. 6, pp. 606-619. ISSN Print: 0042-1324

Johnson D.C., Ballard I.M. & Barnham K.W.J., Connolly J.P., Mazzer M., Bessière A., Calder C., Hill G., Roberts J.S. (2007). Observation of the photon recycling in strain-balanced quantum well solar cell. *Applied Physics Letters,* Vol. 90, No.21, (May 2007), pp.213505-1-3, ISSN 0003-6951.

Klaers J., Schmitt J., & Vewinger F., Weitz M. (2009). Bose–Einstein condensation of photons in an optical microcavity. *Nature,* Vol. 468, pp. 545-548, doi:10.1038/nature09567. ISSN: 0028-0836.

Kompaneets A.S., (1957). The establishment of thermal equilibrium between quanta and electrons, *Soviet Physics - JETP,* Vol. 4, No. 5, pp. 730-740, ISSN: 0038-5646.

Kondepudi D. & Prigogin I. (1998). *Modern Thermodynamics: From Heat Engines to Dissipative Structure,* John Wiley & Sons, Inc., ISBN 5-03-765432-1, New York, USA.

Kozar T. V., Karapuzova N. A. & Laptev G. V., Laptev V. I., Khlyap G. M., Demicheva O. V., Tomishko A. G., Alekseev A. M. (2010). Silicon Solar Cells: Electrical Properties of Copper Nanoclusters Positioned in Micropores of Silver Stripe-Geometry Elements, *Nanotechnologies in Russia,* Vol. 5, № 7-8, p.549-553, DOI: 10.1134/S1995078010070165, ISSN: 1995-0780 (print), ISSN: 1995-0799 (online).

Kuz'min V.A. & Shaposhnikov M.E. (1978). Condensation of photons in the hot universe and longitudinal relict radiation, *JETP Lett.,* Vol. 27, No. 11, pp.628-631, ISSN: 0370-274X.

Landsberg, P.T. & Leff, H. (1989). Thermodynamic cycles with nearly universal maximum-work efficiencies. *Journal of Physics A*, Vol. 22, No.18, (September 1989), pp.4019-4026. ISSN 1751-8113(print).

Landsberg P.T. & Tonge G. (1980). Thermodynamic energy conversion efficiency, *Journal of Applied Physics*, Vol. 51, No. 7, (July 1980), pp. R1-R20, ISSN 0021-8979.

Landsberg P.T. (1978). *Thermodynamics and statistical mechanics*. Oxford University Press, ISBN 0-486-66493-7, Oxford, Great Britain.

Landsberg P.T. (1977). A note on the thermodynamics of energy conversion in plants, *Photochemistry and photobiology*, Vol. 26, Issue 3, pp. 313-314, Online ISSN: 1751-1097.

Laptev V.I. (2010). Chemical Potential and Thermodynamic Functions of Thermal Radiation, *Russian Journal of Physical Chemistry A*, Vol. 84, No. 2, pp. 158–162, ISSN 0036-0244.

Laptev V.I. (2009). Thermodynamic Scale of the Efficiency of Chemical Action of Solar Radiation. *Doklady Physical Chemistry*, Vol. 429, Part 2, pp. 243–245, ISSN 0012-5016.

Laptev V.I. (2008). Solar and heat engines: thermodynamic distinguish as a key to the high efficiency solar cells, In: *Solar Cell Research Progress*, Carson J.A. (Ed.), pp. 131–179, Nova Sci. Publ., ISBN 978-1-60456-030-5, New York, USA.

Laptev V.I. & Khlyap H. (2008). High-Effective Solar Energy Conversion: Thermodynamics, Crystallography and Clusters, In: *Solar Cell Research Progress*, Carson J.A. (Ed.), pp. 181–204, Nova Sci. Publ., ISBN 978-1-60456-030-5, New York, USA.

Laptev V.I. (2006). The Special Features of Heat Conversion into Work in Solar Cell Energy Reemission. *Russian Journal of Physical Chemistry*, Vol. 80, No. 7, pp. 1011–1015, ISSN 0036-0244.

Laptev V.I. (2005). Conversion of solar heat into work: A supplement to the actual thermodynamic description, *J.Appl. Phys.*, Vol. 98, 124905, DOI: 10.1063/1.2149189, ISSN 0021-8979(print), 1089-7550 (online).

Leff H. (1987). Thermal efficiency at maximum work output: New results for old heat engines. *American Journal of Physics*, Vol. 55, Issue 7, July 1987, pp.602-610, ISSN 0002-9505.

Lehninger A.L., Nelson D.L. & Cox M.M. (2008). *Lehninger Principles of Biochemistry*, 5th ed., Freeman, ISBN: 1572599316, ISBN-13: 9781572599314, New York, USA.

Lukash V.N, Rubakov V.A. (2008). Dark energy: myths and reality. Phys. Usp. Vol. 51, No. 3. pp. 283–289. ISSN: 1063-7869(Print), 1468-4780(Online).

Luque A. & Marti A. (2003). In: *Handbook of Photovoltaic Science and Engineering*, Luque A. & Hegedus S.(Eds.), John Wiley and Sons Ltd., ISBN: 978-0-471-49196-5. pp. 113-151, New York, USA.

Mazenko G.F. (2000). *Equilibrium Statistical Mechanics*, Wiley & Sons, Inc, ISBN 0471328391, New York, USA.

Muenster A. (1970). *Classical Thermodynamics*, Wiley-Interscience, ISBN: 0471624306, ISBN-13: 9780471624301, New York, USA.

Novikov I. (1958). The efficiency of atomic power stations. *Journal of Nuclear Energy*, Vol. 7, No. 1-2, (August 1958), pp.125-128.

Odum E.P. (1983). *Basic Ecology*, CBS College Publ., ISBN: 0030584140, ISBN-13: 9780030584145 New York, USA.

Pechurkin N.S. (1988). *Energiya i zhizn'* (Energy and Life), Nauka, Novosibirsk, Russia.

Peleshchak R.M. & Yatsyshyn V.P. (1996). About effect of inhomogeneous deformation on electron work function of metals, *Physics of Metals and Metallography*, MAIK Nauka Publishers – Springer, vol. 82, No. 3, pp.18-26, ISSN Print: 0031-918X, ISSN Online: 1555-6190.

Pillai S , Catchpole K.R. & Trupke T., Green M.A. (2007). Surface plasmon enhanced silicon solar cells, *Journal of Applied Physics*, Vol. 101, No.9, (May 2007), pp. 093105-1-8, doi:10.1063/1.2734885, ISSN: 0021-8979 (print), 1089-7550 (online)

Raven P.H.; Evert, R.F. & Eichhorn S.E. (1999). *Biology of Plants*. 6nd ed., Worth Publiscers, Inc., ISBN: 1572590416, USA.

Rhoderick E. H., (1978). *Metal-semiconductor contacts*. Clarendon Press, ISBN 0198593236, Oxford, Great Britain.

Rubin M. (1979). Optimal configuration of a class of irreversible heat engines. I. *Physical Review A*, Vol. 19, No. 3, (March 1979), pp.1272-1276. ISSN 1094-1622 (online), 1050-2947 (print).

Semenchenko V.K. (1966). *Selected Chapters of Theoretical Physics*, Education, Moscow

Shockley W.; Queisser H., (1961). Detailed Balance Limit of Efficiency of *p-n* Junction Solar Cells, *J. Appl. Phys.*, , 32, 510-519. ISSN 0021-8979.

Slaoui A. & Collins R.T. (2007). Advanced Inorganic Materials for Photovoltaics. *MRS Bulletin*, Vol. 32, No.3, pp.211-214, ISSN: 0883-7694.

Spavieri G. & Rodrigues M. (2007). Photon mass and quantum effects of the Aharonov-Bohm type, *Phys. Rev. A*, Vol. 75, 05211, ISSN 1050-2947 (print) 1094-1622 (online).

Steinfeld A. & Palumbo R. (2001). *Encyclopedia of Physical Science & Technology*. R.A. Mayers (Editor)Vol. 15, pp. 237–256, Academic Press, ISBN: 0122269152, ISBN-13: 9780122269158, New York.

Sze S.M. & Ng, K.K. (2007). *Physics of semiconductor devices*, J. Wiley & Sons, Inc., ISBN 0-471-14323-5, Hoboken, New Jersey, USA.

Voet D.J., Voet J.G. & Pratt C.W. (2008). *Principles of Biochemistry*, 3d ed., John Wiley & Sons Ltd, ISBN:0470233966, ISBN-13: 9780470233962, New York,USA.

Wegh R.T., Donker, H. & Oskam K.D., Meijerink A. (1999). Visible Quantum Cutting in LiGdF$_4$:Eu^{3+} Through Downconversion, *Science*, 283, 663-666, *DOI:10.1126/science.283.5402.663*, ISSN 0036-8075 (print), 1095-9203 (online).

Werner J.; Kolodinski S. & Queisser H. (1994). Novel optimization principles and efficiency limits for semiconductor solar cells. *Physical Review Letters*, vol.72, No.24 (June 1994), p.3851-3854. ISSN 0031-9007 (print), 1079-7114 (online)

Weymann R. (1965). Diffusion Approximation for a Photon Gas Interacting with a Plasma via the Compton Effect , *Phys. Fluids*, Vol. 8, No. 11, pp. 2112-2114, Print: ISSN 1070-6631 Online: ISSN 1089-7666.

Wuerfel P. (2005). *Physics of Solar Cells*, WILEY-VCH Verlag GmbH and Co. KGaA, ISBN 978-3527408573, Wienheim, Germany.

Yungman V.S. & Glushko (Eds). (1999). *Thermal Constant of Substances*, 8 volume set, Vol. 1, John Wiley & Sons, ISBN: 0471318558 New York, USA.

Zel'dovich Ya.B. & Syunyaev R.A. (1972). Shock wave structure in the radiation spectrum during bose condensation of photons, *Soviet Physics - JETP*, Vol. 35, No. 1, pp. 81-85, ISSN: 0038-5646.

7

Hybrid Solar Cells Based on Silicon

Hossein Movla[1], Foozieh Sohrabi[1],
Arash Nikniazi[1], Mohammad Soltanpour[3] and Khadije Khalili[2]
[1]Faculty of Physics, University of Tabriz
[2]Research Institute for Applied Physics and Astronomy (RIAPA), University of Tabriz
[3]Faculty of Humanities and Social Sciences, University of Tabriz
Iran

1. Introduction

Human need for renewable energy resources leads to invention of renewable energy sources such as Solar Cells (SCs). Historically, the first SCs were built from inorganic materials. Although the efficiency of such conventional solar cells is high, very expensive materials and energy intensive processing techniques are required. In comparison with the conventional scheme, the hybrid Si-based SC system has advantages such as; (1) Higher charging current and longer timescale, which make the hybrid system have improved performances and be able to full-charge a storage battery with larger capacity during a daytime so as to power the load for a longer time; (2) much more cost effective, which makes the cost for the hybrid PV system reduced by at least 15%(Wu et al., 2005). Thus, hybrid SCs can be a cheap alternative for conventional SCs.

One type of hybrid SCs is a combination of both organic and inorganic materials which combines the unique properties of inorganic semiconductors with the film forming properties of conjugated polymers. Organic materials are inexpensive, easily processable, enabling lightweight devices and their functionality can be tailored by molecular design and chemical synthesis. On the other hand, inorganic semiconductors can be manufactured as nanoparticles and inorganic semiconductor nanoparticles offer the advantage of having high absorption coefficients, size tenability and stability. By varying the size of nanoparticles the bandgap can be tuned therefore the absorption range can be tailored (Günes & Sariciftci, 2008). These kinds of hybrid SCs based on organic-inorganic materials are fabricated by using different concepts such as solid state dye-sensitized SCs and hybrid SCs using Bulk Heterojunction (BHJ) concept such as TiO_x(Hal et al., 2003), ZnO (Beek et al., 2006), CdSe (Alivisatos, 1996; Huynh et al., 2002), Cds (Greenham et al., 1996), PbS (McDonald et al.,2005), and $CuInS_2$.

Another generation of hybrid SCs are silicon-based modules due to the direct bandgap and high efficiency of Si. This system includes SC module consisting of crystalline and amorphous silicon-based SCs. The methods for enhancing the efficiencies in these types of hybrid SCs such as applying textured structures for front and back contacts as well as implementing an intermediate reflecting layer (IRL) between the individual cells of the tandem will be discussed (Meillaud et al., 2011). This chapter brings out an overview of principle and working of hybrid SCs consisting of HJ SCs which is itself devided into two groups, first organic-inorganic module and second, HJ SCs based on single crystalline, amorphous and microcrystalline Si and SCs in dye-sensitized configuration. Afterward, material characterization of these kinds of SCs will be investigated. Precisely, Crystalline Si thin film SCs and later amorphous and microcrystalline Si SCs and the recent works are discussed.

2. Principle and working of hybrid solar cells

One of the methods to build hybrid SCs is Bulk Hetrojunction (BHJ) SCs, composed of two semiconductors. Excitons created upon photoexcitation are separated into free charge carriers at interfaces between two semiconductors in a composite thin film such as a conjugated polymer and fullerene mixtures. One of these materials of an HJ obviously must be an absorber. The other may be an absorber, too, or it may be a window material; i.e., a wider–gap semiconductor that contributes little or nothing to light absorption but is used to create the HJ and to support carrier transport. Window materials collect holes and electrons, which function as majority-carrier transport layers, and can separate the absorber material from deleterious recombination at contacts. The interface they form with the absorber is also used for exciton dissociation in cells where absorption is by exciton formation. Absorber and window materials may be inorganic semiconductors, organic semiconductors, or mixtures (Fonash, 2010, as cited in Khalili et al. 2010; Sohrabi et al. 2011). For applying HJ structure (HJS) for hybrid SCs, the blends of inorganic nanocrystals with semiconductive polymers as a photovoltaic layer should be employed.

Schematically, the HJ hybrid SCs consist of at least four distinct layers, excluding the substrate, which may be glass. These our layers are anode, cathode, hole transport layer and active layer. Induim tin oxide (ITO) is a popular anodic material due to its transparency and glass substrate coated with ITO is commercially available. A layer of the conductive polymer mixture PEDOT:PSS may be applied between anode and the active layer. The PEDOT:PSS layer serves several functions. It not only serves as a hole transporter and exciton blocker, but it also smoothens out the ITO surface, seals the active layer from oxygen, and keeps the anode material from diffusing into the active layer, which can lead to unwanted trap sites. Next, on the top of the PEDOT:PSS, layer deposited is the active layer. The active layer is responsible for light absorption, exciton generation/dissociation and charge carrier diffusion (Chandrasekaran et al., 2010). The so-called two materials are inserted in active layer namely donor and acceptor. Polymers are the common donors whereas nanoparticles act as common acceptors. On the top of active layer is cathode, typically made of Al, Ca, Ag and Au (Chandrasekaran et al., 2010).

BHJ hybrid SCs attracts much interest due to these features:

a. HJs allow the use of semiconductors that can only be doped either n-type or p-type and yet have attractive properties which may conclude their absorption length, cost, and environmental impact. The existence of concentration gradient of the n-type nanoparticles within the p-type polymer matrix may allow optimization of the topology of the HJ network.

b. HJs allow the exploitation of effective forces.

c. HJs of window-absorber type can be used to form structures that shield carriers from top-surface or back-surface recombination sinks (Fonash , 2010).

d. The affinity steps at HJ interfaces can be used to dissociate excitons into free electrons and holes.

e. HJs can also permit open-circuit voltages that can be larger than the built-in electrostatic potential.

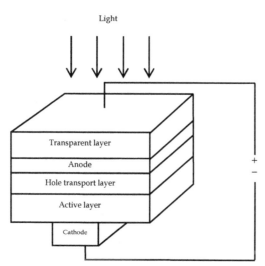

Fig. 1. Structure of HJ hybrid SCs

f. Inorganic semiconductor materials can have high absorption coefficients and photoconductivity as many organic semiconductor materials (Günes & Sariciftci, 2008). Typically, inorganic semiconductors in macroscopic dimensions, irrespective of their size, will absorb all electromagnetic radiation with energy greater than the bandgap. However, if the particles become smaller than that of the exciton in the bulk semiconductor (typically about 10 nm), their electronic structure has changed. The electronic properties of such small particles will depend not only on the material of which they are composed, but also on their size, the so-called quantum confinement effect (Arici et al., 2004, as cited in Weller, 1993; Steigerwald & Brus, 1990; Alivisatos, 1996; Empedocles & Bawendi, 1999; Murphy & Coffer, 2002; Movla et al. 2010a). The lowest energy of optical transition, among others, will increase significantly due to the quantum confinement with decreasing size of the inorganic clusters. Since the energy levels of the polymers can be tuned by chemical modification of the backbone chain and the energy levels of the nanoparticles can be tuned through the size-dependent quantum confinement effects, blends of the two materials offer the possibility of tailoring optimal conditions for a solar cell, including energy gain from charge transfer for the efficient charge separation and the spectral range of the absorbing light (Arici et al., 2004). Therefore, in order to obtain hybrid polymer SCs with high current and fill factor, both electron and hole mobilities must be optimized and most importantly balanced (Chandrasekaran et al., 2010). However, diffusion of nanoparticles into the polymer matrix takes place with the penetration depth controlled by temperature, swelling of the polymer layer, and not at least by the size and shape of the nanocrystals.

Another module of HJ hybrid SCs consists of crystalline and amorphous silicon-based SCs which is the main discussion in this chapter. The present PV market is dominated by three kinds of Si-based solar cells, that is, single-, multi-crystalline or amorphous Si-based solar cells (for short, marked hereafter as Sc-Si, Mc-Si and a-Si solar cells, respectively). The conventional PV system in general uses Sc-Si or Mc-Si solar cell module as the element for solar energy conversion, which have comparatively higher conversion efficiency. However, it is not only the module efficiency that decides whether a PV system is cost effective but also the timescale during which the module works efficiently in a daytime of use and the cost the module itself requires. At this point, a-Si solar cell comes with its advantages of broader timescale and lower cost (Wu et al., 2005, as cited in Goetzberger et al., 2003).

The broader timescale merit of a-Si solar cell arises from its high absorption of light with wavelength around 300–800nm, no matter if it is scattered or not, and no matter if it is weak or blazing. The Sc-, Mc- and a-Si solar cells, therefore, reinforce each other in performances, which could be exploited to construct a hybrid PV system with lower cost in view of the well balanced set of system performance (Wu et al., 2005). The last efficiencies reported for c-Si, Mc-Si and a-Si are approximately 25%, 20% and 10%, respectively (Green et al., 2011).

The newest configuration for hybrid SCs is dye-sensitized SC developed by O'Reagan and Graetzel in 1991.This class of cell has reached efficiencies of over 11%. The basic structure of a dye-sensitized SC involves a transparent (wide-band-gap) n-type semiconductor configured optimally in a nanoscale network of columns, touching nanoparticles, or coral-like protrusions. The surface area of the network is covered everywhere with a monolayer of a dye or a coating of quantum dots, which functions as the dye (Fonash, 2010). A monolayer of dye on a flat surface can only harvest a negligibly small fraction of incoming light. In this case it is useful to enlarge this interface between the semiconductor oxide and the dye. As mentioned above, it is achieved by introducing a nanoparticle based electrode construction which enhances the photoactive interface by orders of magnitude (Grätzel, 2004). The dye sensitizer is the absorber. An electrolyte is then used to permeate the resulting coated network structure to set up a conduit between the dye and the anode. The dye absorbs light, producing excitons, which dissociate at the dye-semiconductor interface, resulting in photogenerated electrons for the semiconductor and oxidized dye molecules that must be reduced and thereby regenerated by the electrolyte (Fonash, 2010).

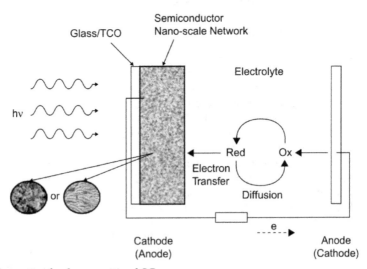

Fig. 2. Schematic of a dye-sensitized SC

A dye-sensitized SC of Graetzel type comprises of several different materials such as nanoporous TiO_2 electrodes, organic or inorganic dyes, inorganic salts and metallic catalysts (Grätzel, 2004, 2005, as cited in Nogueira et al., 2004; Mohammadpour et al., 2010) which will be discussed more in next section. The dye sensitization of the large band-gap semiconductor electrodes is achieved by covering the internal surfaces of porous TiO$_2$ electrode with special dye molecules which absorb the incoming photons (Halme, 2002).

Alternatives to the liquid electrolyte in dye-sensitized SCs are a polymer gel electrolyte or solid state dye-sensitized SCs which can contain organic hole conductor materials, inorganic p-type semiconductors or conjugated polymers (Fonash, 2010). The impetus for this effort is the increased practicality of an all-solid-state device and the avoidance of chemical irreversibility originating from ionic discharging and the formation of active species (Fonash, 2010, as cited in Tennakone et al., 2000). And also this method avoids problems such as leakage of liquid electrolyte. In solid state dye-sensitized SCs, the sensitizer dye is regenerated by the electron donation from the hole conductor (Wang et al., 2006). The hole conductor must be able to transfer holes from the sensitizing dye after the dye has injected electrons into the TiO$_2$. Furthermore, hole conductors have to be deposited within the porous nanocrystalline (nc) layer penetrating into the pores of the nanoparticle and finally it must be transparent in the visible spectrum, or, if it absorbs light, it must be as efficient in electron injection as the dye. (Günes & Sariciftci, 2008)

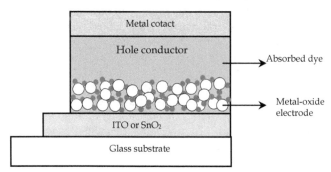

Fig. 3. Schematic of solid state dye-sensitized SC

Other quasi-dye sensitized SCs are nanoparticle sensitized SCs and extremely thin absorber (ETA) SCs. Nanoparticle sensitized SCs are prepared by replacing the dye with inorganic nanoparticles or quantum dots. They can be adsorbed from a colloidal quantum dot solution (Zaban et al., 1998; as cited in Günes et al., 2006; Guenes et al., 2007) or produced in situ (Liu & Kamat, 1993; as cited in Hoyer & Könenkamp, 1995). Inorganic nanocrystals instead of organic dyes could imply tunability of the band-gap and thereby the absorption range. Nanocrystals have large extinction coefficients due to quantum confinement and intrinsic dipole moments, leading to rapid charge separation and are relatively stable inorganic materials (Alivisatos, 1996). To embed the particles into porous TiO$_2$ films and to use those modified layers as light converting electrodes, the incorporated nanoparticles need to be much smaller than the pore sizes of the nanoporous TiO$_2$ electrodes (Shen, 2004).

ETA SCs are conceptually close to the solid state dye-sensitized solar cells. In the ETA SCs, an extremely thin layer of a semiconductor such as CuInS$_2$ or CdTe or CuSCN replaces the dye in TiO$_2$ based SCs. The ETA SCs has the advantage of enhanced light harvesting due to the surface enlargement and multiple scattering. Similar to the solid state dye sensitized Scs, the operation of the ETA SC is also based on a heterojunction with an extremely large interface (Nanu et al., 2005).

3. Material characterization of hybrid solar cells

Relating to the configuration of hybrid SCs like HJ SCs or dye-sensitized SCs, various materials have been suggested by research groups. The BHJ devices were characterized by an interpenetrating network of donor and acceptor materials, providing a large interface area where photo-induced excitons could efficiently dissociate into separated electrons and holes. However, the interpenetrating network cannot be easily formed in the blended mixture. In addition, the organic materials are not good in carrier transport. Thus, the power conversion efficiency is still limited by the low dissociation probability of excitons and the inefficient hopping carrier transport (Huang et al., 2009, as cited in Sirringhaus et al.,1999; Shaw et al.,2008). Semiconductor nanostructures are used to be combined with the organic materials to provide not only a large interface area between organic and inorganic components for exciton dissociation but also fast electron transport in semiconductors. Therefore, many research groups combined organic materials with semiconductor nanostructures to overcome the drawbacks of the organic solar cells. Many inorganic nanowire (NW) had been experimented for this purpose, including CdTe, CdS, CdSe, ZnO, and TiO_2 NWs (Huang et al., 2009, as cited in Kang & Kim, 2006). Totally, BHJ hybrid SCs itself have been demonstrated in various inorganic materials such as CdSe nanodots, nanorods and tetrapods, TiO_2, ZnO, ZnS nanoparticles, $CuInS_2$, $CuInSe_2$, CuPc, CdS, SnS, CIS , PbSe or PbS nanocrystals , HgTe ncs, Si NWs, Si ncs (Chandrasekaran et al., 2010 as cited in Kwong et al.,2004, Arici et al., 2004 Greenham et al., 1996 Qi et al., 2005; Choudhury et al., 2005), etc. which act as acceptors and polymer materials acting as donors are P3HT, PPHT, P3OT, P3BT, P3MeT (Lin et al., 2006), MDMO-PPV, MEH-PPV, MOPPV, etc. (Chandrasekaran et al., 2010). However,CdTe, CdS,and CdSe materials are harmful to the environment, while ZnO and TiO_2 have a bandgap higher than 3eV and so cannot effectively absorb the solar spectrum. To overcome this, SiNWs are suitable for this application because they are environmental friendly and have high absorption coefficient in the infrare dregion (Huang et al., 2009).

J.Haung et al. (2009) reported the fabrication of the SiNW/P3HT:PCBM blend hybrid SCs using the SiNW transfer technique.

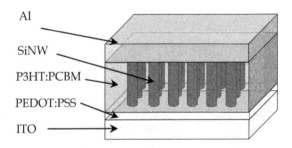

Fig. 4. A schematic of the hybrid SC using SiNWs and P3HT:PCBM blend

Their investigation showed that after introducing the SiNWs, the J_{sc} increases from 7.17 to 11.61 mA/cm^2 and η increases from 1.21% to 1.91% (Haung et al., 2009). This increase is due to this fact that the NWs act as a direct path for transport of charge without the presence of grain boundaries.(Movla et al., 2010b).

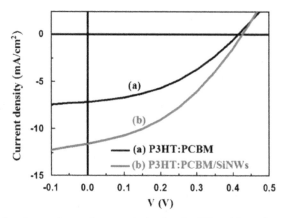

Fig. 5. The current density–voltage characteristics for the SCs with and without the SiNWs under simulated AM1.5 illumination. Reprinted with permission from Solar Energy Materials & Solar Cells Vol.93, Huang, J. et al. Well-aligned single-crystalline silicon nanowire hybrid solar cells on glass, pp. 621–624 © 2009, Elsevier.

More precisely, the results clearly indicate that combination of the SiNWs and P3HT:PCBM blend is an attractive route to obtain high J_{sc} and efficiencies by improving the optical absorption, dissociation of excitons, and the electron transport. Silicon wafer is commercially available and cheap. SiNWs can be fabricated at low temperature from solution processing without any vacuum equipment or high-temperature processing. In addition, this transfer method for SiNWs is simple and fast. It is not a laborious way.This method is suitable for plastic SCs because it can be processed fast, is cheap and simple (Haung et al., 2009).

Similar work was done by G. Kalita et al. (2009) for demonstrating hybrid SCs using Si NWs and polymer incorporating MWNTS. This fabricated device with the structure of Au/P3OT+O-MWNTS/n-Si NWs marked a conversion efficiency of 0.61% (Bredol et al., 2009). Another study was done by C. Y. Liu et al. (2009) about fabricating the hybrid SCs on blends of Si ncs and P3HT (Liu et al., 2009). Also, V. Svrcek et al. (2009), investigated the photoelectric property of BHJ SC based on Si-ncs and P3HT. They came into conclusion that I-V characteristic enhanced when BHJ was introduced into TiO_2 nanotube (nt). The arrangement of Si-ncs/P3HT BHJ within ordered TiO_2 nt perpendicular to the contact facilitated excition separation and charge transfer along nts (Chandrasekaran et al., 2010, as cited in Svrcek et al., 2009).

A new approach for hybrid metal-insulator-semiconductor (MIS) Si solar cells is adopted by the Institute of Fundamental Problems for High Technology, Ukrainian Academy of Sciences. In this technique, the porous silicon layers are created on both sides of single crystal wafers by chemical etching before an improved MIS cell preparation process. The porous Si exhibits unique properties. It works like a sunlight concentrator, light scattering diffuser and reemitter of sunlight as well as an electrical isolator in the multilayer Si structure. The most important advantage of using porous Si in SCs is its band gap which behaves as a direct band gap semiconductor with large quantum efficiency and may be adjusted for optimum sunlight absorption. Employing a specific surface modification, porous Si improves the PV efficiency in UV and NIR regions of solar spectra (Tuzun et al.,

2006, as cited in Tiris et al., 2003). In this approach, due to high quality starting materials and rapid low-temperature (<800 ⁰C) processing a high minority carrier life time is attainable; this, in turn, gives rise to a high photogenerated current collection. Therefore, the SCs with efficiencies above 15% have been obtained under AM1.5 condition (under 100 mW/cm² illumination at 25 ⁰C) (Tuzun et al., 2006).

Another study was reported by J. Ackermann et al. (2002) about the growth of quaterthiophene (4T), a linear conjugated oligomer of thiophene behaving as a p-type semiconductor, on n-doped GaAs and Si substrates to form hybrid HJ SCs. This study shows that in the case of Si as substrate there is almost defect-free high ordered films with grain sizes of several micrometers up to a film thickness of 250 mm (Ackermann et al., 2002). K.Yamamoto et al. (2001) investigated a-Si/poly-Si hybrid (stacked) SC paying attention to the stabilized efficiency, since a-Si has photo degradation, while a poly-Si is stable. This tandem cell exhibited a stabilized efficiency of 11.3%. Also, this research group prepared three-stacked cell of a-Si:H/poly-Si/poly-Si(triple), which will be less sensitive to degradation by using the thinner a-Si. This triple cell showed a stabilized efficiency of 12%(Yamamoto et al., 2001) . Their reasons for applying poly-Si are (i) high growth rate (ii) large area and uniform deposition at the same time and (iii) monolithic series interconnection. In addition, for enhancing the absorption, they suggested natural surface texture and a back reflector (See Fig .6.) (Yamamoto et al., 2001).

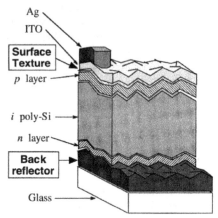

Fig. 6. Schematic view of thin film poly-Si solar cell with natural surface texture and enhanced absorption with back reflector structure. Reprinted with permission from Solar Energy Materials & Solar Cells Vol.93, Huang, J. et al. Well-aligned single-crystalline silicon nanowire hybrid solar cells on glass, pp. 621–624 © 2009, Elsevier.

Apart from BHJ SCs, solid state dye-sensitized SCs can experience various materials. CuI, CuBr, CuSCN, MgO (Tennakone et al.,2001) can be replacements for liquid crystals as inorganic p-type semiconductors. Also, 2,2',7,7'-tetranis(N,N-di-p-methoxyphenyl-amine)9,9'-spirobifluorene (OMETAD) can replace the liquid crystals as organic p-type semiconductor due to their low cost processability (Bach et al.,1995). Poly (3 alkylthiophenes) were used to replace the liquid electrolyte by Sicot et al. (Sicot et al., 1991) and Gebeyehu et al. (Gebeyenha et al.,2002a,2002b) as conjugated polymers although high molecular weight polymers cast from solution, do not penetrate into the pores of the nanoparticles. A polymeric gel electrolyte is considered as a compromise between liquid electrolytes and hole conductors in quasi solid state dye-sensitized SCs (Günes & Sariciftci,2008, as cited in Nogueira et al.,2004;Murphy,1998; Megahed & Scosati,1995) . A mixture of NaI, ethylene carbonate, propylene carbonate and polyacrylonitrile was reported by Cao et al. [55]. Poly (vinylidenefluoride- co-hexafluoropropylene) (PVDF-HFP) used to solidify 3-methoxypropionitrile (MPN) was utilized by Wang et al. (Wang et al.,2004). The last efficiency for dye-sensitized reported by Sharp Corporation is about 10.4 which stands in lower rank in comparison with crystalline Si showing efficiency of approximately 25% (Green et al., 2011).

3.1 Crystalline silicon thin film solar cells

Silicon is the leading material used in microelectronic technology and shows novel photoelectrochemical properties in electrolyte solutions (Wang et al.,2010). Now and before Si-based cells especially crystalline Si has shown higher efficiencies. The last efficieny reported by UNSW PERL is 25% which exceeds other types of Si-based SCs (Green et al., 2011, Zhao et al., 1998).

Crystalline silicon (c-Si) is an extremely well suited material for terrestrial photovoltaics (PV). It is non-toxic and abundant (25% of the Earth's crust), has excellent electronic, chemical and mechanical properties, forms a simple monoelemental semiconductor that has an almost ideal bandgap (1.1 eV) for terrestrial PV, and gives long-term stable SCs and modules. Furthermore, it is the material of choice in the microelectronics industry, ensuring that a large range of processing equipment exists and is readily available. Given these properties of c-Si, it is not surprising that almost all (>90%) terrestrial PV modules sold today use Si wafer SCs. However, the fabrication of Si wafers is both material and energy intensive. Therefore, there is a need for a less material intensive c-Si technology calling a thin-film technology. Besides cost savings on the materials side, thin-film technologies offer the additional benefits of large-area processing (unit size about 1 m^2) on a supporting material, enabling monolithic construction and cell interconnection. The recent c-Si thin-film PV approaches can broadly be classified as follows: (i) fabrication of thin, long stripes of c-Si material from thick single crystalline Si wafers ("ultrathin slicing"); (ii) growth of c-Si thin-films on native or foreign supporting materials (Aberle, 2006).

Recently photoelectrochemical (PEC) SCs based on 1D single crystalline semiconductor micro/nanostructures have attracted intense attention as they may rival the nanocrystalline dye-sensitized SCs (Wang et al., 2010, as cited in Law et al., 2005; Peng et al., 2008; Baxter & Aydil, 2005; Jiang et al., 2008; Mor et al., 2006; Hwang et al., 2009; Dalchiele et al., 2009; Goodey et al., 2007; Maiolo et al., 2007). Therefore, X.Wang et al. (2010) proposed single crystalline ordered silicon wire/Pt nanoparticle hybrids for solar energy harvesting. In this configuration, wafer-scale Si wire arrays are fabricated by the combination of ultraviolet lithography (UVL) and metal-assisted etching. This method emphasizes that the etching technology forms SiNWs non-contaminated by catalyst material while SiNWs by vapor-liquid-solid (VLS) is generally contaminated by gold (Wang et al., 2010, as cited in Allen et al., 2008). It is found that PtNPs modified Si wire electrochemical PV cell generated significantly enhanced photocurrents and larger fill factors. The overall conversion efficiency of PtNPs modified Si wire PEC solar cell is up to 4.7%. Array of p-type Si wire modified with PtNPs also shows significant improvement for water splitting (Wang et al., 2010).

3.2 Amorphous (protocrystalline) and microcrystalline silicon solar cells

The use of thin-film silicon for SCs is one of the most promising approaches to realize both high performance and low cost due to its low material cost, ease of manufacturing and high efficiency. Microcrystalline silicon (μc) SCs as a family of thin film SCs formed by plasma CVD at low temperature are assumed to have a shorter carrier lifetime than single-crystal cells, and it is common to employ a p–i–n structure including an internal electric field in the same way as an amorphous SC. These cells can be divided into p–i- n and n–i–p types according to the film deposition order, although the window layer of the SC is the p-type layer in both cases. A large difference is that the underlying layer of a p–i–n cell is the transparent p-type electrode, whereas the underlying layer of an n–i–p cell is the n-type back electrode. Light-trapping techniques are a way of increasing the performance of mc-SCs. This is a core technique for cells made from μc-silicon because – unlike a-silicon – it is essentially an indirect absorber with a low absorption coefficient. That is, the thickness of the Si film that forms the active layer in a mc-silicon SC is just a few μm, so it is not able to absorb enough incident light compared with SCs using ordinary crystalline substrates. As a result, it is difficult to obtain a high photoelectric current. Light trapping technology provides a means of extending the optical path of the incident light inside the SC by causing multiple reflections, thereby improving the light absorption in the active layer (Yamamoto et al., 2004). Light trapping in this method of categorizing the SCs according to p-i-n or n-i-p types, can be achieved in two ways: (1) by introducing a highly reflective layer at the back surface to reflect the incident light without absorption loss, and (2) by introducing a textured structure at the back surface of the thin-film Si SC (see Fig.7.) (Komatsu et al.,2002; Yamamoto et al.,2004).

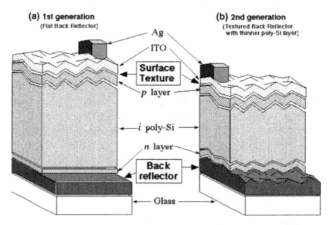

Fig. 7. Cross-sections through light-trapping μc-silicon SC devices: (a) first generation (flat back reflector); (b) second generation (textured back reflector, thinner polycrystalline silicon layer). Reprinted with permission from Solar Energy Vol. 77, Kenji Yamamoto et al., A high efficiency thin film silicon solar cell and module, pp. 939–949 © 2004, Elsevier.

Although the μc-silicon cells formed at low temperature have a potential for high efficiently, their efficiency in single-cell structures is currently only about 10%, which is much lower than that of bulk polycrystalline cells. In order to achieve high efficiencies, Yamamoto et al.(2004) investigated the use of two-and three-stacked (hybrid) structures in which multiple cells with different light absorption characteristics are stacked together. This approach allows better characteristics to be obtained with existing materials and processes. The advantages of using a layered structure include the following: (1) it is possible to receive light by partitioning it over a wider spectral region, thereby using the light more effectively; (2) it is possible to obtain a higher open-circuit voltage; and (3) it is possible to suppress to some extent the rate of reduction in cell performances caused by photo-degradation phenomena that are observed when using a-silicon based materials. Therefore, they have engaged in thin film amorphous and microcrystalline (a-Si/µc-Si) stacked solar cell (Yamamoto et al., 2004).

The advantage of a high Jsc for our µc-Si Sc as mentioned before was applied to the stacked cell with the combination of a-Si cell to gain stabilized efficiency as the study done by K.Yamamoto et al. (2001) since a-Si has a photo-degradation while a µc-Si cell is stable. They have also prepared three stacked cell of a-Si:H/µc-Si/c-Si (triple), which will be less sensitive to degradation by using the thinner a-Si and they have investigated the stability of a-Si:H/µc-Si/µc-Si (triple) cell, too (Yamamoto et al., 2001, 2004). Some other three stacked Si-based SCs can be named such as a-Si/a-SiGe/a-SiGe(tandem) and a-Si/nc-Si/nc-Si (tandem) SCs. The last efficiency reported for a-Si/a-SiGe/a-SiGe(tandem) is about 10.4% and for a-Si/nc-Si/nc-Si (tandem) is approximately 12.5% (Green et al., 2011).

As a next generation of further high efficiency of SC, the new stacked thin film Si SC is proposed where the transparent inter-layer was inserted between a-Si and µc-Si layer to enhance a partial reflection of light back into the a-Si top cell (see fig.8.). This structure is called as internal light trapping enabling the increase of current of top cell without increasing the thickness of top cell, which leads less photo-degradation of stacked cell (Yamamoto et al., 2004).

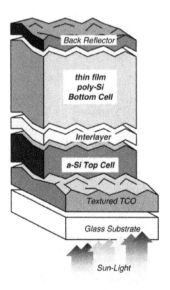

Fig. 8. Schematic view of a-Si/poly-Si (µc-Si) stacked cells with an interlayer. Reprinted with permission from Solar Energy Vol. 77, Kenji Yamamoto et al., A high efficiency thin film silicon solar cell and module, pp. 939-949 © 2004, Elsevier.

By introduction of this interlayer, a partial reflection of light back into the a-Si top cell can be achieved. The reflection effect results from the difference in index of refraction between the interlayer and the surrounding silicon layers. If n is the refractive index and d is the thickness of inter-layer, the product of $\Delta n \times d$ determine the ability of partial reflection. Namely, the light trapping is occurred between the front and back electrode without inter-layer, while with inter-layer, it is also occurred between inter-layer and back electrode. This could reduce the absorption loss of TCO and a-Si:H (see Fig.9.) (Yamamoto et al., 2004).

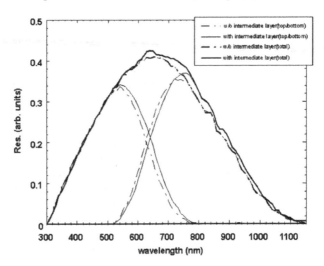

Fig. 9. Spectral-response of the cell with and without interlayer. Bold and normal line shows the spectral-response of the cell with and without inter-layer. Reprinted with permission from Solar Energy Vol. 77, Kenji Yamamoto et al., A high efficiency thin film silicon solar cell and module, pp. 939–949 © 2004, Elsevier.

Some of the advantages of thin film SCs are being characterized to low temperature coefficient, the design flexibility with a variety of voltage and cost potential. Therefore, the thin film Si SCs can be used for the PV systems on the roof of private houses as seen in Fig.10. (Yamamoto et al., 2004).

Another similar work was done by F. Meillaud et al.(2010). They investigated the high efficiency (amorphous/microcrystalline) "micromorph" tandem silicon SCs on glass and plastic substrates. High conversion efficiency for micromorph tandem SCs as mentioned before requires both a dedicated light management, to keep the absorber layers as thin as possible, and optimized growth conditions of the μc-silicon(μc-Si:H) material. Efficient light trapping is achieved in their work by use of textured front and back contacts as well as by implementing an intermediate reflecting layer (IRL)between the individual cells of the tandem.The latest developments of IRLs at IMT Neuchˆatel are: SiOx based for micromorphs on glass and ZnO based IRLs for micromorphs on flexible substrates successfully incorporated in micromorph tandem cells leading to high,matched, current above 13.8mA/cm² for p-i-n tandems. In n-i-p configuration, asymmetric intermediate reflectors (AIRs) were employed to achieve currents of up to 12.5mA/cm² (see fig.11.) (Meillaud et al., 2011).

Fig. 10. KW system at the top of the building Osaka (HYBRID modules). Note that modules are installed at low angle (5 degree off from horizontal).

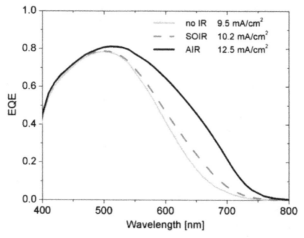

Fig. 11. External quantum efficiency of the top cell in an n-i-p/n-i-p micromorph on flexible substrate (i-layer thickness:200nm). Reprinted with permission from Solar Energy Vol. 95, Meillaud et al., Realization of high efficiency micromorph tandem silicon solar cells on glass and plastic substrates: Issues and potential, pp. 127-130 © 2010, Elsevier.

The last efficiency reported by Oerlikon Solar Lab, Neuchatel for a-Si/μc-Si (thin film cell) is about 11.9% (Green et al., 2011, as cited in Bailat et al., 2010). More precisely talk about tandem cells will be done in following sections.

3.3 Tandem cell

The tandem junction cell is a high-performance silicon solar cell, which is best suited for terrestrial solar power systems. The most distinctive design feature of this device is the use of only back contacts to eliminate the metal shadowing effects because of lower conversion efficiency and steady-state bias requirement. Here we discuss tandem devices consisting of the a-si:H with other forms of silicon:

3.3.1 Tandem cell with multicrystalline Si

In the tandem cell configuration in 1983, Hamakawa reported the structure that a-Si: H/μc-Si heterojunction cell has been investigated as a bottom cell (Hamakawa et al. 1982). Its advantage is that it does not require high temperature processing for junction formation, and the top a-Si:H cell can be fabricated continuously. A heterojunction cell that is reported in 1994 by Matsuyama consisting of a-Si:H as p-type and μc-Si as n-type, with a 10 pm thick μc-Si film fabricated by solid phase crystallization yielded an efficiency of 8.5%. By using a p-μc-Si: C/n-μc-Si/n-pc-Si heterojunction bottom solar cell a conversion efficiency of 20.3% and good stability can be gained. At least in this type of devices the highest efficiency reported up to now is 20.4% (Green et al., 2011).

3.3.2 Tandem cell with microcrystalline Si

Microcrystalline silicon (μc-Si) has been studied extensively for three decades and has been used for doped layers in a-Si solar cells for over 15 years. Microcrystalline silicon is a complex material consisting of conglomerates of silicon nanocrystallites embedded into amorphous silicon. It can be more easily doped than a-Si:H; but, on the other hand, it is also more sensitive to contaminants than a-Si:H. The nucleation and growth of μc-Si:H are determinant for device quality; a certain amount of amorphous material is needed for the passivation of the nanocrystallites and for the reduction of defect related absorption. During the growth of the layer, the formation of crystallites starts with a nucleation phase after an amorphous incubation phase. During continued layer deposition, clusters of crystallites grow (crystallization phase) until a saturated crystalline fraction is reached. These processes are very much dependent on the deposition conditions. In general, crystalline growth is enhanced by the presence of atomic hydrogen, which chemically interacts with the growing surface.

Much research effort has been put worldwide into the development of both fundamental knowledge and technological skills that are needed to improve thin film silicon multijunction solar cells. The research challenges are:

1. To enhance the network ordering of amorphous semiconductors (leading to protocrystalline networks), mainly for improving the stability;
2. To increase the deposition rate, in particular for microcrystalline silicon;
3. To develop thin doped layers, compatible with the new, fast deposition techniques;
4. To design light-trapping configurations, by utilizing textured surfaces and dielectric mirrors.

The first report on practical microcrystalline cells is given in 1992 by Faraji that reported a thin film silicon solar cell with a μc-Si: H: O i-layer by using VHF PECVD and the first solar cell with a μc-Si: H i-layer was fabricated with 4.6% power efficiency (Meier et al., 1994).

In 1996, Fischer reported that microcrystalline (pc) Si: H p-i-n junctions have an extended infrared response, and are entirely stable under photoexcitation with an efficiency of 7.7% (Fischer et al., 1996). The tandem arrangement of an a-Si: H solar cell with a μc-Si: H solar cell also appears promising a remaining problem of the insufficient deposition rates for the μc-Si: H layers. In 2002, Meier published a micromorph tandem cell with efficiency of 10.8% in which the bottom cell was deposited at a rate of R_d =0.5 nm/s with the thickness of 2 μm (Meier et al.,2002). At least in this type of devices the highest efficiencies reported to date are 15.4% (tandem cell consisting of microcrystalline silicon cell and amorphous silicon cell) (Yan et al., 2010).

Further development and optimization of a-Si: H/μc-Si: H tandems will remain very important because it is expected that in the near future, its market share can be considerable. For example, in the European Roadmap for PV R&D, it is predicted that in 2020, the European market share for thin film silicon (most probably a-Si: H/μc-Si: H tandems) will be 30%. This shows the importance of thin film multibandgap cells as second-generation solar cells.

4. Conclusions and outlook

Although conventional SCs based on inorganic materials specially Si exhibit high efficiency, very expensive materials and energy intensive processing techniques are required. In comparison with the conventional scheme, the hybrid Si-based SC system has advantages such as; (1) Higher charging current and longer timescale, which make the hybrid system have improved performances and be able to full-charge a storage battery with larger capacity during a daytime so as to power the load for a longer time; (2) much more cost effective, which makes the cost for the hybrid PV system reduced by at least 15% (Wu et al., 2005). Therefore, hybrid SCs can be suitable alternative for conventional SCs. Among hybrid SCs which can be divided into two main groups including HJ hybrid SCs and dye-sensitized hybrid SCs, HJ hybrid SCs based on Si demonstrate the highest efficiency. Thus, the combination of a-Si/μc-Si has been investigated. These configurations of SCs can compensate the imperfection of each other. For example, a-Si has a photo-degradation while a μc-Si cell is stable so the combination is well stabilized. Furthermore, applying textured structures for front and back contacts and implementing an IRL between the individual cells of the tandem will be beneficial to enhancement of the efficiencies in these types of hybrid SCs. Due to recent studies; a-Si/μc-Si (thin film cell) has an efficiency of about 11.9%. Another study is done over three stacked cell of a-Si:H/μc-Si/c-Si (triple), which will be less sensitive to degradation by using the thinner a-Si. The last efficiency reported for a-Si/a-SiGe/a-SiGe(tandem) is about 10.4% and for a-Si/nc-Si/nc-Si (tandem) is approximately 12.5%.

Furthermore, Si based SC systems are being characterized to low temperature coefficient, the design flexibility with a variety of voltage and cost potential, so it can be utilized in large scale. In near future, it will be feasible to see roofs of many private houses constructed by thin film Si solar tiles. Although hybrid SCs are suitable replacements for conventional SCs, these kinds of SCs based on inorganic semiconductor nanoparticles are dependent on the synthesis routes and the reproducibility of such nanoparticle synthesis routes. The surfactant which prevents the particles from further growth is, on the other hand, an insulating layer which blocks the electrical transport between nanoparticles for hybrid SCs son such surfactants should be tailored considering the device requirements. Therefore, there is an increased demand for more studies in the field of hybrid SCs to find solutions to overcome these weak points.

5. Acknowledgment

The authors would like to express their thanks to Prof. Dr. Ali Rostami from Photonic and Nanocrystal Research Lab (PNRL) and School of Engineering Emerging Technologies at the University of Tabriz, for grateful helps to prepare this chapter. The corresponding author would like to acknowledge financial support of Iran Nanotechnology Initiative Council.

6. References

Aberle , A.G. (2006). Fabrication and characterisation of crystalline silicon thin-film materials for solar cells. *Thin Solid Films*, Vol. 511 – 512, pp. 26 – 34.

Ackermann, J.; Videlot ,C. & El Kassmi, A. (2002). Growth of organic semiconductors for hybrid solar cell application, *Thin Solid Films*, Vol. 403 –404 , pp. 157-161

Arici,E.; Serdar Sariciftci, N. & Meissner, D. (2004). Hybrid Solar Cell. *Encyclopedia of Nanoscience and Nanotechnology*, Nalwa, H.S., (Ed.), pp. 929-944, American Scientific Publishers, ISBN: I -58883-059-4

Chandrasekaran, J.; Nithyaprakash, D.; Ajjan, K.B.; Maruthamuthu, M.; Manoharan, D.& Kumar, S. (2011). Hybrid Solar Cell Based on Blending of Organic and Inorganic Materials — An Overview. *Renewable and Sustainable Energy Reviews*, Vol.15, Issue 2, pp. 1228-1238

Eshaghi G. N.; Movla, H.; Sohrabi, F.; Hosseinpour, A.; Rezaei, M. & Babaei, H. (2010). The effects of recombination lifetime on efficiency and J-V characteristics of InxGa1_xN/GaN quantum dot intermediate band solar cell. Physica E, Vol.42, pp. 2353-2357

Fischer, D.; Dubail, S.; Selvan, J. A. A.; Vaucher, N. P.; Platz, R.; Hof, Ch.; Kroll, U.;   Meier, J.; Torres, P.; Keppner, H.; Wyrsch, N.; Goetz, M.; Shah, A. and Ufert, K.-D. (1996). The "micromorph" solar cell: extending a-Si:H technology towards thin film crystalline silicon, *Photovoltaic Specialists Conference, 1996., Conference Record of the Twenty Fifth IEEE*, Washington, DC , USA, pp. 1053 – 1056

Fonash, S.J. (2010). *Solar Cell Device Physics* (2nd edition), Academic Press (Elsevier), ISBN 978-0-12-374774-7, United State of America

Green, M.A.; Emery,K.; Hishikawa, Y. & Wilhelm, W. (2011). Solar Cell Efficiency Tables (version 37). *Prog. Photovolt: Res. Appl.*, Vol.19, No.9, pp. 84-92

Günes, S. & Serdar Saiciftci, N. (2008). Hybrid Solar Cells, *Inorganica Chimica Acta*, Vol. 361, pp. 581-588

Hamakawa Y. (1982), *Amorphous Semiconductor, Technologies & Devices*, Elsevier, ISBN: 978-0-44-487977-6, United State of America

Huang, J.; Hsiao, C.; Syu, S.; Chao, J. & Lin, C. (2009). Well-aligned single-crystalline silicon nanowire hybrid solar cells on glass, *Solar Energy Materials & Solar Cells* , Vol.93 , pp. 621-624

Khalili Kh.; Asgari A.; Movla H.; Mottaghizadeh A. & Najafabadi H. A. (2011). Effect   of interface recombination on the performance of SWCNT\GaAs heterojunction solar cell. *Procedia Engineering, Physics Procedia*, Vol.8, pp. 275–279

Komatsu, Y.; Koide, N.; Yang, M.; Nakano, T.; Nagano, Y.; Igarashi, K.; Yoshida, K.; Yano, K.; Hayakawa, T.; Taniguchi, H.; Shimizu, M. & Takiguchi, H. (2002). a-Si/mc-Si hybrid solar cell using silicon sheet substrate, *Solar Energy Materials & Solar Cells*, Vol.74, pp. 513–518

Meier, J.; Flückiger R.; Keppner H.; and A. Shah (1994). Complete microcrystalline p-i-n solar cell—Crystalline or amorphous cell behavior?, *Applied Physics Letters*, Vol.65, pp. 860-862

Meier, J.; Dubail, S. ; Golay, S. ; Kroll, U.; Faÿ, S.; Vallat-Sauvain, E.; Feitknecht, L. ; Dubail, J.; Shah, A (2002). Microcrystalline Silicon and the Impact on Micromorph Tandem Solar Cells, *Solar Energy Materials and Solar Cells*, Vol. 74, num. 1-4, pp. 457-467

Meillaud, F; Feltrin, A; Despeisse, M.; Haug, FJ.; Domine, D.; Python, M.; Söderström, T.; Cuony, P.; Boccard, M.; Nicolay, S. & Ballif, C. (2011). Realization of high efficiency micromorph tandem silicon solar cells on glass and plastic substrates: Issues and potential. *Solar Energy Materials & Solar Cells*, Vol.95, No.1, pp. 127-130

Mohammadpour, A. & Shankar, K. (2010). Anodic TiO_2 nanotube arrays with optical wavelength-sized apertures, *Journal of Materials Chemistry*, Vol. 20, pp. 8474–8477

Movla, H.; Gorji N. E.; Sohrabi, F.; Hosseinpour, A. & Babaei H. (2010). Application of nanostructure materials in solar cells, *Proceeding of The 2th International Conference on Nuclear and Renewable Energy Resources, (NURER 2010)*, Ankara, Turkey 4-7 July, 2010.

Movla, H.; Sohrabi, F.; Fathi J.; Nikniazi A.; Babaei H. & Gorji N. E. (2010). Photocurrent and Surface Recombination Mechanisms in the InxGa1-xN\GaN Different-sized Quantum Dot Solar Cells, *Turkish Journal of Physics*, Vol. 34, pp. 97-106.

Sohrabi, F.; Movla H.; Khalili Kh.; Najafabadi H. A.; Nikniazi A. & Fathi J. (2011). J-V Characteristics of Heterojunction Solar Cell Based on Carbon Nanotube-Silicon, *proceeding of 2nd Iranian Conference on Optics & Laser Engineering*, Malek Ashtar University of Technology, Isfahan, Iran, 19-20 May, 2011.

Tuzun, O.; Oktik, S.; Altindal, S. & Mammadov, T.S. (2006). Electrical characterization of novel Si solar cells, *Thin Solid Films*, Vol. 511– 512, pp. 258-264

Wang, X.; Peng, K-Q.; Wu, X-L. & Lee, S-T. (2010). Single crystalline ordered silicon wire/Pt nanoparticle hybrids for solar energy harvesting. *Electrochemistry Communications*, Vol. 12, pp. 509–512

Wu, L.; Tian ,W. & Jiang, X. (2005). Silicon-based solar cell system with a hybrid PV module, *Solar Energy Materials & Solar Cells*, Vol.87, pp. 637-645

Yamamoto, K.; Yoshimi, M.; Tawada, Y.; Okamoto, Y. & Nakajima, A. (2001). Cost effective and high-performance thin Film Si solar cell towards the 21st century, *Solar Energy Materials & Solar Cells*, Vol.66, pp. 117-125

Yamamoto, K.; Nakajima, A.; Yoshimi, M.; Sawada, T.; Fukuda, S.; Suezaki, T; Ichikawa, M.; Koi, Y.; Goto, M.; Meguro, T.; Matsuda, T.; Kondo, M.; Sasaki, T.; Tawada, Y. (2004). A high efficiency thin film silicon solar cell and module. *Solar Energy*, Vol.77, No.6, pp. 939–949

Yan, B.; Yue, G.; Xu, X.; Yang, J.; and Guha, S. (2010). High efficiency amorphous and nanocrystalline silicon solar cells, Physica Status Solidi A, Vol.207, No. 3, pp.671–677

8

AlSb Compound Semiconductor as Absorber Layer in Thin Film Solar Cells

Rabin Dhakal, Yung Huh, David Galipeau and Xingzhong Yan
Department of Electrical Engineering and Computer Science,
Department of Physics, South Dakota State University, Brookings SD 57007,
USA

1. Introduction

Since industrial revolution by the end of nineteenth century, the consumption of fossil fuels to drive the economy has grown exponentially causing three primary global problems: depletion of fossil fuels, environmental pollution, and climate change (Andreev and Grilikhes, 1997). The population has quadrupled and our energy demand went up by 16 times in the 20th century exhausting the fossil fuel supply at an alarming rate (Bartlett, 1986; Wesiz, 2004). By the end of 2035, about 739 quadrillion Btu of energy (1 Btu = 0.2930711 W-hr) of energy would be required to sustain current lifestyle of 6.5 billion people worldwide (US energy information administration, 2010). The increasing oil and gas prices, gives us enough region to shift from burning fossil fuels to using clean, safe and environmentally friendly technologies to produce electricity from renewable energy sources such as solar, wind, geothermal, tidal waves etc (Kamat, 2007). Photovoltaic (PV) technologies, which convert solar energy directly into electricity, are playing an ever increasing role in electricity production worldwide. Solar radiation strikes the earth with 1.366 KWm^{-2} of solar irradiance, which amounts to about 120,000 TW of power (Kamat 2007). Total global energy needs could thus be met, if we cover 0.1% of the earth's surface with solar cell module with an area 1 m^2 producing 1KWh per day (Messenger and Ventre, 2004).

There are several primary competing PV technologies, which includes: (a) crystalline (c-Si), (b) thin film (a-Si, CdTe, CIGS), (c) organic and (d) concentrators in the market. Conventional crystalline silicon solar cells, also called first generation solar cells, with efficiency in the range of 15 - 21 %, holds about 85 % of share of the PV market (Carabe and Gandia, 2004). The cost of the electricity generation estimates to about \$4/W which is much higher in comparison to \$0.33/W for traditional fossil fuels (Noufi and Zweibel, 2006). The reason behind high cost of these solar cells is the use of high grade silicon and high vacuum technology for the production of solar cells. Second generation, thin film solar cells have the lowest per watt installation cost of about \$1/W, but their struggle to increase the market share is hindered mainly due to low module efficiency in the range of 8-11% ((Noufi and Zweibel, 2006; Bagnall and Boreland, 2008). Increasing materials cost, with price of Indium more than \$700/kg (Metal-pages, n.d.), and requirements for high vacuum processing have kept the cost/efficiency ratio too high to make these technologies the primary player in PV market (Alsema, 2000). Third generation technologies can broadly be divided in two categories: devices achieving high efficiency using novel approaches like concentrating and tandem solar cells and moderately efficient organic based photovoltaic solar cells (Sean and Ghassan, 2005; Currie et al., 2008). The technology and science for third generation solar cells are still immature and subject of widespread research area in PV.

1.1 Thin film solar cells

Second Generation thin film solar cells (TFSC) are a promising approach for both the terrestrial and space PV application and offer a wide variety of choices in both device design and fabrication. With respect to single crystal silicon technology, the most important factor in determining the cost of production is the cost of 250-300 micron thick Si wafer (Chopra et al., 2004). Thin-film technologies allow for significant reduction in semiconductor thickness because of the capacity of certain materials for absorbing most of the incident sunlight within a few microns of thickness, in comparison to the several hundred microns needed in the crystalline silicon technology (Carabe and Gandia, 2004). In addition, thin-film technology has an enormous potential in cost reduction, based on the easiness to make robust, large-area monolithic modules with a fully automatic fabrication procedure. Rapid progress is thus made with inorganic thin-film PV technologies, both in the laboratory and in industry (Aberle, 2009).

Amorphous silicon based PV modules have been around for more than 20 years Chithik et al. first deposited amorphous silicon from a silane discharge in 1969 (Chittik et al., 1969) but its use in PV was not much progress, until Clarson found out a method to dope it n or p type in 1976 (Clarson, n.d.) Also, it was found that the band gap of amorphous silicon can be modified by changing the hydrogen incorporation during fabrication or by alloying a-Si with Ge or C (Zanzucchi et al., 1977; Tawada et al. 1981). This introduction of a-Si:C:H alloys as p-layer and building a hetero-structure device led to an increase of the open-circuit voltage into the 800 mV range and to an increased short-circuit current due to the "window" effect of the wideband gap p layer increasing efficiency up to 7.1% (Tawada et al. 1981, 1982). Combined with the use of textured substrates to enhance optical absorption by the "light trapping" effect, the first a-Si:H based solar cell with more than 10% conversion efficiency was presented in 1982 (Catalakro et al., 1982). However, there exists two primary reasons due to which a-Si:H has not been able to conquer a significant share of the global PV market. First is the low stable average efficiency of 6% or less of large-area single-junction PV modules due to "Staebler–Wronski effect", i.e. the light-induced degradation of the initial module efficiency to the stabilized module efficiency (Lechner and Schad, 2002; Staebler and Wronski, 1977). Second reason is the manufacturing related issues associated with the processing of large (>1 m²) substrates, including spatial non-uniformities in the Si film and the transparent conductive oxide (TCO) layer (Poowalla and Bonnet, 2007).

Cadmium Telluride (CdTe) solar cell modules have commercial efficiency up to 10-11% and are very stable compound (Staebler and Wronski, 1977). CdTe has the efficient light absorption and is easy to deposit. In 2001, researches at National Renewable Energy Laboratory (NREL) reported an efficiency of 16.5% for these cells using chemical bath deposition and antireflective coating on the borosilicate glass substrate from $CdSnO_4$ (Wu et al., 2001). Although there has been promising laboratory result and some progress with commercialization of this PV technology in recent years (First Solar, n.d.), it is questionable whether the production and deployment of toxic Cd-based modules is sufficiently benign environmentally to justify their use. Furthermore, Te is a scarce element and hence, even if most of the annual global Te production is used for PV, CdTe PV module production seems limited to levels of a few GW per year (Aberle, 2009).

The CIGS thin film belongs to the multinary Cu-chalcopyrite system, where the bandgap can be modified by varying the Group III (on the Periodic Table) cations among In, Ga, and Al and the anions between Se and S (Rau and Schock, 1999). This imposes significant challenges for the realization of uniform film properties across large-area substrates using high-throughput equipment and thereby affects the yield and cost. Although CIGS technology is a star performer in laboratory, with confirmed efficiencies of up to 19.9% for small cells (Powalla and Bonnet, 2007) however the best commercial modules are presently 11–13% efficient (Green et al. 2008). Also there are issues regarding use of toxic element cadmium and scarcity of indium associated with this technology. Estimates indicate that all known reserves of indium would only be sufficient for the production of a few GW of CIGS PV modules (Aberle, 2009).

This has prompted researchers to look for new sources of well abundant, non toxic and inexpensive materials suitable for thin film technology. Binary and ternary compounds of group III-V and II-VI are of immediate concern when we look for alternatives. AlSb a group III-V binary compound is one of the most suitable alternatives for thin film solar cells fabrication because of its suitable optical and electrical properties (Armantrout et al., 1977). The crystalline AlSb film has theoretical conversion efficiency more than 27% as suggested in literature (Zheng et al., 2009).

1.2 Aluminum antimony thin films

Aluminum Antimony is a binary compound semiconductor material with indirect band gap of 1.62 eV thus ideal for solar spectrum absorption (Chandra et al., 1988). This also has become the material of interest due to relatively easy abundance and low cost of Al and Sb. AlSb single crystal has been fabricated from the Czochralski process but the AlSb thin film was prepared by Johnson et al. by co-evaporation of Al and Sb. They also studied the material properties to find out its donor and acceptor density and energy levels (Johnson, 1965). Francombe et al. observed the strong photovoltaic response in vacuum deposited AlSb for the first time in 1976 (Francombe et al., 1976). Number of research groups around the world prepared thin AlSb thin film and studied its electrical and optical properties by vacuum and non vacuum technique. These include, Leroux et al. deposited AlSb films on number of insulating substrate by MOCVD deposition technique in 1979 (Leroux et al., 1980). Dasilva et al. deposited AlSb film by molecular beam epitaxy and studied its oxidation by Auger and electron loss spectroscopy in 1991 (Dasilva et al., 1991). Similarly, AlSb film was grown by hot wall epitaxy and their electrical and optical properties was studied by Singh and Bedi in 1998 (Singh and Bedi, 1998). Chen et al. prepared the AlSb thin film by dc magnetron sputtering and studied its electrical and optical properties (Chen et al. 2008) in 2007. Gandhi et al. deposited AlSb thin film on by the alternating electrical pulses from the ionic solution of $AlCl_3$ and $SbCl_3$ in EMIC (1-methyl-3-ethylimidazolium chloride) and studied its electrical and optical characters (Gandhi et al. 2008). However, AlSb thin film had never been successfully employed as an absorber material in photovoltaic cells. We, electro-deposited AlSb thin film on the TiO_2 substrate by using the similar technique as Gandhi et al. and had observed some photovoltaic response in 2009 (Dhakal et al., 2009). Al and Sb ions are extremely corrosive and easily react with air and moisture. Thus it becomes very difficult to control the stoichiometry of compound while electroplating. Thus, vacuum deposition techniques become the first choice to prepare AlSb thin films for the solar cell applications. In this work, we fabricated AlSb thin film co-sputtering of Al and Sb target. The film with optimized band-gap was used to fabricate p-n and p-i-n device structures. The photovoltaic response of the devices was investigated.

2. Fabrication of AlSb thin films

AlSb film was prepared from dc magnetron sputtering of Al and Sb target (Kurt J. Lesker, Materials Group, PA) simultaneously in a sputtering chamber. Sputtering is a physical vapor deposition process whereby atoms are ejected from a solid target material due to bombardment of the target by plasma, a flow of positive ions and electrons in a quasi-neutral electrical state (Ohring, 2002). Sputtering process begins when the ion impact establishes a train of collision events in the target, leading to the ejection of a matrix atom (Ohring, 2002). The exact processes occurring at the target surface is depends on the energy of the incoming ions. Fig. 1 shows the schematic diagram of sputtering using DC and RF power. DC sputtering is achieved by applying large (~2000) DC voltages to the target (cathode) which establishes a plasma discharge as Ar⁺ ions will be attracted to and impact the target. The impact cause sputtering off target atoms to substrates.

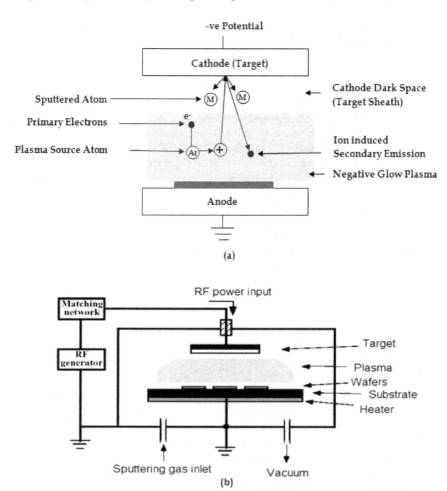

Fig. 1. Schematic Diagrams of (a) DC sputtering and (b) RF sputtering (Ohring, 2002)

In DC sputtering, the target must be electrically conductive otherwise the target surface will charge up with the collection of Ar⁺ ions and repel other argon ions, halting the process. RF Sputtering - Radio Frequency (RF) sputtering will allow the sputtering of targets that are electrical insulators (SiO_2, etc). The target attracts Argon ions during one half of the cycle and electrons during the other half cycle. The electrons are more mobile and build up a negative charge called self bias that aids in attracting the Argon ions which does the sputtering. In magnetron sputtering, the plasma density is confined to the target area to increase sputtering yield by using an array of permanent magnets placed behind the sputtering source. The magnets are placed in such a way that one pole is positioned at the central axis of the target, and the second pole is placed in a ring around the outer edge of the target (Ohring, 2002). This configuration creates crossed E and B fields, where electrons drift perpendicular to both E and B. If the magnets are arranged in such a way that they create closed drift region, electrons are trapped, and relies on collisions to escape. By trapping the electrons, and thus the ions to keep quasi neutrality of plasma, the probability for ionization is increased by orders of magnitudes. This creates dense plasma, which in turn leads to an increased ion bombardment of the target, giving higher sputtering rates and, therefore, higher deposition rates at the substrate.

We employed dc magnetron sputtering to deposit AlSb thin films. Fig. 2 shows the schematic diagram of Meivac Inc sputtering system. Al and Sb targets were placed in gun 1 and 2 while the third gun was covered by shutter. Both Al and Sb used were purchase from Kurt J. Lesker and is 99.99% pure circular target with diameter of 2.0 inches and thickness of 0.250 inches.

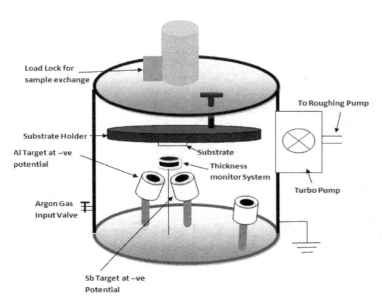

Fig. 2. Schematic diagram of dc magtron sputtering of Al and Sb targets.

Firstly, a separate experiment was conducted to determine the deposition rate of aluminum and antimony and the associated sputtering powers. Al requires more sputtering power than Sb does for depositing the film at same rates. Next Al and Sb was co-sputtered to produce 1 micron AlSb film in different deposition ratio for Al:Sb. The film was annealed at 200 C in vacuum for 2 hrs and cooled down naturally. Table 1 summarizes the deposition parameters of different AlSb films.

Al: Sb Ratio	Deposition Rate (Å/s)		Sputtering Power (W)		Ar Gas Pressure (mTorr)	Film Thickenss (kÅ)
	Al	Sb	Al	Sb		
1:3	2	6	104	37	20.1	10
2:5	2	5	104	33	20.1	10
3:7	3	7	150	42	20.1	10
1:1	1	1	150	24	20.1	10
7:3	7	3	261	24	20.1	10

Table 1. Deposition Parameters of Different AlSb films.

The film was deposited on glass slides for electrical and optical characterization. The microscopic glass substrate (1 cm x 1 cm) was cleaned using standard substrate cleaning procedure as follows: soaked in a solution of 90% boiling DI and 10% dishwashing liquid for five minutes, followed by soaking in hot DI (nearly boiled) water for five minutes. The substrate was then ultra sonicated, first in Acetone (Fisher Scientific) and then isopropyl alcohol (Fisher Scientific) for 10 minutes each. The substrate was then blown dry with nitrogen.

The morphology of the AlSb film was checked by SEM and was used to validate the grain size and crystalline nature of AlSb particles and shown in Fig. 3.

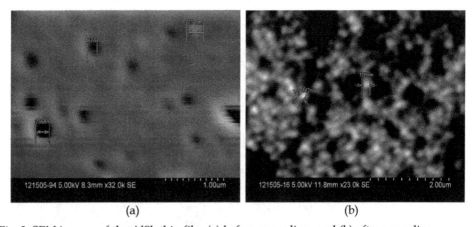

(a)   (b)

Fig. 3. SEM images of the AlSb thin film (a) before annealing, and (b) after annealing.

The AlSb grains were found to have been developed after annealing of the film due to proper diffusion and bonding of Al and Sb. Only low magnified image could be produced before annealing the film and holes were seen on the surface. The AlSb microcrystal is formed with an average grain size of 200 nm. Also seen are holes in the film which are primarily the defect area, which could act as the recombination centers. Better quality AlSb film could be produced if proper heating of the substrate is employed during deposition process.

3. Optical characterization of AlSb thin film

The transmittance in the thin film can be expressed as (Baban et al. 2006):

$$T = \frac{I_T}{I_0} = (1 - R1)(1 - R2)(1 - R3)(1 - S)e^{-\alpha d} \tag{1}$$

Where, a is the absorption coefficient antd d is the thickness of the semiconductor film. $R1$, $R2$ and $R3$ are the Fresnel power reflection coefficient and the Fresnel reflection coefficient at semiconductor - substrate and substrate – air interface. S measures the scattering coefficient of the surface.

UV Visible Spectrophotometer (Lambda 850) was used to measure the absorption and transmission data. This system covered the ultraviolet-visible range in 200 – 800 nm. The procedures in the Lambda 850 manual were followed. Figure 4 shows the transmittance spectra of the AlSb thin films. The films have an strong absorption in the visible spectral range up to 550 nm for film with Al:Sb ratio 2:5. Similarly for films with Al:Sb in the ratio of 1:3, 1:1 and 3:7 have strong absorption up to 700 nm. The films were transparent beyond these levels.

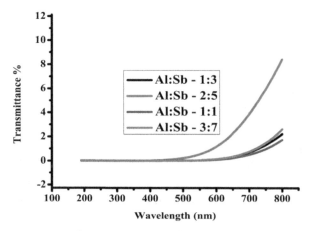

Fig. 4. Transmittance Spectra of AlSb films with different Al:Sb growth ratios.

The film with Al:Sb ratio of 7:3 didn't have a clear transmittance spectra and thus not shown in the figure. This was because the increasing the content of aluminum would make the film metallic thus absorbing most of the light in visible spectrum.

Absorption coefficient of a film can be determined by solving equation 1 for absorption and normalizing the Transmittance in the transparent region as (Baban et al. 2006):

$$\alpha = -\frac{1}{d}\ln\left(T_{normalized}\right) \tag{2}$$

Optical band gap of the film was calculated with the help of transmission spectra and reflectance spectra by famous using Tauc relation (Tauc, 1974)

$$\alpha h v = d\left(h v - E_g\right)^n \tag{3}$$

Where Eg is optical band gap and the constant n is $1/2$ for direct band gap material and n is 2 for indirect band gap. The value of the optical band gap, Eg, can be determined form the intercept of $(\alpha hv)^{1/2}$ Vs Photon energy, hv, at $(\alpha hv)^{1/2} = 0$.

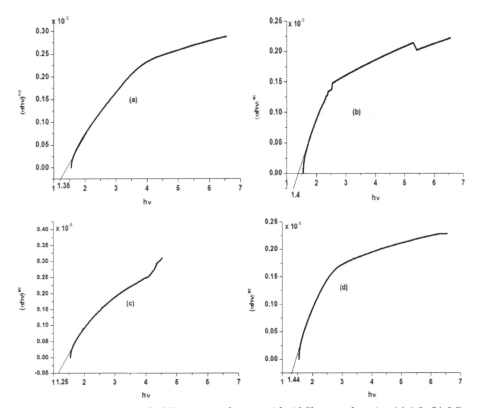

Fig. 5. Bandgap estimation of AlSb semiconductor with Al:Sb growth ratios (a) 1:3, (b) 2:5, (c) 1:1 and (d) 3:7.

The optical absorption coefficient of all the films was calculated from the transmittance spectra and was found in the range of 10^5 cm^{-1} for photon energy range greater than 1.2 eV. Fig. 5 shows the square root of the product of the absorption coefficient and photon energy (hv) as a function of the photon energy. The band gap of the film was then estimated by extrapolating the straight line part of the $(\alpha hv)^{1/2}$ vs hv curve to the intercept of horizontal axis.

This band gap for Al:Sb growth ratio 1:3, 2:5, 1:1 and 3:7 was found out to be 1.35 eV, 1.4 eV, 1.25 eV and 1.44 eV respectively. Since the ideal band gap of AlSb semiconductor is 1.6 eV we have taken the Al:Sb growth ratio to be 3:7 to characterize the film and fabricate the solar cells.

4. Electrical characterization

Material's sheet resistivity, ρ, can be measured using the four point probe method as show in Fig. 6. A high impedance current source is used to supply current (I) through the outer two probes and a voltmeter measures the voltage (V) across the inner two probes.

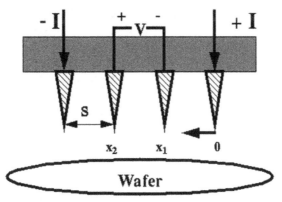

Fig. 6. Schematic diagram of Four point probe configuration.

The sheet resistivity of a thin sheet is given by (Chu et al. 2001):

$$\rho = RCF \frac{V_{measured}}{I_{measured}} \quad (4)$$

Where, RCF is the resistivity correction factor and given by $RCF = \dfrac{\pi \alpha}{\ln 2}$ The sheet resistance, R_s, could be thus be calculated as $R_s = \rho/d$ and measured in ohms per square. Conductivity (σ) is measured as reciprocal of resistivity and could be related to the activation energy as (Chu et al., 2001):

$$\sigma = \sigma_0 e^{\frac{-\Delta E}{k_b T}} \quad (5)$$

Where, ΔE is the activation energy. This describes the temperature dependence of carrier mobility. The dark conductivity of AlSb film measured as a function of temperature and is shown in Fig. 7.

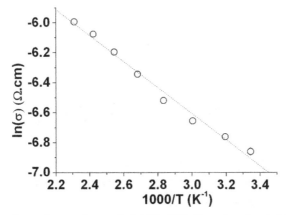

Fig. 7. Temperature dependence of annealed AlSb (3:7) film when heated from 26 - 240 °C (dot line for guiding the eyes).

The annealed film shows a linear $\ln\sigma$ vs $1/T$ relationship. The activation energy of the dark conductivity was estimated to be 0.68 eV from the temperature dependence of the conductivity curve for AlSb film. This value is in good agreement with work done by Chen et al. (Chen et al., 2008). This curve also confirms the semiconducting property of the AlSb (3:7) film because the conductivity of the film was seen to be increasing with increasing the excitation.

5. Simulation of solar cell

AMPS 1D beta version (Penn State Univ.) was used to simulate the current voltage characteristics of p-i-n junction AlSb solar cells. The physics of solar cell is governed by three equations: Poisson's equation (links free carrier populations, trapped charge populations, and ionized dopant populations to the electrostatic field present in a material system), the continuity equations (keeps track of the conduction band electrons and valence band holes) for free holes and free electrons. AMPS has been used to solve these three coupled non-linear differential equations subject to appropriate boundary conditions. Following simulation parameters was used for the different layers of films.

Contact Interface		
Barrier Height (eV)	0.1 (E_C-E_F)	0.3 (E_F-E_V)
S_e (cm/s)	1.00×10^8	1.00×10^8
S_h (cm/s)	1.00×10^8	1.00×10^8

*S surface recombination velocity of electrons or holes.

Semiconductor Layers				
	CuSCN	AlSb	ZnO	TCO
Thicknesses, $\mathbf{d}$ (nm)	100	1000	45	200
Permittivity, $\varepsilon/\varepsilon_0$	10	9.4	10	9
Band gap, E_g (eV)	3.6	1.6	2.4	3.6
Density of electrons on conduction band, N_C (cm-3)	1.80×10^{18}	7.80×10^{17}	2.22×10^{18}	2.22×10^{18}
Density of holes on valence band, N_V (cm-3)	2.20×10^{19}	1.80×10^{19}	1.80×10^{19}	1.80×10^{19}
Electron mobility, μ_e (cm2/Vs)	100	80	100	100
Hole mobility, μ_p (cm2/Vs)	25	420	25	25
Acceptor or donor density, N_A or N_D (cm-3)	$N_A =$ 1×10^{18}	$N_A =$ 1×10^{14}	$N_D =$ 1.1×10^{18}	$N_D =$ 1×10^{18}
Electron affinity, X (eV)			4.5	4.5

* $\varepsilon_0 = 8.85 \times 10^{-12}$ F/m electric constant; TCO is In_2O_3: SnO_2.

Gaussian Midgap defect states				
N_{DG}, N_{AG} (cm-3)	$A = 1\times$ 10^{19}	$D = 9\times$ 10^{10}	$A = 1\times$ 10^{19}	$D = 1\times$ 10^{16}
W_G (eV)	0.1	0.1	0.1	0.1
σ_e (cm2)	1.00×10^{-13}	1.00×10^{-8}	1.00×10^{-16}	1.00×10^{-11}
σ_p (cm2)	1.00×10^{-13}	1.00×10^{-11}	1.00×10^{-13}	1.00×10^{-14}

* $N_{DG/AG}$ the donor-like or acceptor-like defect density, W_G the energy width of the Gaussian distribution for the defect states, τ carrier lifetime, and σ capture cross section of electrons (σ_e) or holes (σ_p).

Table 2. Parameters of the simulating the IV behavior or p-i-n junction solar cells.

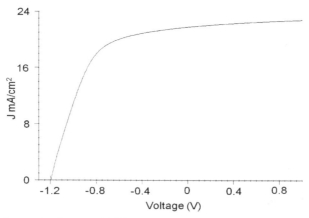

Fig. 8. Current-voltage simulation of AlSb p-i-n junction structure in AMPS 1D software.

Fig. 8 shows the current voltage simulation curve of pin junction solar cell - CuSCN/AlSb/ZnO with AlSb as an intrinsic layer. CuSCN was used as a p layer and ZnO as a n layer. The cell was illuminated under one sun at standard AM 1.5 spectrum.

The simulation result shows that the solar cell has the FF of 55.5% and efficiency of 14.41%. The short circuit current for the cell was observed to be 21.7 mA/cm^2 and the open circuit voltage was observed to be 1.19 V. AlSb is thus the promising solar cell material for thin film solar cells. The efficiency of the same cell structure could be seen increased up to 19% by doubling the thickness of AlSb layer to 2 micron.

6. Solar cell fabrication

Both p-n and p-*i*-n junction solar cells were designed and fabricated in 1cm x 2 cm substrate with AlSb as a p type and an absorber material respectively. Variety of n type materials including TiO$_2$ and ZnO were used to check the photovoltaic response of AlSb thin film. Fig. 9 shows the p-n and p-*i*-n based solar cell design with ZnO and TiO$_2$ are an n-type layer and CuSCN as a p-type layer.

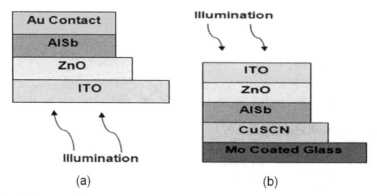

Fig. 9. Solar Cell Design (a) p-n and (b) p-*i*-n structure.

ZnO thin film was prepared by RF sputtering of 99.999% pure ZnO target (Kurt J. Lesker, PA, diameter 2 inches and thickness 0.25 inches). ZnO intrinsic film was deposited by RF power of 100 W at 0.7 Å/s and subsequently annealed in air at 150 °C. ITO film was also prepared from 99.99% pure ITO target (Kurt J. Lesker, PA, diameter 2 inches and thickness 0.25 inches) on the similar fashion using dc magnetron sputtering. Transparent ITO film was deposited at plasma pressure of 4.5 mTorr. The sputtering power of 20 W yields deposition rate 0.3 Å/s. The film was then annealed at 150 °C in air for 1 hour. The highly ordered mesoporous TiO$_2$ was deposited by sol gel technique as described by Tian et al. (Tian et al. 2005). CuSCN thin film was prepared by spin coating the saturated solution of CuSCN in dipropyl sulphide and dried in vacuum oven at 80 °C (Li et al., 2011). The thickness of all three films ZnO, ITO and TiO$_2$ film was about 100 nm and the thickness of AlSb layer is ~1 micron. The active layer was annealed.

7. I-V Characterization of solar cell

Current voltage measurement of the solar cells was carried out using Agilent 4155c (Agilent, Santa Clara, CA) semiconductor parameter analyzer equipped with solar cell simulator in SDSU. Fig. 10 shows the experimental set up used for measuring I-V response of the solar cells, where 2 SMUs (source measurement unit) were used. The SMU s could operate as a voltage source (constant sweep voltage) or a current source and it could measure voltage and current at the same time. The SMUs could measure from 10^{-12} A to 1 A and -10 V to 10 V. SMU 1 was set as voltage sweep mode from -1 V to 1 V with steps of 0.01 V, and SMU 2 was set to measure current of the solar cell during IV measurement. IV responses were measured under both dark and illuminated condition. During illumination, the intensity of the simulated light was 100 mW cm^{-2} and calibrated using the NREL calibrated standard cell.

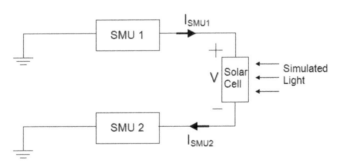

Fig. 10. Experimental set up for measuring IV response of a solar cell.

Table 3 shows the current voltage characteristics of p-n junction solar cells with structures AlSb/TiO$_2$, AlSb/ZnO. The active cell area was 0.16 cm^2 and fabricated on ITO coated glass surface.

The V_{oc} of the best cell with ZnO as an n type layer was found out to be 120 mV and I_{sc} to be 76 uA. The FF of the cell was calculated to be 0.24 and the efficiency was 0.009%. The cell with TiO$_2$ as an n layer has even lower V_{OC} and I_{sc}. TiO$_2$ is less suitable n-type layer for making junction with AlSb than ZnO because it is far more conductive than TiO$_2$. A number of reasons may be attributed for this low efficiency. First, is due to small electric field at the junction between AlSb and the n type material (ZnO or TiO$_2$). This severely limits the charge separation at the junction and decreases V_{OC} of the device. A better material needs to be explored to dope AlSb n type to increase the built in field. The field could also be extended using the p-*i*-n structure to design the solar cells.

Cell	V_{OC} (mV)	I_{SC} (mA)	FF	Efficiency %
AlSb/TiO2	80	12×10^{-3}	0.23	0.001
AlSb/ZnO	120	76×10^{-3}	0.24	0.009

Table 3. Current-voltage characteristics of p-n junction solar cells

Interesting results were obtained with a p-*i*-n junction, CuSCN/AlSb/ZnO. The used cell has an active cell area of this cell was 0.36 cm^2 and fabricated on Mo coated glass surface. Charge was collected from the silver epoxy fingers casted on top of ITO surface and Mo back contact. The cell showed a V_{OC} of ~ 500 mV and a J_{SC} of 1.5 mA/cm^2. With a FF value of 0.5, the efficiency of this cell was calculated to be 0.32%. This observation may be attributed to the more efficient charge separation than that in the p-n junction devices due to a strong build-in field. However, the efficiency of the p-*i*-n junction device is very low in comparison to other available thin film solar cells devices. There are still many unknown factors including the interfaces in the junction. Such a low efficiency could be attributed to the defects along the AlSb interface with both the p- and n-type of layers. Interfaces between AlSb and other layers needed to be optimized for a better performance.

8. Summary

AlSb thin film has been prepared by co-sputtering aluminum and antimony. The deposition rate of Al:Sb was required to be 3:7 to produce the stoichiometric AlSb film with optical band gap of 1.44 eV. After annealing the film at 200 ^{0}C in vacuum for two hours, the film likely formed crystalline structures with a size of ~200 nm and has strong absorption coefficient in the range of 10^5 cm^{-1} in the visible light. p-n and p-*i*-n heterojunction solar cells were designed and fabricated with AlSb as a p-type material and an intrinsic absorber layer. The simulation of the p-i-n junction solar cell with CuSCN/AlSb/ZnO using AMPS at AM1.5 illumination shows efficiency of 14% when setting ~1 μm-thick absorber layer. The p-n junction solar cells were fabricated with different types of n layers shows the photovoltaic responses. The p-*i*-n showed better photovoltaic performance than that of p-n junction cells. All the preliminary results have demonstrated that AlSb is promising photovoltaic material. This work is at the early stage. More experiment is needed for the understanding of the crystallization and properties of the AlSb films and the interface behaviors in the junctions.

9. Acknowledgments

Support for this project was from NSF-EPSCoR Grant No. 0554609, NASA-EPSCoR Grant NNX09AU83A, and the State of South Dakota. Simulation was carried out using AMPS 1D beta version (Penn State University). Dr. Huh appreciates AMES Lab for providing sputtering facility. We appreciate AMPS 1D beta version (Penn State Univ.)

10. References

Aberle, A. G. (2009). Thin Film Solar cells, *Thin Solid Films*, Vol. 517, pp. (4706-4710).

Alsema, E. A. (2000). Energy Pay-back Time and CO_2 Emissions of PV Systems, *Prog. Photovolt. Res. Appl.*, Vol. 8, pp. 17-25.

Andreev, V. M., & Grilikhes, V. A. (1997), *Photovoltaic conversion of concentrated sunlight*, John Wiley and sons, Inc., England.

Armantrout, G. A., Swierkowski, S. P., Sherohman, J. W., &Yee, J. H., 1977, *IEEE Trans. Nucl. Sci.*, NS-24, Vol. 121.

Baban, C., Carman, M., & Rusu, G. I. (2006) Electronic transport and photoconductivity of polycrystalline CdSe thin films, *J. Opto. Elec. and Adv. Materials*, Vol. 8, No. 3, pp. 917-921.

Bagnall, D. M., & Boreland, M. (2008). Photovoltaic Technologies, *Energy Policy*, Vol. 36, No. 12, pp. 4390-4396.

Bartlett, A. A. (1986). Sustained availability: A management program for nonrenewable resources, *Am. J. Phys.*, 54, pp (398-402).

Carabe, J., & Gandia, J. (2004), Thin – film - silicon solar cells, *Optoelectronics Review*, Vol. 12, No. 1, pp. 1-6.

Catalakro, A. (1982). Investigation on High efficiency thin film solar cells, *Proc. of 16th IEEE Photovoltaic Specialist Conf.*, San Diego (CA), pp. 1421-1422.

Chandra, Khare, S., & Upadhaya, H. M. (1988). Photo-electrochemical solar cells using electrodeposited GaAs and AISb semiconductor films, *Bulletin of Material Sciences*, Vol. 10, No. 4, pp. 323-332.

Chen, W., Feng, L., Lei, Z., Zhang, J., Yao, F., Cai, W., Cai, Y.,Li, W.,Wu, L., Li, B., & Zheng, J. (2008). AlSb thin films prepared by DC magnetron sputtering and annealing, *International Journal of Modern Physics B, Vol. 22*, No. 14, pp. 2275-2283.

Chittik, R. C., Alexander, J. H., & Sterling, H. E. (1968). The preparation and properties of amorphous Silicon, *Journal of Electrochemical Society*, Vol. 116, No. 1, pp. (77-81).

Chopra, K. L., Paulson, P. D., & Dutta, V. (2004). Thin Film Solar cells: An Overview, *Prog. Photovolt. Res. Appl.*, Vol. 12, pp. (69-92).

Chu, D. P., McGregor, B. M., & Migliorato, P. (2001), Temperature dependence of the ohmic conductivity and activation energy of $Pb_{1+}y(Zr0_{.3}Ti_{0.7})O_3$ thin films, *Appl. Phys. Letter*, Vol. 79, No. 4, pp. 518-520.

Clarson, D. E. (1977). Semiconductor device having a body of amorphous silicon, *US patent* no. 4064521.

Currie, M. J., Mapel, J. K., Heidel, T. D., Goffri, S., & Baldo, M. A. (2008). High-Efficiency Organic Solar Concentrators for Photovoltaics, *Science*, Vol. 321, No. 5886, pp. 226-228.

Dasilva, F. W. O., Raisian, C., Nonaouara, M., & Lassabatere, L. (1991). Auger and electron energy loss spectroscopies study of the oxidation of AlSb(001) thin films grown by molecular beam epitaxy, *Thin Solid Films*, Vol. 200 pp. 33-48.

Dhakal, R., Kafford, J., Logue, B., Roop, M., Galipeau, D., & Yan, X. (2009). Electrodeposited AlSb compound semiconductor for thin film solar cells, *34th IEEE Photovoltaic Specialists Conference (PVSC)*, Philadelphia, pp. 001699 – 001701.

First Solar, n.d. Available from <http://www.firstsolar.com/company_overview.php>

Francombe, M. H., Noreika, A. J., & Zietman, S. A. (1976). Growth and properties of vacuum-deposited films of AlSb, AlAs and AlP, *Thin Solid Films*, Vol. 32, No. 2, pp. 269-272.

Gandhi, T., Raja, K. S., & Mishra, M. (2008). Room temperature Electro-deposition of aluminum antimonide compound semiconductor, *Electrochemica Acta*, Vol. 53, No. 24, pp. 7331-7337.

Green, M. A., Emery, K., Hishikawa, Y., & Warta, W. (2008). Solar Cell Efficiency Tables (Version 32), *Prog.Photovolt:Res.Appl.*Vol. 16, pp. 61-67.

Johnson, J. E. (1965). Aluminum Antimonide Thin Films by co-evaporation of elements, *Journal of Applied Physics*, Vol. 36, pp. 3193-3196.

Kamat, P. V. (2007). Meeting the Clean Energy Demand: Nanostructure Architecture for Solar energy conversion, *J. Phys. Chem. C*, Vol. 111, pp. 2834-2860.

Lechner, P., & Schade, H. (2002). Photovoltaic thin-film technology based on hydrogenated amorphous silicon, *Prog. Photovolt.*, Vol. 10, pp. 85-97.

Leroux, M. (1980). Growth of AlSb on insulating substrates by metal organic chemical vapor depositon, *Journal of crystal growth*, Vol. 48, pp. 367-378.

Li, Y. Yan, M., Jiang, M., Dhakal, R., Thapaliya, P. S., & Yan, X. (2011), Organic inorganic hybrid solar cells made from hyperbranched phthalocyanines, *J of Photonics Energy*, Vol. 1, No. 011115.

Messenger, R. A., & Ventre, J. (2004). *Photovoltaic systems engineering*, CRC Press, pp (2).

Metal Pages, n.d. Available from <http://www.metal-pages.com/metalprices/indium/>

Noufi, R., & Zweibel, K. (2006). High efficiency CdTe and CIGS solar cells, *Photovoltaic Energy Conversion Conference*, pp. 317-320.

Ohring, M. (2002). Material Science of Thin Films, *Academic Press*, pp. 95-273.

Powalla, M., & Bonnet, D. (2007). Thin-Film Solar Cells Based on the Polycrystalline Compound Semiconductors CIS and CdTe, *Advances in OptoElectronics 2007*, article ID 97545, 6 pages.

Rau, U., & Schock, H. W. (1999). Electronic properties of Cu(In,Ga)Se$_2$ hetero-junction solar cells – recent achievements, current understanding, and future challenges, *Appl. Phys., A Mater. Sci. Process.*, Vol. 69, No. 2, pp. 131-147.

Sean, E., & Ghassan, E. (2005). Organic-Based Photovoltaics: Towards Low Cost Power Generation, *MRS bulletin*, Vol. 30, No. 1, pp. 10-15.

Singh, T., & Bedi, R. D. (1998). Growth and properties of aluminium antimonide films produced by hot wall epitaxy on single-crystal KCl, *Thin Solid Films*, Vol. 312, pp. 111-115.

Staebler, D. L., & Wronski, C. R. (1977). Estimation of the degradation of amorphous silicon solar cells, *Appl. Phys. Lett.*, Vol. 31, pp. 292-294.

Tauc, J. (1974). Amorphous and Liquid Semiconductors, *New-York: Plenum*, pp. (159).

Tawada, Okamoto, Y., H., & Hamakawa, Y. (1981). a-Si:C:H/a-Si:H heterojunction solar cells having more than 7.1% conversion efficiency, *Applied Physics Letters*, Vol. 39, pp. 237-239.

Tawada, Y., Kondo, M., Okamoto, Y., H., & Hamakawa, Y. (1982). Hydrogenated amorphous silicon carbide as a window material for high efficiency a-Si solar cells, *Solar Energy Mater.* Vol. 6, pp. 299-315.

Tian, B., Li, F., Bian, Z., Zhao, D., & Huang, C. (2005). Highly crystallized mesoporous TiO$_2$ films and their applications in dye sensitized solar cells, *J. Mater. Chem.*, Vol. 15, pp. 2414-2424.

US energy information administration (2010). Available from <http://www.eia.doe.gov/oiaf/ieo/world.html>.

Wesiz, P. B. (2004). *Basic choices and constraints on long-term energy supplies*, Physics Today.

Wu, X. (2001). 16.5% CdS/CdTe polycrystalline thin film solar cell, *Proceedings of 17th European Photovoltaic Solar energy Conference*, Munich, pp. 995-1000.

Zanzucchi, P., Wornski, C. R., & Clarson, D. E. (1977). Optical photoconductivity properties of discharge produced a-Si, *J. Appl. Phys.*, Vol. 48, No. 12, pp. 5227-5236.

Zheng, H., Wu, L. Li, B., Hao, X., He, J., Feng, L. Li, W., Zhang, J., & Cai, Y. (2009). The electrical, optical properties of AlSb polycrystalline thin films deposited by magnetron co-sputtering without annealing, *Chin. Phys. B*, Vol. 19, No. 12, pp. 127204-127207.

9

Organic Bulk Heterojunction Solar Cells Based on Poly(p-Phenylene-Vinylene) Derivatives

Cigdem Yumusak[1,2] and Daniel A. M. Egbe[2]
[1]*Department of Physics, Faculty of Arts and Sciences, Yildiz Technical University,*
Davutpasa Campus, Esenler, Istanbul,
[2]*Linz Institute for Organic Solar Cells (LIOS), Physical Chemistry,*
Johannes Kepler University of Linz, Linz,
[1]*Turkey*
[2]*Austria*

1. Introduction

Since the discovery of electrical conductivity in chemically doped polyacetylene (Shirakawa et al., 1977; Chiang et al., 1977; Chiang et al., 1978), enormous progress has been made in the design, synthesis and detailed studies of the properties and applications of π-conjugated polymers (Yu et al., 1998; Skotheim et al., 1998; Hadziioannou et al., 1998). The award of the Nobel prize in Chemistry three decades later in the year 2000 to Alan J. Heeger, Alan G. MacDiarmid and Hideki Shirakawa for the abovementioned discovery and development of semiconducting polymers, was greeted worldwide among researchers as a recognition for the intensified research, which has been going on in the field of organic π-conjugated polymers (Shirakawa, 2001). Such polymers are advantageous compared to inorganic semiconductors due to their low production cost, ease of processability, flexibility as well as tenability of their optical and electronic properties through chemical modifications. These outstanding properties make them attractive candidates as advanced materials in the field of photonics and electronics (Forrest, 2004; Klauk, 2006; Bao & Locklin, 2007; Sun & Dalton, 2008; Moliton, 2006; Hadziioannou & Mallarias, 2007; Shinar & Shinar, 2009; Nalwa, 2008).

Among the most used polymers in optoelectronic devices are the poly(p-phenylene-vinylene)s (PPV), polyfluorenes, polythiophenes and their derivatives. The insertion of side-chains in these polymers reduces the rigidity of the backbone, increases their solubility and enables the preparation of films through inexpensive, solution-based methods, such as spin-coating (Akcelrud, 2003). Besides, these ramifications can also be used to tune the photophysical and electrochemical properties of these polymers using a variety of routes.

Solar cells based on solution-processable organic semiconductors have shown a considerable performance increase in recent years, and a lot of progress has been made in the understanding of the elementary processes of photogeneration (Hoppe & Sariciftci, 2004; Mozer & Sariciftci, 2006; Günes et al., 2007). Recently, organic bulk heterojunction solar cells with almost 100% internal quantum yield were presented, resulting in up to almost 8% power conversion efficiency (Park et al., 2009; Green et al., 2010). This device concept has been shown to be compatible with solution-processing at room temperature, for instance, by high-throughput printing techniques. Processing on flexible substrates is possible, thus allowing for roll-to-roll manufacturing as well as influencing the properties of the finished electronic devices. The recent considerable achievements in terms of power conversion efficiency have been made possible now by more than 15 year long research and development on solution-processed organic solar cells. Nevertheless, in order to let the scientific progress be followed by a commercial success, further improvements in term of efficiency and device lifetime have to be made.

In this chapter, we will briefly introduce the basic working principles of organic solar cells and present an overview of the most often studied PPV-type materials as applied within the photoactive layer.

2. Organic solar cells

2.1 A brief history

The first organic solar cells consisted of a single layer of photoactive material sandwiched between two electrodes of different work functions (Chamberlain, 1983; Wohrle & Meissner, 1991). However, due to the high binding energy of the primary photoexcitations, the separation of the photogenerated charge carriers was so inefficient that far below 1% power conversion efficiency could be achieved.

The next breakthrough was achieved in 1986 by introducing the bilayer heterojunction concept, in which two organic layers with specific electron or hole transporting properties were sandwiched between the electrodes (Tang, 1986). In this organic bilayer solar cell were consisting of a light-absorbing copper phthalocyanine layer in conjunction with an electronegative perylene carboxylic derivative. The differing electron affinities between these two materials created an energy offset at their interface, thereby driving exciton dissociation.

The efficiencies of the first organic solar cells reported in the 1980s were about 1% at best at that time. Primarily, this is due to the fact that absorption of light in organic materials almost always results in the production of a mobile excited state, rather than free electron-hole pairs as produced in inorganic solar cells. This occurs because in organic materials the weak intermolecular forces localize the exciton on the molecules. Since the exciton diffusion lengths in organic materials are usually around 5-15 nm (Haugeneder et al., 1999), much shorter than the device thicknesses, exciton diffusion limits charge-carrier generation in these devices because most of them are lost through recombination. Photogeneration is therefore a function of the available mechanisms for excitons dissociation.

The discovery of ultrafast photoinduced electron transfer (Sariciftci et al., 1992) from a conjugated polymer to buckminsterfullerene (C_{60}) and the consequent enhancement in charge photogeneration provided a molecular approach to achieving higher performances from solution-processed systems. In 1995 the first organic bulk heterojunction organic solar cell was fabricated based on a mixture of soluble p-phenylene-vinylene (PPV) derivative with a fullerene acceptor (Yu et al., 1995). In 2001, Shaheen et al. obtained the first truly promising results for bulk heterojunction organic solar cells when mixing the conjugated polymer poly(2-methoxy-5-(3',7'-dimethyl-octyloxy)-p-phenylene vinylene) (MDMO-PPV) and methanofullerene [6,6]-phenyl C_{61}-butyric acid methyl ester (PCBM) yielding a power conversion efficiency of 2.5% (Shaheen et al., 2001).

Padinger *et al.* (Padinger et al., 2003) presented a further increase in the power conversion efficiency by using a blend, which is nowadays the best investigated organic solar cell system: a poly(3-hexyl thiophene) donor (P3HT) in conjunction with PCBM. It was shown that annealing at a temperature above the glass transition of the polymer enabled an enhancement of the efficiency from 0.4% to 3.5%.

In the following years, the power conversion efficiency could be increased steadily. This is, to a large fraction, due to the considerable amount of time that has been spent by many laboratories around the world on the optimization of bulk heterojunction solar cells — many of them using P3HT:PCBM — but also by new approaches. Additives have been used in order to allow an increased control of the phase segregation during film formation of a copolymer–fullerene blend (Park et al., 2009; Peet et al., 2007), thus yielding efficiencies of up to 6%. The process additive is a solvent for the fullerene, but not the polymer, thus allowing the PCBM an extended time for self-organization during the drying process. A positive effect by heating the solvent before the film application could also be shown (Bertho et al., 2009). Today, up to 8% power conversion efficiency are reported in this kind of organic solar cells (Park et al., 2009; Green et al., 2010).

2.2 Organic bulk heterojunction solar cells

The sequential process involved in the light into electricity conversion can be summarized by the following steps: First, incident light is absorbed within the photoactive layer leading to the created of a bound electron-hole pairs (singlet excitons); the created excitons start to diffuse within the donor phase leading to charge separation; the separated charge carriers are transported to the corresponding electrodes.

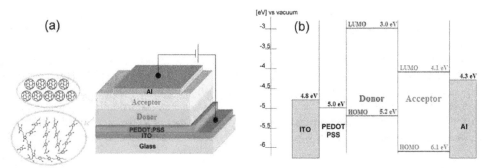

Fig. 1. (a) Schematic device structure and (b) energy diagram for an organic bilayer solar cell

Figure 1 (a) shows the simplest structure of an organic bilayer solar cell appears to be the superposition of donor and acceptor materials on top of each other, providing the interface needed to ensure the charge transfer. The schematic energy diagram of such an organic bilayer solar cell is depicted in Figure 1 (b). The excitons photogenerated in the donor or in the acceptor can diffuse to the interface where they are dissociated. According to the Onsager theory (Onsager, 1938) that can be invoked as a first approximation in organic semiconductors, photoexcited electrons and holes, by virtue of the low dielectric constant intrinsic to conjugated polymers, are coulombically bound. Due to the related exciton binding energy, which with around 0.5 eV is much larger than the thermal energy, the photogenerated excitons are not easily separated. Once excitons have been generated by the absorption photons, they can diffuse over a length of approximately 5-15 nm (Haugeneder et al., 1999). Since the exciton diffusion lengths in conjugated polymers are less than the photon absorption length, the efficiency of a bilayer cell is limited by the number of photons that can be absorbed within the effective exciton diffusion range at the polymer/electron interface. This limits drastically the photocurrent and hence the overall efficiency of the organic bilayer solar cells. To overcome this limitation, the surface area of the donor/acceptor interface needs to be increased. This can be achieved by creating a mixture of donor and acceptor materials with a nanoscale phase separation resulting in a three-dimensional interpenetrating network: the "bulk heterojunction solar cells" (Figure 2).

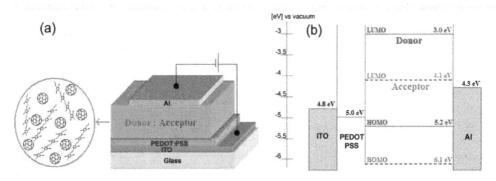

Fig. 2. (a) Schematic device structure and (b) energy diagram for an organic bulk heterojunction solar cell

The discovery of 1-(3-methoxycarbonyl)propyl-1-phenyl[6]C_{61} (PCBM) (Hummelen et al., 1995), a soluble and processable derivative of fullerene C_{60}, allowed the realization of the first organic bulk heterojunction solar cell by blending it with poly(2-methoxy-5-(2'-ethyl-hexoxy)-1,4-phenylene-vinylene) (MEH-PPV) (Yu et. Al., 1995). Figure 2(b) demonstrates the schematic energy diagram of an organic bulk heterojunction solar cell. Contrary to Figure (1b), excitons experience dissociation wherever they are generated within the bulk. Indeed, the next interface between donor and acceptor phases is present within the exciton diffusion length everywhere in the device. After having been generated throughout the bulk, the free carriers have to diffuse and/or be driven to the respective electrodes (Dennler & Sariciftci, 2005).

2.3 Characteristics of bulk heterojunction solar cells

Conjugated polymer thin films sandwiched between two metal electrode are usually described using a metal-insulator-metal (MIM) picture (Parker, 1994). The different operating regimes the MIM device due to externally applied voltages is shown in Figure 3.
As illustrated in Figure 3(a), the vacuum levels (E_{vac}) of the stacked materials shall align themselves (Shottky-Mott model).
Figure 3(a) indicates the energy diagram of a bulk heterojunction solar cell in open circuit condition. The E_{vac} of the different materials are aligned as explained above, and no electrical field is present within the device. Figure 3 (b) represents the short circuit condition. The Fermi levels of the two electrodes align themselves and a built-in field appears in the bulk, resulting in a constant slope for the HOMO and LUMO levels of the donor and acceptor (respectively, HD, LD, HA, and LA) and for the E_{vac}.

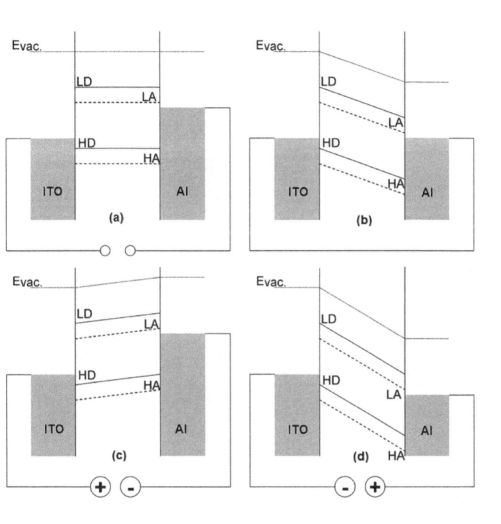

Fig. 3. MIM picture for a polymer diode under different operating modes. (a) open circuit condition, (b) short circuit condition, (c) forward bias, (d) reverse bias.

When polarized in the forward direction (high work function electrode (ITO) connected to (+) and low work function electrode (Al) connected to (-)) as in Figure 3 (c), electrons can be injected from the Al electrode to ITO electrode and holes from ITO electrode to Al electrode. The effective field in the device will ensure the drift of electrons from Al electrode to ITO electrode and hole from ITO electrode to Al electrode. Finally, when the device is polarized in the reverse direction (ITO connected to (-) and Al connected to (+)) (Figure 3 (d), charge injection is hindered by the field present in the device (Dennler & Sariciftci, 2005).

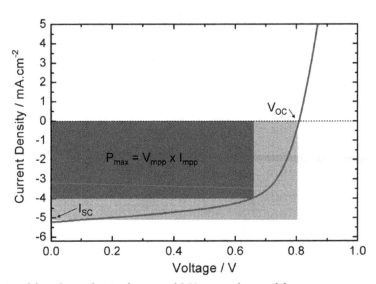

Fig. 4. First and fourth quadrant of a typical J-V curve observed for a Glass/ITO/PEDOT:PSS/MDMO-PPV:PCBM(1:4)/Al solar cell. Shown are the short circuit current (I_{SC}), the open circuit voltage (V_{OC}), the current (I_{mpp}) and voltage (V_{mpp}) at the maximum power point (P_{max})

Solar cells are operated between open circuit and short circuit condition (fourth quadrant in the current-voltage characteristics), as shown in Figure 4. In the dark, there is almost no current flowing, until the contacts start to inject heavily at forward bias for voltages larger than the open circuit voltage. Under illumination, the current flows in the opposite direction than the injected currents. The overall efficiency of a solar cell can be expressed by the following formula:

$$\eta = \frac{V_{OC}.I_{SC}.FF}{P_{in}} \tag{1}$$

where V_{OC} is the open circuit voltage, I_{SC} is the short circuit current, and P_{in} is the incident light power. The fill factor (FF) is given by

$$FF = \frac{I_{mpp}.V_{mpp}}{V_{OC}.I_{SC}} \tag{2}$$

where I_{mpp} and V_{mpp} represent the current and voltage at the maximum power point (P_{max}) in the fourth quadrant, respectively (Figure 4).

3. p-phenylene-vinylene based conjugated polymers

3.1 Poly(p-phenylene-vinylene) and its derivatives
Poly(p-phenylene-vinylene)s (PPVs) and its derivatives are one of the most promising classes of conjugated polymers for organic solar cells due to their ease of processability as

well as tunability of their optical and electronic properties through chemical modifications. Since the first report of electroluminescence from PPV, a great research attention has been focused on these types conjugated polymers (Burroughes et al., 1990). This focus was moreover up heaved after the discovery of an ultrafast photoinduced charge transfer from alkoxy-substituted PPV to the buckminsterfullerene (Sariciftci et al., 1992). PPV and its derivatives remain the most popular conjugated polymers for this application and continue to generate considerable interest and much research for photovoltaic applications (Cheng et al., 2009).

The pure PPV is insoluble, intractable, and infusible and therefore difficult to process. Solution processability is desirable as it allows polymeric materials to be solution cast as thin films for various applications. A general methodology to overcome this problem is to develop a synthetic route that involves a solution-processable polymer precursor. First synthetic route for high quality PPV films with high molecular weights was first introduced by Wessling (Figure 5) allowed the synthesis of soluble precursors, which can be processed into thin films prior to thermal conversion to PPV (Wessling & Zimmerman, 1968; Wessling, 1985). A potential drawback of the precursor routes is the limited control over poly-dispersity and molecular weight of the resulting polymer.

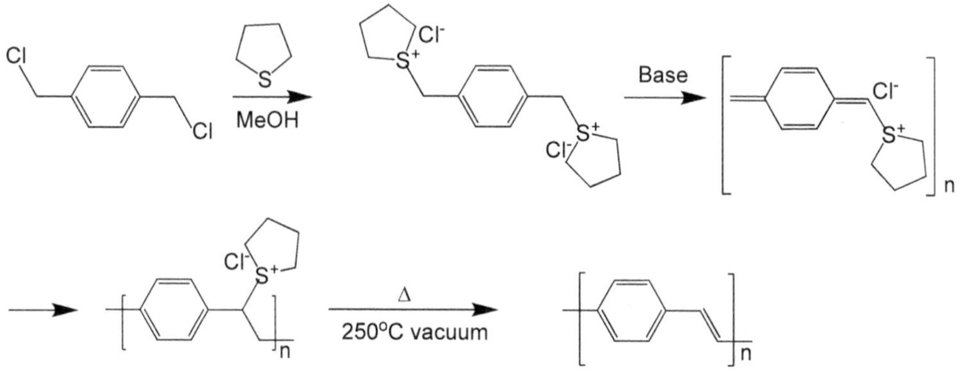

Fig. 5. Synthesis of PPV **(P1)** via the Wessling Route

Some of the drawbacks of this precursor approach include the generation of toxic side products during the solid state elimination process, structural defects arising from incomplete thermal conversion or oxidation, and undefined molecular weights and distribution (Papadimitrakopoulos et al., 1994). By incorporating long alkyl or alkoxy chains into the phenylene ring before polymerization to ensure the solubility, a one step approach can be applied to make processable PPV derivatives which can then be cast into thin films directly without conversion for device fabrication (Braun & Heeger, 1991). To date, the most widely used method for the preparation of PPV derivatives is the Gilch route (Gilch, 1966). A typical Gilch route to the synthesis of a representative solution-processable poly(2-methoxy-5-((2'-ethylhexyl)oxy)-1,4-phenylene-vinylene) (MEH-PPV, **P2**) is represented in Figure 6 (Neef & Ferraris, 2000). By following the same synthetic route, poly(2-methoxy-5-((3',7'-dimethyloctyl)oxy)-1,4-phenylenevinylene) (MDMO-PPV, **P3**) can also be synthesized

(Figure 7). This route involves mild polymerization conditions, and the molecular weights of the polymers obtained are generally high (Cheng et al., 2009).

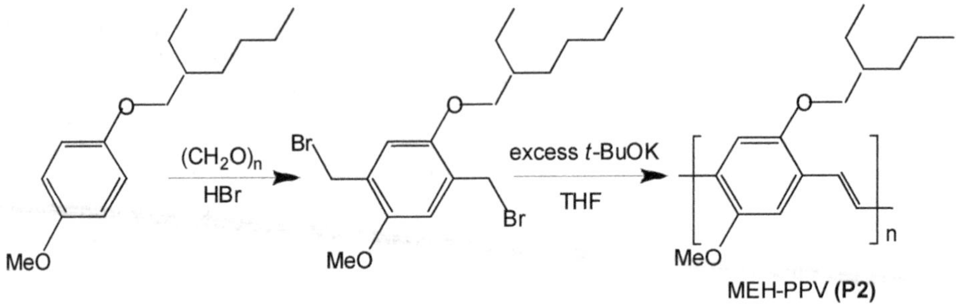

Fig. 6. Synthesis of MEH-PPV (**P2**) via the Gilch Route

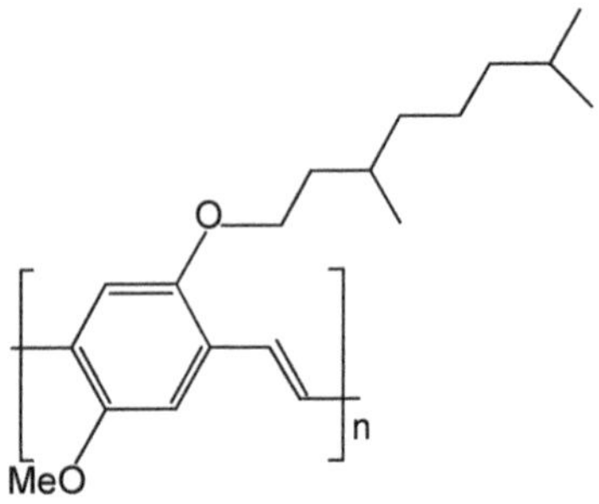

MDMO-PPV (**P3**)

Fig. 7. Chemical structure of MDMO-PPV (**P3**)

Solution-processable PPV derivatives were first reported by Wudl et al. (Wudl et al., 1991; Braun & Heeger, 1991) and by Ohnishi et al. (Doi et al., 1993). The solubility of the materials was achieved by grafting of long alkoxy chains, which cause some conformational mobility of the polymers. Consequently the soluble derivatives have lower glass transition temperatures than pure PPV. Poly[(2,5-dialkoxy-1,4-phenylene)-vinylene]s including long alkoxy side chains are soluble in common organic solvents such as chloroform, toluene, chlorobenzene, dichloromethane, tetrahydrofuran. Fig. 8 shows the chemical structures of PPV and its various alkoxy-substituted derivatives. The solubility increases from left to right, whereby the solubility of the much studied MEH-PPV and MDMO-PPV is enhanced by the branched nature of their side chains (Braun & Heeger, 1991).

PPV derivatives are predominantly hole conducting materials with high-lying LUMO (lowest unoccupied molecular orbital) levels. The unbalanced charge carrier transport properties and the relatively high barrier for electron injection from electrode metals such as aluminium limit the efficiency of the photovoltaic devices.

Several approaches have been explored to improve the electron affinity of PPVs. The insertion of weak electron-withdrawing triple bonds (—C≡C—) within the PPV backbone can be regarded as one of the most original approach of enhancing the electron affinity of PPVs Brizius et al., 2000; Egbe et al., 2001).

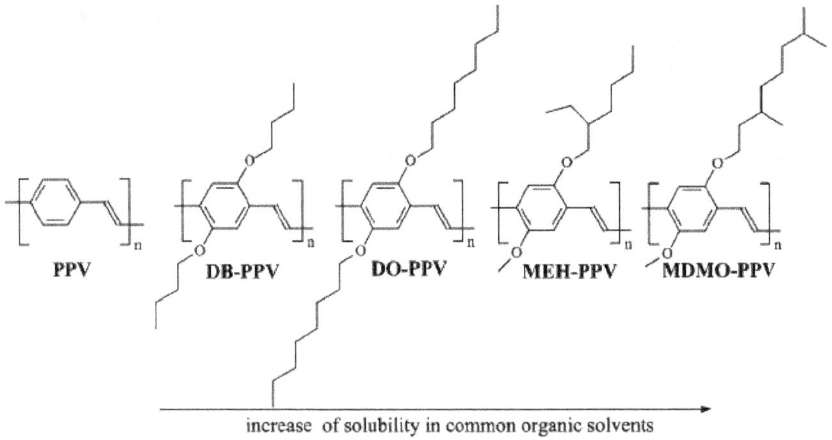

increase of solubility in common organic solvents

Fig. 8. Chemical structures of PPV and its different alkoxy-substituted derivatives

The substituent on the phenylene ring of PPV allows for the tuning of the band gap. Enhanced electron affinity of substituted PPVs is reflected by higher absorption coefficients, lower-lying HOMO (highest occupied molecular orbital) levels, and higher open circuit voltages from fabricated organic solar cells. The HOMO and LUMO levels of unsubstituted PPV are reported to be ca. -5.1 and -2.7 eV, respectively, with a band gap around 2.4 eV. Introducing two alkoxy groups into the phenylene ring to perturb the molecular orbitals lowers the LUMO to -2.9 eV with an essentially unchanged HOMO level. Hence, the band gap is reduced to 2.2 eV (Alam & Jenekhe, 2002). As a consequence, PPV emits a green-yellow light, while MEH-PPV exhibits a yellow-orange emission. Because PPV derivatives are the earliest conjugated polymers developed for organic electronics application, they were also frequently used as the active materials in organic bilayer solar cells before the concept of bulk heterojunction configuration was widely accepted. For organic bilayer solar cells, PPV and MEH-PPV serve as the electron donor in conjunction with a poly(benzimidazobenzophenanthroline ladder) (BBL, **P4**) as the electron acceptor (Figure 9). The photovoltaic parameters of the bilayer device with the configuration ITO/PPV/**P4**/Al showed a J_{sc} of 2.3 mA/cm², a V_{oc} of 1.06 V, an FF of 47%, and power conversion efficiency (PCE) value of 1.4%. In case of using MEH-PPV as the electron donor, the device reached a PCE of 0.8% under the same conditions (Alam & Jenekhe, 2004). The better photovoltaic performance of PPV over MEH-PPV can be accredited to the greater crystallinity and structural order of the PPV main chain compared to alkylated MEH-PPV.

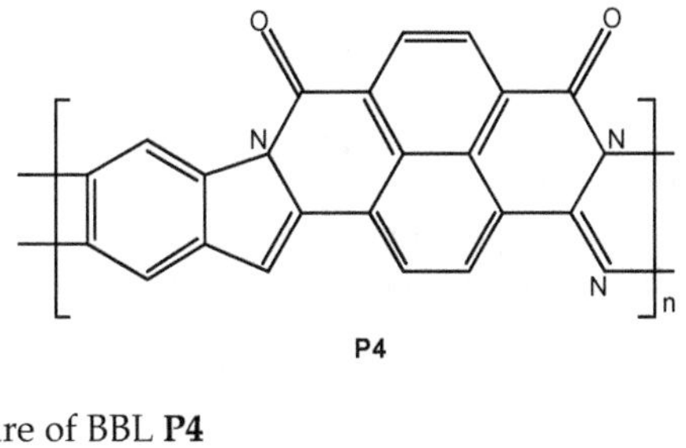

P4

Fig. 9. Chemical structure of BBL **P4**

Soluble MEH-PPV was also combined with PCBM for organic solar cell applications. The organic bilayer solar cell with the configuration ITO/PEDOT:PSS/MEH-PPV/PCBM/Al showed a J_{sc} of 2.1 mA/cm², a V_{oc} of 0.75 V, an FF of 23%, and a PCE value of 0.46% (Zhang et al., 2002). An organic bulk heterojunction solar cell including MEH-PPV/PCBM as the photoactive layer showed better PCE values in the range 1.1-1.3% than bilayer solar cells. Furthermore, by stacking two independent single organic solar cells together with the help of the transparent cathode LiF/Al/Au, the PCE of the multiple-device stacked structure can be dramatically improved to 2.6% (Shrotriya et al., 2006). The devices can be stacked together and connected either in parallel or in series, resulting in doubled J_{sc} or V_{oc}, respectively, compared to those of a single device (Cheng et al., 2009).

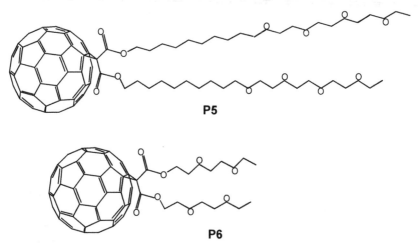

Fig. 10. Chemical structures of C_{60} derivatives **P5** and **P6**

Two alternative soluble methanofullerene derivatives **P5** and **P6** have been developed to serve as electron acceptors and combined with MEH-PPV to produce organic solar cells. The chemical structures of the C_{60} derivatives are shown in Figure 10. Due to a better compatibility of **P5** with MEH-PPV, the MEH-PPV/**P5** system shows a better device PCE of 0.49% than the MEH-PPV/**P6** system, which has a PCE of 0.22% (Li et al., 2002). In addition to C_{60} derivatives, different types of titanium oxide (TiO_2) were blended with MEH-PPV for photovoltaic device applications (Breeze et al., 2001; Song et al., 2005; Wei et al., 2006; Neyshtadt et al., 2008; Shim et al., 2008). However, their device performances were generally low, with PCE values lower than 0.5% (Cheng et al., 2009).

In organic bulk heterojunction solar cells, MDMO-PPV is the most widely used PPV derivative to serve as the electron donor in combination with C_{60} electron acceptor derivatives. Organic solar cells based on combined MDMO-PPV:PCBM (1:4, w/w) were fabricated by Shaheen and co-workers (Shaheen et al., 2001). It was found that when chlorobenzene was used as the casting solvent instead of toluene to deposit the active layer, an optimal morphology with suppressed phase segregation and enhanced microstructure was obtained, resulting in increased charge carrier mobility for both holes and electrons in the active layer. And this device achieved a J_{sc} of 5.23 mA/cm², a V_{oc} of 0.82 V, and a high PCE of 2.5% (Shaheen et al., 2001).

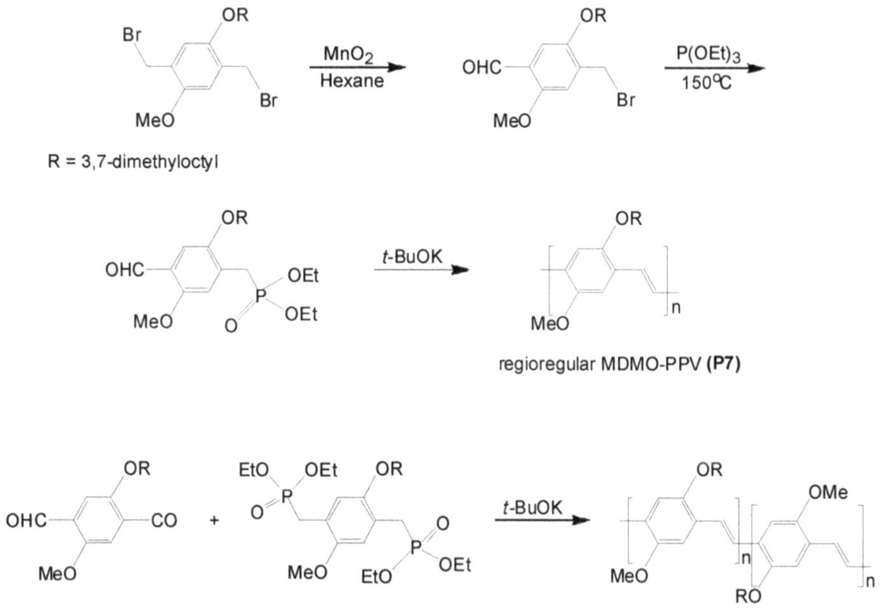

Fig. 11. Synthesis of regioregular and regiorandom MDMO-PPVs ((**P7**) and (**P8**))

Regioregularity in MDMO-PPV also plays an important role in determining the photovoltaic device performance. A fully regioregular MDMO-PPV (**P7**) by the Wittig-Hornor reaction of a single monomer comprised of aldehyde and phosphonate functionalities was synthesized (Tajima et al., 2008). Regiorandom MDMO-PPV (**P8**), from dialdehyde and diphosphnate monomers, was also prepared for comparison (Figure 11). The device achieved a PCE of 3.1%, a J_{sc} of 6.2 mA/cm^2, a V_{oc} of 0.71 V, and an FF of 70% with regioregular MDMO-PPV (**P7**). This is the highest efficiency reported for the PPV:PCBM system so far. But the device based on regiorandom MDMO-PPV (**P8**)/PCBM only achieved a PCE of 1.7%. It is concluded that higher crystallinity of the polymer for higher hole mobility and better mixing morphology between the polymer and PCBM contribute to the improvement of photovoltaic device performance with regioregular MDMO-PPV (Cheng et al., 2009).

Miscellaneous physics and engineering aspects have been investigated for devices based on the MDMO-PPV/PCBM bulk heterojunction active layer system: photooxidation (Pacios et al., 2006), stacked cells (Kawano et al., 2006), active layer thickness (Lenes et al., 2006), NMR morphology studies (Mens et al., 2008), and insertion of a hole-transporting layer between PEDOT and the active layer (Park et al., 2007). Besides the PCBM organic acceptor, inorganic electron acceptors (van Hal et al., 2003; Beek et al., 2005; Boucle et al., 2007; Sun et al., 2003) such as metal oxides or quantum dots are also under active development and have been combined with MDMO-PPV to prepare hybrid (organic-inorganic) bulk heterojunction solar cells. Optimized photovoltaic devices using blends of MDMO-PPV:ZnO (Beek et al., 2005) or MDMO-PPV:cadmium selenide (Sun et al., 2003) showed moderate PCE values of 1.6% and 1.8%, respectively.

3.2 Cyano-substituted poly(*p*-phenylene-vinylene)s

Cyano-substituted poly(*p*-phenylene-vinylene)s (CN-PPV) with electron deficient cyano groups on the vinyl units are synthesized by Knoevenagel polycondensation polymerization of terephthaldehyde and 1,4-bis(cyanomethyl)benzene in the presence of the base *t*-BuOK (Figure 12). Hence, the LUMO and HOMO levels of PPV derivatives can also be tuned by incorporating electronic substituent into the vinylene bridges (Cheng et al., 2009).

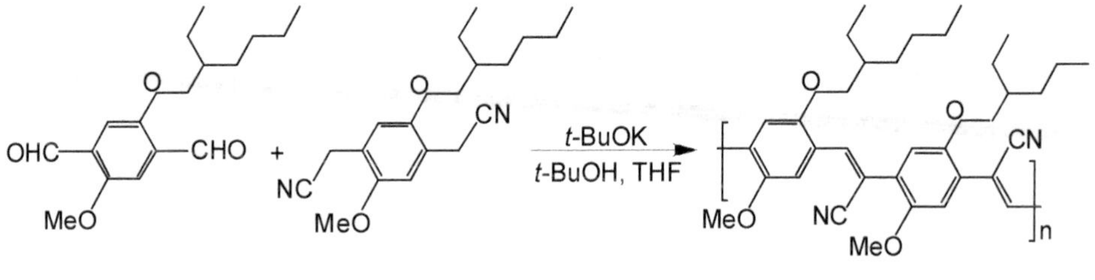

P9

Fig. 12. Synthesis of CN-PPV copolymer **P9** via a Knoevenagel Polycondensation

CN-PPVs show high electron affinity to reduce the barrier to electron injection and good electron-transport properties as a result of the electron-withdrawing effect of the cyano side group and suitable electron acceptors in organic photovoltaic devices (Granström et al., 1998; Halls et al., 1995; Gupta et al., 2007). To effectively reduce the band gap of CN-PPV below 2 eV, electron-rich thiophene units with lower aromaticities have been incorporated into the main chain to form a D-A arrangement. A series of copolymers based on the bis(1-cyano-2-thienylvinylene)pheniylene structures with different alkyl or alkoxy side chains on the thiophene rings were reported by Vanderzande et al. (Colladet et al., 2007) (Figure 13).

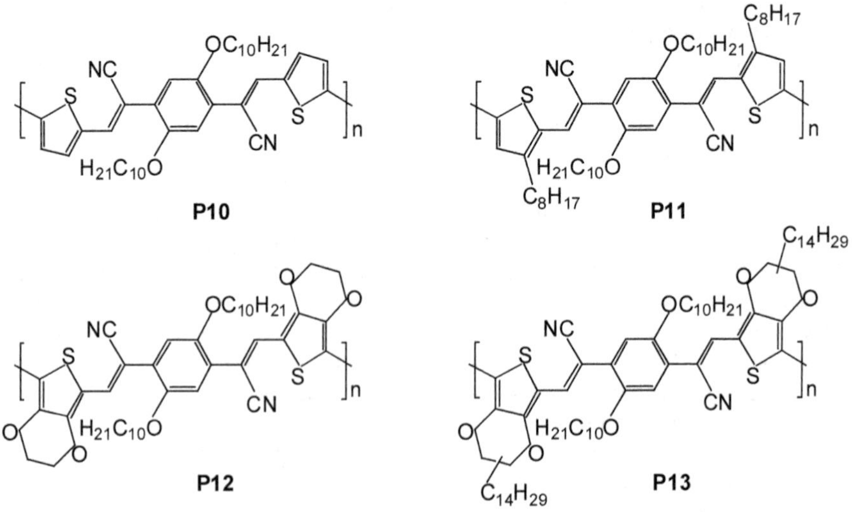

P10

P11

P12

P13

Fig. 13. Chemical structures of **P10-P13**

These monomers were all prepared by Knoevenagel condensations to construct cyanovinylene linkages. The electron-rich nature of thiophene units in these polymers makes them good candidates to serve as electron donors in organic bulk heterojunction solar cells. For example, both **P11**/PCBM- and **P12**/PCBM-based solar cells achieved a PCE of around 0.14%. Optimization of these devices by thermal annealing showed a slight increase of PCE to 0.19%. Reynold et al. also reported synthesizing a range of CN-PPV derivatives (**P14-P17**) containing dioxythiophene moieties in the polymer main chain (Figure 14) (Thompson et al., 2005; Thompson et al., 2006). The best photovoltaic device based on these CN-PPV derivatives with PCBM as the active layer achieved a PCE of 0.4%.

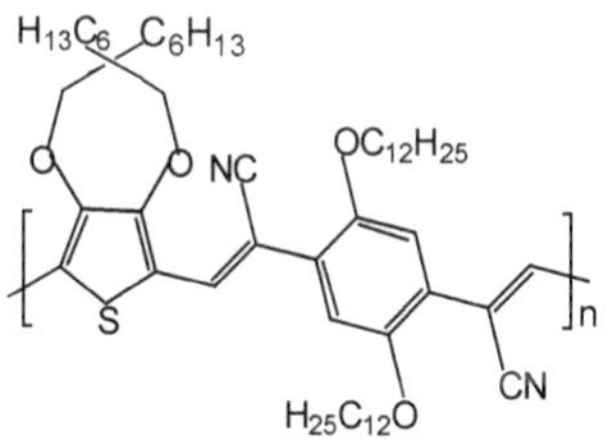

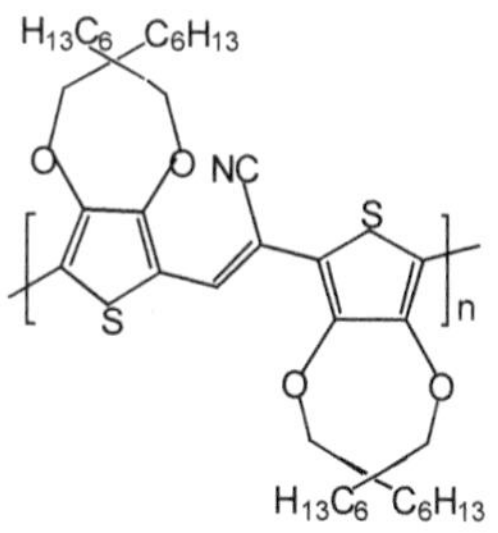

Fig. 14. Chemical structures of **P14-P17**

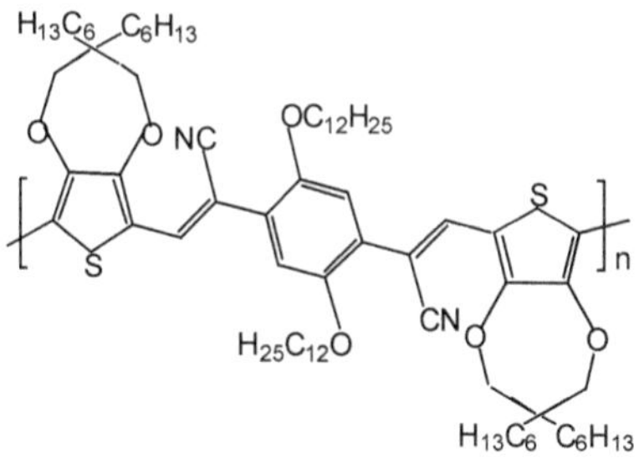

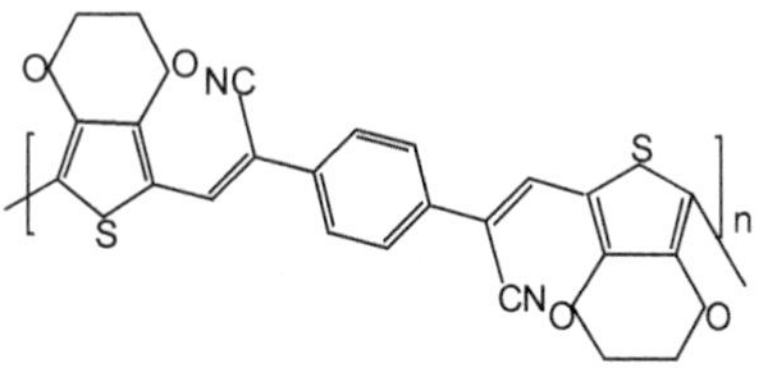

3.3 Acetylene-substituted poly(p-phenylene-vinylene)s

Acetylene-substituted PPV derivatives can be synthesized via the Wittig – Horner Reaction (Figure 15). The coplanar and rigid nature of the acetylene moiety in the polymer chain may have the potential to obtain a higher degree of packing and thus improve the photovoltaic performance of such devices. Having coplanar electron-rich anthracene units and triple bond bridges, **P18** exhibits broader absorption, a lower HOMO level, and a smaller optical band gap of 1.9 eV, compared to MDMO-PPV. A device with the configuration of ITO/PEDOT/**P18**:PCBM (1:2, w/w)/LiF/Al, achieved a PCE value of up to 2% with a high V_{oc} of 0.81 V. Figure 16 shows the chemical structures of a series of acetylene-substituted PPV derivatives synthesized by similar procedures. For polymers **P21** and **P22** (Egbe et al., 2007), the introduction of a thiophene ring into the polymer backbone showed an improvement in the PCE ranging from 1.2% to 1.7%. This is higher than those for **P19** and **P20** (Egbe et al., 2005), based on the same device configuration of ITO/PEDOT/polymer:PCBM (1:3, wt%)/LiF/Al (Cheng et al., 2009).

P21
P22
t-BuOK
P18

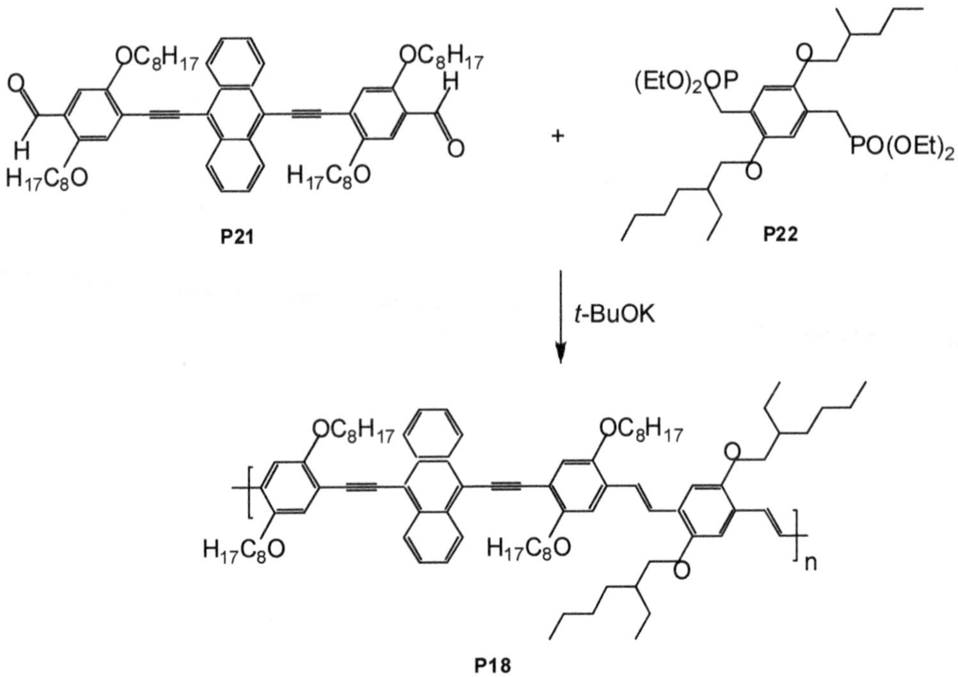

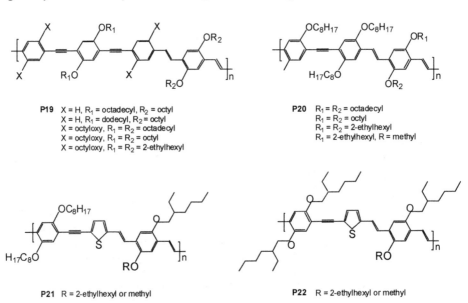

Fig. 15. Synthesis of acetylene-containing **P18** via a Wittig - Horner reaction

P19 X = H, R1 = octadecyl, R2 = octyl
X = H, R1 = dodecyl, R2 = octyl
X = octyloxy, R1 = R2 = octadecyl
X = octyloxy, R1 = R2 = octyl
X = octyloxy, R1 = R2 = 2-ethylhexyl
P20 R1 = R2 = octadecyl
R1 = R2 = octyl
R1 = R2 = 2-ethylhexyl
R1 = 2-ethylhexyl, R = methyl
P21 R = 2-ethylhexyl or methyl
P22 R = 2-ethylhexyl or methyl

Fig. 16. Chemical structures of acetylene-containing PPV derivatives **P19-P22**

4. Conclusion

The efficiency of organic solar cells is increasing steadily by means of interdisciplinary approach. Extensive efforts are currently carry out by chemists in order to create new low bandgap materials to harvest more photons and increase the power conversion efficiency. Furthermore, processability of conjugated polymers that can be deposited from liquid solutions at low temperature make them suitable for large scale production on flexible substrates at low cost roll-to-roll process. To integrate new advanced device concepts and the nanostructure engineering of the morphology are also important in bringing high efficiency and low cost organic solar cells one step closer to successful commercialization.

5. References

Akcelrud, L. (2003). Electroluminescent polymers. *Prog. Polym. Sci.*, Vol.28, pp.875-962.

Alam, M.M.; Jenekhe, S.A. (2002). Polybenzobisazoles are efficient electron transport materials for improving the performance and stability of polymer light-emitting diodes. *Chem. Mater.*, Vol.14, pp.4775-4780.

Alam, M.M.; Jenekhe, S.A. (2004). Efficient solar cells from layered nanostructures of donor and acceptor conjugated polymers. *Chem. Mater.*, Vol.16, pp.4647-4656.

Bao, Z.; Locklin, J. (2007). Organic Field Effect Transistors., Taylor & Francis, Boca Raton, FL.

Beek, W.J.E.; Wienk, M.M.; Kemerink, M.; Yang, X.; Janssen, R.A.J. (2005). Hybrid zinc oxide conjugated polymer bulk heterojunction solar cells. *J. Phys. Chem. B*, Vol.109, pp.9505-9516.

Bertho, S.; Oosterbaan W.D.; Vrindts, V.; DHaen, J.; Cleij, T.J.; Lutsen, L.; Manca, J.; Vanderzande, D. (2009). Controlling the morphology of nanofiber-P3HT:PCBM blends for organic bulk heterojunction solar cells. *Org. Electron.*, Vol.10, pp.1248-1251.

Boucle´, J.; Ravirajan, P.; Nelson, J. (2007). Hybrid polymer-metal oxide thin films for photovoltaic applications. *J. Mater. Chem.*, Vol.17, pp.3141-3153.

Breeze, A.J.; Schlesinger, Z.; Carter, S.A.; Brock, P. (2001). Charge transport in TiO_2/MEH-PPV polymer photovoltaics. *J. Phys. Rev.B*, Vol.64, pp.125205-1 – 125205-9.

Braun, D.; Heeger, A.J. (1991). Visible-light emission from semiconducting polymer diodes. *Appl. Phys. Lett.*, Vol.58, pp.1982.

Burroughes, J.H.; Bradley, D.D.C.; Brown, A.R.; Marks, R.N.; Mackay, K.; Friend, R.H.; Burns, P.L.; Holmes, A.B. (1990). Light-emitting diodes based on conjugated polymers. *Nature*, Vol.347, pp.539-541.

Chamberlain, G.A. (1983). Organic solar cells: a review. *Solar Cells*, Vol.8, pp.47-83.

Cheng, Y.J.; Yang, S.H.; Hsu, C.S. (2009). Synthesis of conjugated polymers for organic solar cell applications. *Chem. Rev.*,Vol.109, pp.5868-5923.

Chiang, C.K.; Fischer, C.R.; Park, Y.W.; Heeger, A.J.; Shirakawa, H.; Louis, E.J.; Gau, S.C.; MacDiarmid, A.G. (1977). Electrical conductivity in doped polyacetylene. *Phys. Rev. Lett.*, Vol.39, pp.1098-1101.

Chiang, C.K.; Druy, M.A.; Gau, S.C.; Heeger, A.J.; Louis, E.J.; MacDiarmid, A.G.; Park, Y.W.; Shirakawa, H. (1978). Synthesis og highly conducting films of derivatives of polyacetylene, (CH). *J. Am. Chem. Soc.*, Vol.100, pp.1013-1015.

Colladet, K.; Fourier, S.; Cleij, T.J.; Lutsen, L.; Gelan, J.; Vanderzande, D.; Nguyen, L.H.; Neugebauer, H.; Sariciftci, S.; Aguirre, A.; Janssen, G.; Goovaerts, E. (2007). Low band gap donor-acceptor conjugated polymers toward organic solar cells applications. *Macromolecules*, Vol.40, pp.65-72.

Dennler, G.; Sariciftci, N.S. (2005). Flexible conjugated polymer-based plastic solar cells: from basics to applications. *Proceedings of The IEEE*, Vol.93, pp.1429-1439.

Egbe, D.A.M.; Nguyen, L.H.; Schmidtke, K.; Wild, A.; Sieber, C.; Guenes, S.; Sariciftci, N.S. (2007). Combined effects of conjugation pattern and alkoxy side chains on the photovoltaic properties of thiophene-containing PPE-PPVs. *J. Polym. Sci., Part A: Polym. Chem.*, Vol.45, pp.1619-1631.

Egbe, D.A.M.; Nguyen, L.H.; Hoppe, H.; Mühlbacher, D.; Sariciftci, N.S. (2005). Side chain influence on electrochemical and photovoltaic properties of yne-containing poly(phenylene vinylene)s. *Macromol. Rapid Commun.*, Vol.26, pp.1389-1394.

Forrest, S.R. (2004). The path to ubiquitous and low-cost organic electronic appliances on plastic. *Nature*, Vol.428, pp.911-918.

Gilch, H.G.; Wheelwright, W.L. (1966). Polymerization of α-halogenated p-xylenes with base *J. Polym. Sci., Part A: Polym.Chem.*, Vol.4, pp.1337-1349.

Granström, M.; Petritsch, K.; Arias, A.C.; Lux, A.; Andersson, M.R.; Friend, R.H. (1998). Laminated fabrication of polymeric photovoltaic diodes. *Nature*, Vol.395, pp.257-260.

Green, M.A.; Emery, K.; Hishikawa, Y.; Warta, W. (2010). Solar cell efficiency tables (version 35). *Prog. Photovolt. Res. Appl.*, Vol.18, pp.144-150.

Gunes, S.; Neugebauer, H. Sariciftci, N.S. (2007). Conjugated polymer-based organic solar cells. *Chem. Rev.*, Vol.107, pp.1324-1338.

Gupta, D.; Kabra, D.; Kolishetti, N.; Ramakrishnan, S.; Narayan, K.S. (2007). An efficient bulk-heterojunction photovoltaic cell based on energy transfer in graded-bandgap polymers. *Adv. Funct. Mater.*, Vol.17, pp.226-232.

Hadziioannou, G.; van Hutten, P.F. (2000). Semiconducting polymers : chemistry, physics and engineering, 1st ed., Wiley-VCH, Weinheim, Germany.

Hadziioannou, G.; Mallarias, G.G. (2007). Semiconducting Polymers., Wiley-VCH, Weinheim, Germany.

Halls, J.J.M.; Walsh, C.A.; Greenham, N.C.; Marseglia, E.A.; Friend, R.H.; Moratti, S.C.; Homles, A.B. (1995). Efficient photodiodes from interpenetrating polymer networks. *Nature*, Vol.376, pp.498-500.

Haugeneder, A.; Neges, M.; Kallinger, C.; Spirkl, W.; Lemmer, U.; Feldmann, J.; Scherf, U.; Harth, E.; Gügel, A.; Müllen, K. (1999). Exciton diffusion and dissociation in conjugated polymer/fullerene blends and heterostructures. *Phys. Rev. B.*, Vol.59, pp.15346-15351.

Hoppe, H.; Sariciftci, N.S. (2004). Organic solar cells: an overview. *J. Mater. Res.*, Vol.19, pp.1924-1945.

Hummelen, J.C.; Knight, B.W.; LePeq, F.; Wudl, F.; Yao, J.; Wilkins, C.L. (1995). Preparation and characterization of fulleroid and methanofullerene derivatives. *J. Org. Chem.*, Vol.60, pp.532-538.

Kawano, K.; Ito, N.; Nishimori, T.; Sakai, J. (2006). Open circuit voltage of stacked bulk heterojunction organic solar cells. *Appl. Phys. Lett.*, Vol.88, pp.073514-1 – 073514-3.

Klauk, H. (2006). Organic Electronics., Wiley-VCH, Weinheim, New York.Shirakawa, H.; Louis, E.J.; MacMiarmid, A.G.; Chiang, C.K.; Heeger, A.J. (1977). Synthesis of electrically conducting organic polymers : halogen derivatives of polyacetylene, $(CH)_x$. *J. C. S. Chem. Comm.*, pp.578-580.

Lenes, M.; Koster, L.J.A.; Mihailetchi, V.D.; Blom, P.W.M. (2006). Thickness dependence of the efficiency of polymer : fullerene bulk heterojunction solar cells. *Appl. Phys. Lett.*, Vol.88, pp.243502-1 – 243502-3.

Li, J.; Sun, N.; Guo, Z.X.; Li, C.; Li, Y.; Dai, L.; Zhu, D.; Sun, D.; Cao, Y.; Fan, L. (2002). Photovoltaic devices with methanofullerenes as electron acceptors. *J. Phys. Chem. B*, Vol.106, pp.11509-11514.

Mens, R.; Adriaensens, P.; Lutsen, L.; Swinnen, A.; Bertho, S.; Ruttens, B.; D'Haen, J.; Manca, J.; Cleij, T.; Vanderzande, D.; Gelan, J. (2008). NMR study of the nanomorphology in thin films of polymer blends used in organic PV devices: MDMO-PPV/PCBM. *J. Polym. Sci., Part A: Polym. Chem.*, Vol.46, pp.138-145.

Moliton, A. (2006). Optoelectronics of Molecules and Polymers., Springer-Verlag, Berlin, Germany.

Mozer, A.; Sariciftci, N.S. (2006). Conjugated polymer photovoltaic devices and materials. *Chimie*, Vol.9, pp.568-577.

Nalwa, H.S. (2008). Handbook of Organic Electronics and Photonics., American Scientific Publishers, Valencia, CA.

Neef, C.J.; Ferraris, J.P. (2000). MEH-PPV: Improved synthetic procedure and molecular weight control. *Macromolecules*, Vol.33, pp.2311-2314.

Neyshtadt, S.; Kalina, M.; Frey, G.L. (2008). Self-organized semiconducting polymer-incorporated mesostructured titania for photovoltaic applications. *Adv. Mater.*, Vol.20, pp.2541-2546.

Onsager, L. (1938). Initial recombination of ions. *Phys. Rev.*, Vol.54, pp.554-557.

Pacios, R.; Chatten, A.J.; Kawano, K.; Durrant, J.R.; Bradley, D.D.C.; Nelson, J. (2006). Effects of photo-oxidation on the performance of poly[2-methoxy-5-(3 ',7 '-dimethyloctyloxy)-1,4-phenylene vinylene]:[6,6]-phenyl C-61-butyric acid methyl ester solar cells. *Adv. Funct. Mater.*, Vol.16, pp.2117-2126.

Padinger, F.; Rittberger, R.S.; Sariciftci, N.S. (2003). Effects of postproduction treatment on plastic solar cells. *Adv. Funct. Mater.*, Vol.13, pp.85-88.

Park, J.; Han, S.H.; Senthilarasu, S.; Lee, S.H. (2007). Improvement of device performances by thin interlayer in conjugated polymers/methanofullerene plastic solar cell. *Sol. Energy Mater.Sol. Cells*, Vol.91, pp.751-753.

Park, S.H.; Roy, A.; Beaupre, S.; Cho, S.; Coates, N.; Moon, J.S.; Moses, D.; Leclerc, M.; Lee, K.; Heeger, A.J. (2009). Bulk heterojunction solar cells with internal quantum efficiency approaching 100%. *Nat. Photon.*, Vol.3, pp.297-303.

Parker, I.D. (1994). Carrier tunneling and device characteristics in polymer light-emitting diodes. *J. Appl. Phys.*, Vol.75, pp.1656-1666.

Papadimitrakopoulos, F.; Konstadinidis, K.; Miller, T.M.; Opila, R.; Chandross, E.A.; Galvin, M.E. (1994). The Role of Carbonyl Groups in the Photoluminescence of Poly(p-phenylenevinylene). *Chem. Mater.*, Vol.6, pp.1563-1568.

Peet, J.; Kim, J.Y.; Coates, N.E.; Ma, W.L.; Moses, D.; Heeger, A.J.; Bazan, G.C. (2007). Efficiency enhancement in low-bandgap polymer solar cells by processing with alkane dithiols. *Nat. Mater.*, Vol.&, pp.497-500.

Sariciftci, N.S.; Smilowitz, L.; Heeger, A.J.; Wudl, F. (1992). Photoinduced electron-transfer from a conducting polymer to buckminsterfullerenes. *Science*, Vol.258, pp.1474-1476.

Shaheen, S.E.; Brabec, C.J.; Sariciftci, N.S.; Padinger, F.; Fromherz, T.; Hunnelen, J.C. (2001). 2.5% efficient organic plastic solar cells. *Appl. Phys. Lett.*, Vol.78, pp.841-843.

Shim, H.S.; Na, S.I.; Nam, S.H.; Ahn, H.J.; Kim, H.J.; Kim, D.Y.; Kim, W.B. (2008). Efficient photovoltaic device fashioned of highly aligned multilayers of electrospun TiO2 nanowire array with conjugated polymer. *Appl. Phys. Lett.*, Vol.92, pp.183107-1 – 183107-3.

Shinar, R.; Shinar, J. (2009). Organic Electronics in Sensors and Biotechnology., McGraw-Hill, New York.

Shirakawa, H. (2001). Die Entdeckung der Polyacetylenefilme – der Beginn des Zeitalters leitfähiger Polymere (Nobel – Aufsatz). *Angew. Chem.*, Vol.113, pp.2642-2628.

Shrotriya, V.; Wu, E.H.E.; Li, G.; Yao, Y.; Yang, Y. (2006). Efficient light harvesting in multiple-device stacked structure for polymer solar cells. *Appl. Phys. Lett.*, Vol.88, pp.064104-1 – 064104-3.

Song, M.Y.; Kim, K.J.; Kim, D.Y. (2005). Enhancement of photovoltaic characteristics using a PEDOT interlayer in $TiO2/MEHPPV$ heterojunction devices. *Sol. Energy Mater. Sol. Cells*, Vol.85, pp.31-39.

Skotheim, T.J.; Elsenbaumer, R.L.; Reynolds, J.R. (1998). Handbook of Conducting Polymers, 2nd ed., Marcel Dekker, New York.

Sun, B.; Marx, E.; Greenham, N.C. (2003). Photovoltaic devices using blends of branched CdSe nanoparticles and conjugated polymers. *Nano Lett.*, Vol.3, pp.961-963.

Sun, S.S.; Dalton, L. (2008). Introduction to Organic Electronic and Optoelectronic Materials and Devices., Taylor & Francis, Boca Raton, FL.

Tajima, K.; Suzuki, Y.; Hashimoto, K. (2008). Polymer photovoltaic devices using fully regioregular poly[(2-methoxy-5-(3 ',7 '-dimethyloctyloxy))-1,4-phenylenevinylene]. *J. Phys. Chem. C*, Vol.112, pp.8507-8510.

Tang, C.W. (1986). 2-Layer Organic Photovoltaic Cell. *Appl. Phys. Lett.*, Vol.48, pp.183-185.

Thompson, B.C.; Kim, Y.G.; Reynolds, J.R. (2005). Spectral broadening in MEH-PPV : PCBM-based photovoltaic devices via blending with a narrow band gap cyanovinylene-dioxythiophene polymer. *Macromolecules*, Vol.38, pp.5359-5362.

Thompson, B.C.; Kim, Y.G.; McCarley, T.D.; Reynolds, J R. (2006). Soluble narrow band gap and blue propylenedioxythiophene-cyanovinylene polymers as multifunctional materials for photovoltaic and electrochromic applications. *J. Am. Chem. Soc.*, Vol.128, pp.12714-12725.

van Hal, P.A.; Wienk, M.M.; Kroon, J.M.; Verhees, W.J.H.; Slooff, L.H.; van Gennip, W.J.H.; Jonkheijm, P.; Janssen, R.A.J. (2003). Photoinduced electron transfer and photovoltaic response of a MDMO-PPV:TiO2 bulk-heterojunction. *Adv. Mater.*, Vol.15, pp.118-121.

Wei, Q.; Hirota, K.; Tajima, K.; Hashimoto, K. (2006). Design and Synthesis of TiO_2 Nanorod Assemblies and Their Application for Photovoltaic Devices. *Chem. Mater.*, Vol.18, pp.5080-5087.

Wessling, R.A.; Zimmerman, R.G. (1968). U.S. Patent 3401152.

Wessling, R.A. (1985). The polymerization of xylylene bisdialkyl sulfonium salts. *J. Polym. Sci., Polym. Symp.*, Vol.72, pp.55-66.

Wohrle, D.; Meissner, D. (1991). Organic Solar Cells. *Adv. Mater.*, Vol.3, pp.129-138.

Yu, G.; Gao, J.; Hummelen, J.C.; Wudl, F.; Heeger, A.J. (1995). Polymer photovoltaic cells: Enhanced efficiencies via a network of internal donor-acceptor heterojunction. *Science*, Vol.270, pp.1789-1791.

Yu, G.; Wang, J.; McElvain, J.; Heeger, A.J. (1998). Large-area, full-color image sensors made with semiconducting polymers. *Adv. Mater.*, Vol.10, pp.1431-1434.

Zhang, F.L.; Johansson, M.; Andersson, M.R.; Hummelen, J.C.; Inganäs, O. (2002). Polymer photovoltaic cells with conducting polymer anodes. *Adv. Mater.*, Vol.14, pp.662-665.

10

Towards High-Efficiency Organic Solar Cells: Polymers and Devices Development

Enwei Zhu[1], Linyi Bian[1], Jiefeng Hai[1], Weihua Tang[1,*] and Fujun Zhang[2,*]

[1]Nanjing University of Science and Technology,
[2]Beijing Jiaotong University
People's Republic of China

1. Introduction

The effective conversion of solar energy into electricity has attracted intense scientific interest in solving the rising energy crisis. Organic solar cells (OSCs), a kind of green energy source, show great potential application due to low production costs, mechanical flexibility devices by using simple techniques with low environmental impact and the versatility in organic materials design (Beal, 2010). In the past years, the key parameter, power conversion efficiencies (PCE), is up to 7% under the standard solar spectrum, AM1.5G (Liang et al., 2010). The PCE of solar cells are co-determined by the open circuit voltage (V_{oc}), the fill factor (FF) and the short circuit density (J_{sc}). Researchers have made great efforts in both developing new organic materials with narrow band gap and designing different structural cells for harvesting exciton in the visible light range.

Solution processing of π-conjugated materials (including polymers and oligomers) based OSCs onto flexible plastic substrates represents a potential platform for continuous, large-scale printing of thin-film photovoltaics (Krebs, 2009; Peet, 2009). Rapid development of this technology has led to growing interest in OSCs in academic and industrial laboratories and has been the subject of multiple recent reviews (Cheng, 2009; Dennler, 2009; Krebs, 2009; Tang, 2010). These devices are promising in terms of low-cost power generation, simplicity of fabrication and versatility in structure modification. The structure modification of π-conjugated materials has offered wide possibilities to tune their structural properties (such as rigidity, conjugation length, and molecule-to-molecule interactions) and physical properties (including solubility, molecular weight, band gap and molecular orbital energy levels). This ability to design and synthesize molecules and then integrate them into organic–organic and inorganic–organic composites provides a unique pathway in the design of materials for novel devices. The most common OSCs are fabricated as the bulk-heterojunction (BHJ) devices, where a photoactive layer is casted from a mixture solution of polymeric donors and soluble fullerene-based electron acceptor and sandwiched between two electrodes with different work functions (Yu et al., 1995). When the polymeric donor is excited, the electron promoted to the lowest unoccupied molecular orbital (LUMO) will lower its energy by moving to the LUMO of the acceptor. Under the built-in electric field caused by the contacts, opposite charges in the photoactive layer are separated, with the holes being transported in the donor phase and the electrons in the acceptor. In this way, the blend can be considered as a network of donor–acceptor heterojunctions that allows efficient charge separation and balanced bipolar transport throughout its whole volume. Remarkably, the power conversion efficiency (PCE, defined as the maximum power produced by a photovoltaic cell divided by the power of incident light) of the OSCs has been pushed to more than 7% from 0.1% after a decade's intensive interdisciplinary research. The current workhorse materials employed for PSCs are regioregular poly(3-hexylthiophene) (P3HT) and [6,6]-phenyl-C61-butyric acid methyl ester (PCBM). This material combination has given the highest reported PCE values of 4%~5% (G. Li, 2005). Theoretically, the PCE of polymer solar cells can be further improved (ca 10%) (Scharber et al., 2006) by implementing new materials (Cheng, 2009; Peet, 2009; Tang, 2010) and exploring new device architecture (Dennler, 2008; Ameri, 2009; Dennler, 2009) after addressing several fundamental issues such as bandgap, interfaces and charge transfer (Li, 2005; Chen, 2008; Cheng, 2009).

In this account, we will update the recent 4 years progress in pursuit of high performance BHJ OSCs with newly developed conjugated polymers, especially narrow bandgap polymers from a viewpoint of material chemists. The correlation of polymer chemical structures with their properties including absorption spectra, band gap, energy levels, mobilities, and photovoltaic performance will be elaborated. The analysis of structure-property relationship will provide insight in rational design of polymer structures and reasonable evaluation of their photovoltaic performance.

2. Fluorene-based conjugated polymers

Fluorene (FL) and its derivatives have been extensively investigated for their application in light-emitting diodes due to its rigid planar molecular structure, excellent hole-transporting properties, good solubility, and exceptional chemical stability.

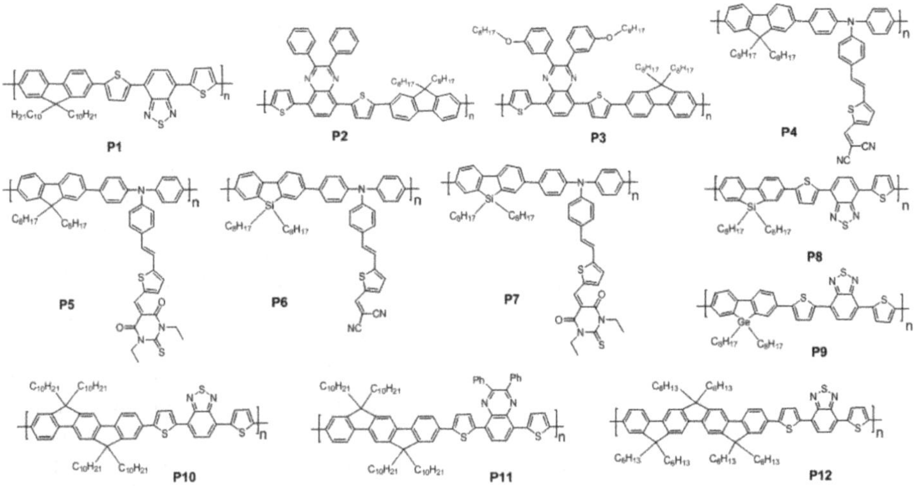

Chart 1. Flourene based narrow band gap polymers.

Along with their low-lying HOMO levels, polyfluorenes (PFs) are expected to achieve higher V_{oc} and J_{sc} in their PSC device, which makes fluorene unit a promising electron-donating moiety in D-A narrow band gap polymers' design. Besides, feasible dialkylation at 9-position and selective bromination at the 2,7-positions of fluorene allow versatile molecular manipulation to achieve good solubility and extended conjugation *via* typical Suzuki or Stille cross-coupling reactions. By using 4,7-dithien-2-yl-2,1,3-benzothiadiazole (DTBT) as electron accepting unit and didecylated FL as donating unit, Slooff (Slooff et al., 2007) developed **P1** (**P1-12** structure in Chart 1) with extended absorption spectrum ranging from 300 to 800nm. Spin-coated from chloroform solution, the device ITO/PEDOT:PSS/**P1**:PCBM(1:4, w/w)/LiF/Al harvested a extremely high PCE of 4.2% (Table 1). An external quantum efficiency (EQE) of 66% was achieved in the active layer with a film thickness up to 140 nm, and further increasing the film thickness did not increase the efficiency due to limitations in charge generation or collection. For 4.2% PCE device, a maximum EQE of about 75% was calculated, indicating efficient charge collection.

By using quinoxaline as electron accepting unit, **P2** was synthesized with an E_g of 1.95eV (Kitazawa et al., 2009). The device performance is dependent upon the ratio of chloroform(CF)/chlorobenzene(CB) in co-solvent for blend film preparation and a maximal J_{sc} is achieved with CF/CB (2:3 v/v) co-solvent. The optimized device showed 5.5% PCE by inserting 0.1nm LiF layer between BHJ active layer and Al cathode with the structure ITO/PEDOT:PSS/**P2**:PC71BM/LiF/Al. Similarly structured **P3** achieved 3.7% PCE by blending with PC71BM (1:3 w/w) (Gadisa et al., 2007).

Polymer	λ_{max}^{abs} nm	E_g eV	μ_h cm^2 V^{-1}s^{-1}	HOMO /LUMO, eV	Polymer: PCBM[b]	J_{sc} mA/cm^2	V_{oc} V	FF	PCE
P1	-	-	-		1:4	7.7	1.0	0.54	4.2
P2	540	1.95	-	-5.37/-	1:4[a]	9.72	0.99	0.57	5.5
P3	542	1.94	-	-6.30/-3.60	1:3	6.00	1.00	0.63	3.7
P4	545	1.87	5.3×10^{-4}	-5.30/-3.43	1:4	9.62	0.99	0.5	4.74
P5	580	1.76	1.2×10^{-3}	-5.26/-3.50	1:4	9.61	0.99	0.46	4.37
P6	541	1.83	1.8×10^{-4}	-5.32/-3.49	1:4[a]	6.69	0.85	0.37	2.50
P7	579	1.74	2.1×10^{-4}	-5.35/-3.61	1:4[a]	6.22	0.90	0.45	3.15
P8	565	1.82	1.0×10^{-3}	-	1:2	9.50	0.90	0.51	5.4
P9	580	1.79	1.1×10^{-4}	-5.58/-3.91	-	6.9	0.79	0.51	2.8
P10	543	1.97	4.2×10^{-3}	-5.47/-3.44	1:3.0	6.10	1.00	0.40	2.44
P11	541	2.00	9.5×10^{-4}	-5.45/-3.36	1:3.5[a]	7.57	1.00	0.40	3.04
P12	531	1.96	9.7×10^{-3}	-5.45/-3.45	1:4.0[a]	10.3	1.04	0.42	4.50

λ_{max}^{abs}: maximum absorption peak in film; E_g: optical band gap; μ_h: hole mobility; J_{sc}: short-circuit current density; V_{oc}: open-circuit voltage; FF: fill factor; PCE: power conversion efficiency; [a]polymer:PC71BM; [b]polymer:PCBM in weight ratio.

Table 1. The optical, electrochemical, hole mobility, and PSC characteristics of **P1-12**

Different from the common linear D-A alternating polymer design, Jen and his coworkers designed a series of novel two-dimensional narrow band gap polymers, whose backbone adopts high hole transporting fluorene-triarylamine copolymer (PFM) and is grafted with malononitrile (**P4**) and diethylthiobarbituric acid (**P5**) through a styrylthiophene π-bridge (Huang et al., 2009). Both of them show two obvious absorption peaks, where the first absorption peaks at ~385 nm are corresponding to the π-π * transition of their conjugated main chains and the others are corresponding to the strong ICT characters of their side chains. Two polymers show narrowed down E_g (<2eV) and present similar HOMO energy level as that of PFM, while these two polymers exhibit much lowered LUMO levels (-3.43 and -3.50eV for **P4** and **P5**, respectively) than PFM,. The devices with structure of ITO/PEDOT:PSS(40nm)/**P4-5**:PC71BM(1:4 w/w, 85nm)/Ca(10nm)/Al(100nm) exhibit a PCE of 4.74% and 4.37% for **P4** and **P5**, respectively (refer to Table 1). Determined by using space-charge-limited-current (SCLC) method, the hole mobility of these two polymer/P71BM blend films was found to be $5.27 \times 10^{-4} cm^2V^{-1}s^{-1}$ for **P4** and $1.16 \times 10^{-3} cm^2V^{-1}s^{-1}$ for **P5**, respectively, which is even higher than that of P3HT with similar device configuration (G. Li, 2005). These two-dimensional polymers are thought to possess better isotropic charge transport ability than linear polymers, which is beneficial for PSC applications. By replacing FL in **P4** and **P5** with silafluorene (2,7-dibenzosilole) unit, similarly structured two-dimensional conjugated polymers **P6** and **P7** were developed (Duan et al., 2010) developed. With an E_g of 1.83 and 1.74eV, **P6** and **P7** OSCs harvest high PCE as 2.5% and 3.51%, respectively, with device configuration ITO/PEDOT:PSS/**P6-7**:PC71BM/Ba/Al. It should be noted that the silofluorene polymers display lower hole transport ability than FL-based **P4** and **P5**.

By alternating silafluorene and DTBT units, copolymer **P8** was synthesized (Boudreault et al., 2007a). With an E_g of 1.82eV, **P8** films presented an absorption spectrum blend covered the range from 350 to 750nm. **P8** device displayed a PCE of 1.6% under AM 1.5 (90 mW/cm²) illumination with the structure ITO/PEDOT:PSS/**P8**:PCBM(1:4)/Al. By using similar polymer, **P8**:PCBM (1:2 w/w) blend film delivered as high as PCE of 5.4% under AM 1.5 (80 mW/cm²) (Zhou et al., 2004). By using DTBT as electron accepting unit, Leclerc (Allard et al., 2010) developed germafluorene based polymer **P9** whose device displayed a good PCE of 2.8% with the structure of ITO/PEDOT-PSS/**P9**:PC71BM/LiF(20nm)/Al.

In order to extend π-conjugation of polymer backbone, the ladder-type oligo-*p*-phenylenes (indenofluorene) consisting of several "linearly overlapping" fluorene was developed. Compared to FL unit, indenofluorene presents a broader more intense absorption band. Solubilizing alkyl chains can be easily introduced into this unique molecular backbone, which may provide a better solution processability of the polymers. Zheng (Zheng et al., 2010) reported three alternating D-A copolymers (**P10-12**) combining indenofluorene as the donor and DTBT or 5,8-dithien-2-yl-2,3-diphenyl quinoxaline (DTQX) as the acceptor unit. By spin-coating from chlorobenzene (CB):o-dichlorobenzene(DCB) (4:1 v/v) co-solvent, **P10** with decylated indenofluorene presents good solubility and its device achieved 2.44% PCE with the configuration ITO/PEDOT:PSS/**P10**:PCBM/Cs₂CO₃/Al (Veldman et al., 2008). In comparison with **P10**, **P12** with one more phenylene group in indenofluorene unit presents greatly improved device PCE (3.67%) with the same configuration, due to great improvement of J_{sc}. By blending with PC71BM in 1:4 weight ratio, **P18** OSC showed significantly improved PCE (4.5%). The DTQX-containing **P11** presents a good PCE of 2.32%, which was attributed to broad absorption spectra and high J_{sc}.

3. Carbazole-based conjugated polymers

Structurally analogous to FL, carbazole has been also known as 9-azafluorene. The central fused pyrrole ring makes tricyclic carbazole fully aromatic and electron-rich with its donating nitrogen. The alkyl chains introduced into nitrogen of carbazole improve the solubility. Carbazole derivatives have been widely used as electron-donor materials due to their excellent thermal and photochemical stabilities as well as relatively high charge mobility and good solubility with 9-position alkylation. When two carbazoles fused together to form indolo[3,2-*b*]carbazole (IC) unit, it exhibits even stronger electron-donating properties, higher hole mobility and better stability than carbazole (Boudreault, 2007b; Li, 2006). With bulky heptadecanyl modified carbazole to improve solubility, Leclerc (Blouin et al., 2007a) developed DTBT-based conjugated poly(N-alkyl-2,7-carbazole) (PC) **P13** (**P13-26** structure in Chart 2), which exhibits excellent thermal stability, relatively high molecular weight and good solubility. Spin-coated from CF solution, **P13**:PCBM(1:4 w/w) OSC delivered a high PCE of 3.6% (Table 2). Such high PCE and excellent stability show its potential application in OSC. By using longer alkyl chain, Hashimoto (Zou et al., 2010) reported a similar PC **P14**, whose device based on **P14**:PCBM (1:3w/w) showed 3.05% PCE, with all photovoltaic parameters similar to those of **P13**. By using 4,7-dithien-2-yl-2,1,3-benzothiadiazoxaline (DTBX) as accepting unit, Leclerc (Blouin et al., 2007b) synthesized **P15**. Due to its symmetrical backbone, **P15** showed good structural organization, which leads to good hole mobility and thus resultant improvement of J_{sc} and FF for its OSC. **P15** based OSC achieved a good PCE of 2.4%.

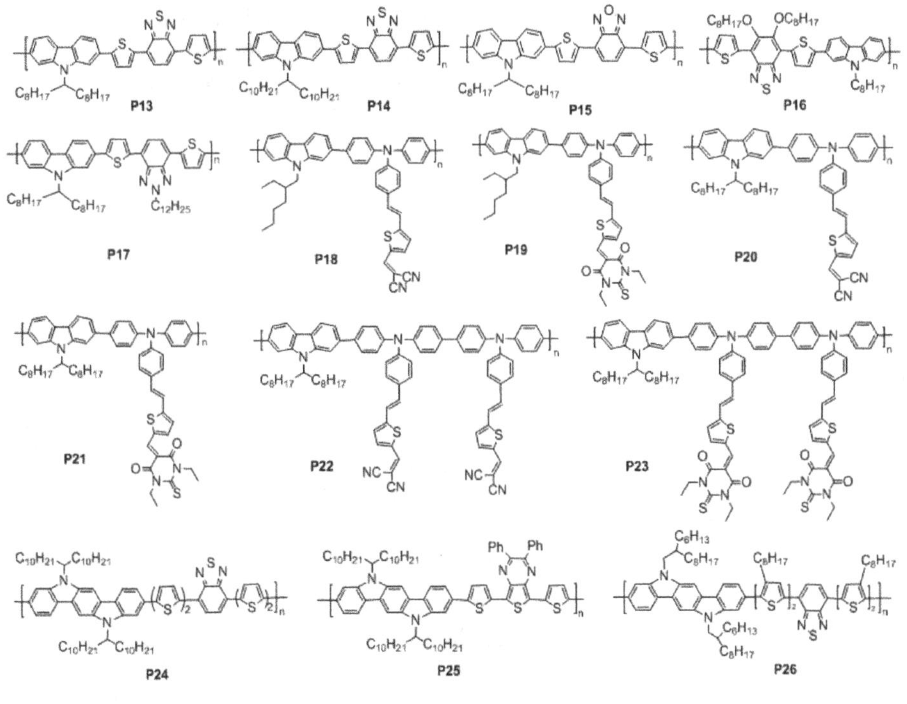

Chart 2. Carbazole based narrow band gap polymers

In order to improve the solubility and close packing of polymer backbone, Zhang (Qin et al., 2009) designed **P16** with planar polymer conformation by introducing two octyloxy chains onto benzothiazole (BT) ring and an octyl chain onto carbazole ring. **P16** showed good solubility at elevated temperature and an E_g of 1.95eV. Spin-coated from DCB mixture solutions with 2.5% 1,8-diiodooctane (DIO), **P16**:PC71BM device showed extremely high

PCE of 5.4% with the structure ITO/PEDOT:PSS/**P16**:PC71BM(1:2.5 w/w)/LiF/Al. By replacing the BT unit with benzotriazole, Cao (Zhang et al., 2010) developed PC **P17**, with an E_g of 2.18eV, which was broader than that of P3HT. The optimized device of **P17** achieved improved PCE from 1.51% to 2.75% by sandwiching a thin layer of PFN between Al cathode and active layer.

On the basis of successful design of two-dimensional PF **P4** and **P5**, Cao [Duan et al., 2010b] developed a series of carbazole-based two-dimensional polymers **P18-23**, with their E_g ranging from 1.7 to 1.9eV. **P18-23** present similar HOMO energy levels as the conjugated backbone while their LUMO levels were mainly determined by narrow band gap side-chains. Among them, **P22** and **P23** exhibited stronger absorption intensity and broader absorption spectra. The photovoltaic performance of **P22** and **P23** devices indicates the 2-ethylhexyl chains had little influence on the V_{oc} of OSC while the bulky heptadecanyl chains on carbazole led to a significant decrease in J_{sc} and FF. In comparison with that of **P4** and **P5**, **P18** devices exhibited excellent PCE of 4.16%. **P18-23** OSCs possessed high V_{oc} (>0.8eV) values, which was relevant with their relatively low HOMO levels. As known, V_{oc} is related to the offset between the HOMO levels of donor materials and the LUMO levels of acceptor materials. The alkyl chains on carbazole unit had important influence on the hole-transporting ability of the resultant polymers: **P18** and **P19** with short side-chains demonstrated hole mobility high as 2.4×10^{-4} and 1.1×10^{-3}cm²V⁻¹s⁻¹, respectively, which are much higher than those of **P20-P23** with bulkier alkyl side-chains. The bulky alkyl chains possibly are thought to result in imbalanced electron transporting and consequently decreased FF of polymer.

Polymer	λ_{max}^{abs} nm	E_g eV	μ_h cm²V⁻¹s⁻¹	HOMO/LUMO, eV	Polymer: PCBM	J_{sc} mA/cm²	V_{oc} V	FF	PCE
P13	576	1.87	-	-5.50/-3.60	1:4	6.92	0.89	0.63	3.6
P14	551	1.72	-	-5.48/-3.46	1:3	7.66	0.80	0.50	3.05
P15	-	1.87	1×10^{-4}	-5.47/-3.65	1:4	3.70	0.96	0.60	2.4
P16	579	1.95	1×10^{-4}	-5.21/-3.35	1:2.5a	9.8	0.81	0.69	5.4
P17	523	2.18	-	-5.54/-3.36	1:2	4.68	0.9	0.65	2.75
P18	548	1.83	2.4×10^{-4}	-5.23/-3.40	1:4a	8.94	0.91	0.51	4.16
P19	586	1.74	1.1×10^{-3}	-5.25/-3.51	1:4a	8.18	0.92	0.47	3.52
P20	545	1.88	5.3×10^{-5}	-5.29/-3.41	1:4a	7.51	0.91	0.44	2.91
P21	575	1.77	1.5×10^{-4}	-5.30/-3.53	1:4a	6.23	0.89	0.38	2.09
P22	546	1.83	8.8×10^{-5}	-5.20/-3.37	1:4a	7.18	0.83	0.39	2.34
P23	598	1.74	1.6×10^{-4}	-5.24/-3.50	1:4a	6.53	0.84	0.40	2.19
P24	551	1.85	-	-5.43/-3.62	1:3	4.83	0.90	0.48	2.07
P25	527	1.91	-	-5.53/-3.54	1:4a	7.38	0.80	0.43	2.54
P26	-	-	-	-5.17/3.15	1:2	9.17	0.69	0.57	3.6

Definitions for all parameters are the same as those in Table 1.

Table 2. The optical, electrochemical, hole mobility, and PSC characteristics of **P13-26**

It is widely accepted that the formation of strong intermolecular π-π stacking is necessary to achieve high charge mobility in OSC (Payne, 2004; Mcculloch, 2006). Rigid coplanar fused aromatic rings can efficiently enhance this interaction and hence improve charge mobilities. Indolo[3,2-*b*]carbazole (IC) has large-size fused aromatic ring, making closer packing of conjugated structure. Hashimoto (Zou et al., 2009) developed narrow band gap (1.90eV) PIC **P24** by alternating IC donating unit and DTBT accepting unit. The optimized **P24** device ITO/PEDOT:PSS/**P24**:PCBM1:3(w/w)/Al demonstrated a PCE of 2.07%. The study reveals that the formation of an interpenetrated network of donors and acceptors inside active layer is critical for obtaining high-performance OSCs. Such interpenetrated networks can be built by tuning device processing parameters such as the ratio of polymer to PCBM in the blend and post-annealing treatment at suitable temperatures. Besides, a simple and effective method to improve the photovoltaic performance of OSCs is to increase the number of unsubstituted thienyl units in D–A copolymers consisting of DTBT as acceptor. The V_{oc} of **P24** is similar with those of some DTBT based narrow band gap polymers with donating units like FL (1.0eV) (Slooff et al., 2007) dibenzosilole (0.9–0.97eV) (Boudreault et al., 2007) or carbazole (0.88 eV) (Blouin et al., 2007) but higher than those of D–A derivatives with donating units such as cyclopenta[2,1-b:3,4-b′]dithiophene (0.6eV), dithienosilole (0.44eV) (Liao et al., 2007) or dithieno[3,2-b:2′,3′-d] pyrrole(0.52eV) (Zhou et al., 2008). By adopting DTTP as accepting unit, the same group developed **P25**, with an E_g of 1.91 Hashimoto (Zhou et al., 2010a). Spin-coating from CF solution, **P25**:PCBM (1:4, w/w) based OSC contributed a PCE of 1.56% and was further improved to 2.54% by using DCB instead of CF.

By using hexylthiophene moiety in DTBT accepting unit and shorter alkyl chains in IC unit, Movileanu (Lu et al., 2008) developed **P24** analogous PIC **P26**. Long solubilizing side-chains generally resulted in lower charge mobility, balance of the alkyl chains in conjugated polymers was important for charge mobility. **P26** film showed pronounced peak broadening and significant red shift (100nm) in absorption spectra than its solution. **P26** device ITO/PEDOT:PSS/**P26**:PCBM(1:2w/w)/LiF/Al achieved a high PCE of 3.6%.

As discussed above, both PC and PF exhibit relatively low-lying HOMO level and thus high V_{oc} for OSCs via copolymerizing with accepting unit including DTBT, DTTP and DTQX. In contrast with other accepting units, DTBT-containing polymers may maintain deep HOMO and narrow the band gap through reducing LUMO, which results in a high V_{oc}. However, PCs devices generally display low J_{sc} and FF, which was similar to PFs.

4. Thiophene-based conjugated polymers

Polythiophenes (PTs) are considered as the most important conjugated polymers for a broad spectrum of optoelectronic applications such as light-emitting diodes, field-effect transistors, and OSCs due to their excellent optical and electrical properties as well as exceptional thermal and chemical stability (Roncali, 1992).

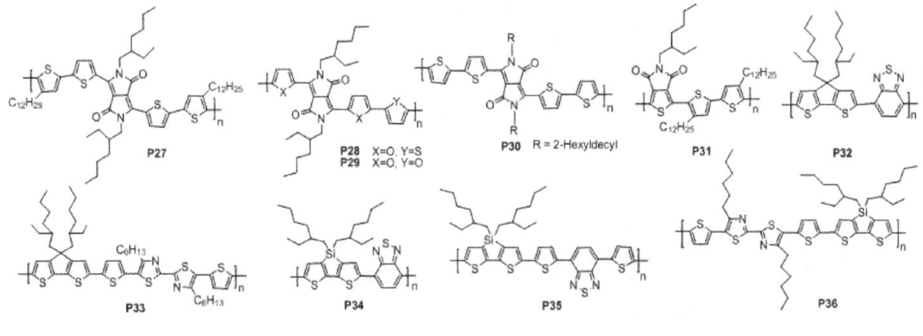

Chart 3. Thiophene-based narrow band gap polymers

By alternating quaterthiophene and diketopyrrolopyrrole (DPP) units, Janssen (Wienk et al., 2008) developed PT **P27** (**P27-36** structure in Chart 3), with an E_g of 1.4eV (Table 3). The absorption spectrum of **P27** in DCB showed stronger red-shift than its solution in CF. Fast evaporation of CF did not allow **P27** to crystallize, which results in poor morphology when spin-coating **P27**:PCBM blend film from CF solution. Certain degree of crystallization could be achieved by heating the film at 130°C. Spin-coated from CF:DCB (4:1 v/v) mixture solution, **P27**:PCBM (1:2 w/w) optimized OSC reached highest PCE of 4%, while OSC achieving only 1.1% PCE using CF as blending solvent and good PCE of 3.2% when prepared from DCB.

Furans have been used as an alternative to thiophenes in organic dyes for dye-sensitized solar cells and have shown very similar optical and electronic properties (R. Li et al., 2009). The first furan-containing DPP-based low band gap polymers **P28-29** were developed (Claire et al., 2010). By spin-coating from CB solution with 1vol% high-boiling-point 1-chloronaphthalene (CN), **P28** device achieved slightly improved PCE from 3.4% to 3.8% with the structure ITO/PEDOT:PSS/**P28**:PCBM(1:3w/w)/LiF/Al, due to optimized blend morphology. The best **P28**:PC71BM (1:3w/w) OSC contributed a PCE of 5.0% with the addition of 9% CN in CB. The device with blend **P29**:PC71BM (1:3 w/w) a PCE of 4.1% with 5% CN addition in CB. The addition of CN led to much finer phase separation between polymer and fullerene to form a fiber-like inter-penetrating morphology at the length scale of 20 nm, which is close to the ideal domain size, assuming an exciton diffusion length of 5-10 nm (Markov et al., 2005; Scully et al., 2006; Shaw et al., 2008). By alternating bithiophene donating unit and DPP accepting unit, Janssen (Johan et al., 2009) developed **P30**, with an E_g of 1.3eV. Terthiophene induced additional planarity, which enhanced packing and charge carrier mobility, as **P30**-based transistors exhibited nearly balanced hole and electron mobilities of 0.04 and 0.01cm²V⁻¹s⁻¹, respectively. The optimized device of **P30**:PCBM (1:2w/w) presented 3.8% PCE by spin-coating from CB solution with a small amount of DIO. The corresponding optimized device of **P30**:PC71BM achieved a PCE of 4.7% by adding 100mg/mL DIO in CF solution, due to low solubility of PC71BM in CF.

Polymer	λ_{max}^{abs} nm	E_g eV	μ_h cm²V⁻¹s⁻¹	HOMO /LUMO, eV	Polymer: PCBM	J_{sc} mA/cm²	V_{oc} V	FF	PCE
P27	650	1.4	-	-	1:2[a]	11.5	0.61	0.58	4.0
P28	789	1.41	-	-5.40/-3.80	1:3[a]	11.2	0.74	0.60	5.0
P29	767	1.35	-	-5.50/-3.80	1:3[a]	9.1	0.73	0.58	4.1
P30	-	1.36	0.04	-5.17/-3.61	1:2[a]	11.8	0.65	0.60	4.7
P31	572	1.82	1.0×10⁻⁴	-5.56/-3.10	1:1.5	8.02	0.95	0.62	4.7
P32	775	1.40	-	-5.30/-3.57	1:3	11.0	0.70	0.47	3.2
P33	560	1.84	-	-5.07/-3.55	1:1	7.70	0.68	0.53	2.79
P34	-	1.45	3×10⁻³	-5.05/-3.27	1:1[a]	12.7	0.68	0.55	5.1
P35	652	1.53	3.0×10⁻⁶	-4.99/-3.17	1:1[a]	9.76	0.60	0.50	2.95
P36	558	1.85	3.1×10⁻⁴	-5.18/-3.09	1:1[a]	7.85	0.68	0.54	2.86

Definitions for all parameters are the same as those in Table 1.

Table 3. The optical, electrochemical, hole mobility, and PSC characteristics of **P27-36**

By alternating bithiophene donating unit and thieno[3,4-c]pyrrole-4,6-dione (TPD) accepting unit, Wei (Yuan et al., 2010) prepared **P31**, with an E_g of 1.82eV. TPD presented strong electron-withdrawing properties, which were attributed to improvement of intramolecular interactions and reduction of E_g as well as relatively low HOMO level due to its symmetric, rigidly fused, coplanar structure (Zou et al., 2010). Controlling the active layer thickness of device ITO/PEDOT:PSS/**P31**:PCBM(1:2 w/w)/Al between 90 and 100nm, the best device presented 4.7% PCE, with relatively high V_{oc} (0.95V) attributed to its deep HOMO level.

By alternating cyclopentadithiophene (CPDT) and BT unit, Mühlbacher (Mühlbacher et al., 2006) developed **P32**, which exhibited an E_g of 1.40eV and an absorption spectrum ranged from 300 to 850nm. The device based on **P32**:PCBM obtained a PCE of 3.2%. By incorporating CPDT with bithiazole unit, Lin (K. Li et al., 2009) developed polymer **P33**, with the best device contributed a PCE of 3.04% with the configuration of ITO/PEDOT:PSS/**P33**:PCBM(1:2 w/w)/Ca/Al.

Recently, dithieno-[3,2-b:2',3'-d]silole-containing polymers have attracted considerable attention due to their potential applications in the field of optoelectronic devices. Many silole-containing polymers have been proven to achieve high V_{oc} by reducing the LUMO levels of polymers (Lu et al., 2008). It is feasible to realize a low-lying HOMO energy level and high V_{oc} simultaneously under the condition of ensuring a sufficient energy level offset (0.3-0.5V) between the LUMO level of silole-containing donor and the LUMO of PCBM. Furthermore, silole-containing polymers applied in OSC have also proved to be effective in improving hole mobility and rendering higher efficiencies compared to CPDT (Wang, 2008; Hou, 2008). By alternating dithieno-[3,2-b:2',3'-d]silole and BT unit, Yang (Hou et al., 2008) developed **P34**, with an E_g low 1.45eV. P40 has a broader absorption spectrum ranging from 350-800nm. Its device ITO/PEDOT/PSS/**P34**:PC70BM/Ca/Al achieved an average PCE of 4.7 % for 100 devices, with the OSC showing a PCE of 5.1%. The same group reported DTBT-based silole-containing polymers **P35** (Huo et al., 2009). The device presented 3.43% PCE with the structure of ITO/PEDOT:PSS/**P35**:PCBM/Ca/Al. By alternating dithieno-[3,2-b:2',3'-d]silole and dithiazole unit, Li (Zhang et al., 2010) designed **P36**, whose device (ITO/PEDOT:PSS/**P36**:PC71BM(1:2 w/w)/Al) achieved a PCE of 2.86%.

Recently, the polymers containing benzo[1,2-b:4,5-b']dithiophene (BDT) have been successfully applied to field-effect transistor (FET) and PSC devices (Huo, 2008; Pan, 2007; Chen, 2009) with a high mobility of 0.15-0.25cm^{2}V^{-1}s^{-1}, owing to the large and planar conjugated structure promoting π-π stacking and thus benefiting charge transport. By incorporating alkoxyl BDT with thieno[3,4-b]thiophene, **P37** (**P37-50** structure in Chart 4) was developed, with an E_g 1.62eV (Liang et al., 2009a). Thieno[3,4-b]thiophene repeating units contributed to the stability of the quinoidal structure of the backbone narrowing the energy gap of resulting polymer (Sotzing & Lee, 2002). Thieno[3,4-b]thiophene with ester substituents in **P37** contributed both good solubility and oxidative stability. The absorption of **P37** coincided with the corresponding maximum photon flux region in solar spectrum. **P37** OSC with the structure of ITO/PEDOT:PSS/**P37**:PCBM(1:1 w/w)/Ca/Al achieved a PCE of 4.76% (Table 4). **P37**:PC71BM presented an improved PCE of 5.30% with the same device structure. The absorption spectra of two devices almost covered the whole visible range. Furthermore, high EQEs were obtained for both **P37**:PCBM and **P37**:PC71BM OSCs. For the former, the maximum EQE was over 60% at 650 nm and was 50% in the range of 550-750 nm; the latter exhibited a higher EQE, almost all over 60% in the range from 400 to 750 nm.

By changing side-chain substituents, **P37** analogous polymers **P38-41** were developed (Liang et al., 2009b). All polymers showed very close between 2.6×10^{-4} and 4.7×10^{-4}cm^{2}V^{-1}s^{-1} and similar absorption spectra, covering the whole visible region.

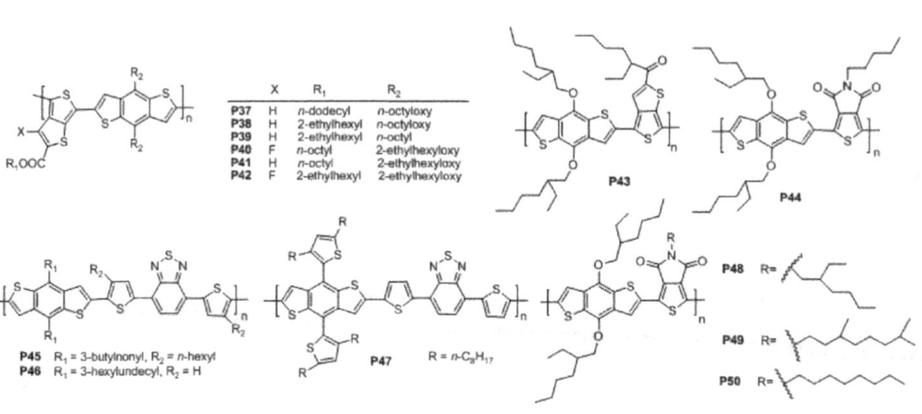

Chart 4. BDT based narrow band gap polymers.

The substitution of octyloxy to octyl group lowered the HOMO energy level of **P38** from -4.94 to -5.04 eV (**P39**). With electron-withdrawing fluorine in polymer backbone, **P40** had significantly lower HOMO level in comparison to **P41**. The devices based on ITO/PEDOT: PSS/**P38-41**:PCBM (1:1 w/w)/Ca/Al exhibited a PCE of 5.10%, 5.53%, 3.20% and 3.02%,respectively, with active layer spin-coated from DCB solution. By using DCB/DIO (97:3 v/v) mixture solvent, PCE of **P39-P41** can be dramatically improved to 5.85%, 5.90% and 4.10%, respectively. By replacing octyl group in **P40** into 2-ethylhexyl, **P52** was further developed (Liang et al., 2010). Spin-coated from DCB solution, **P51**:PC71BM harvested a PCE of 5.74%, 6.22% and 5.58%, respectively at their D/A weight ratio as 1:1, 1:1.5 and 1:2. The best device contributes PCE as high as 7.4% with the mixture solvent of CB/DIO(97:3 v/v), where uniform morphology of blend film was prepared.

Polymer	λ_{max}^{abs} nm	E_g eV	μ_h cm^{2}V^{-1}s^{-1}	HOMO/ LUMO, eV	Polymer: PCBM	J_{sc} mA/cm^2	V_{oc} V	FF	PCE
P37	687	1.62	4.5×10⁻⁴	-4.90/-3.20	1:1.2ᵃ	15.0	0.56	0.63	5.30
P38	683	1.59	4.0×10⁻⁴	-4.94/-3.22	1:1	12.8	0.60	0.66	5.10
P39	682	1.60	7.1×10⁻⁴	-5.04/-3.29	1:1	13.9	0.72	0.58	5.85
P40	682	1.63	7.7×10⁻⁴	-5.12/-3.31	1:1	13.0	0.74	0.61	5.90
P41	677	1.62	4.0×10⁻⁴	-5.01/-3.24	1:1	10.7	0.66	0.58	4.10
P42	675	1.61	5.8×10⁻⁴	-5.15/-3.31	1:1.5ᵃ	14.5	0.74	0.69	7.4
P43	675	1.61	2×10⁻⁴	-5.12/-3.55	1:1.5ᵃ	14.7	0.7	0.64	6.58
P44	610	1.80	-	-5.56/-3.75	1:2ᵃ	9.81	0.85	0.66	5.50
P45	595	1.70	1.6×10⁻⁵	-5.26/-2.96	1:1	9.70	0.81	0.55	4.31
P46	644	1.70	3.8×10⁻⁵	-5.33/-3.17	1:1	7.79	0.83	0.60	3.85
P47	596	1.75	-	-5.31/-3.44	1:2ᵃ	10.7	0.92	0.58	5.66
P48	608	1.75	-	-5.48/-	1:2	8.1	0.87	0.56	4.0
P49	616	1.70	-	-5.57/-	1:1.5	9.7	0.81	0.67	5.7
P50	627	1.73	-	-5.40/-	1:1.5	11.5	0.85	0.68	6.8

Definitions for all parameters are the same as those in Table 1.

Table 4. The optical, electrochemical, hole mobility, and PSC characteristics of **P37-50**

BDT-containing polymers can finely tune the HOMO level and efficiently narrow the band gap (Pan, 2007; Liang, 2008; Liang, 2009). Besides, alkoxy groups have been proved to be beneficial to reduce the HOMO level of polymer due to its stronger electron-donating property (Liang et al., 2009b). By removal of the oxygen atom on ester group in thieno[3,4-b]thiophene unit, ketone-substituted polymer **P43** (Chen et al. 2009) was developed to further reduce the HOMO level. The best OSC based on **P43**:PC71BM showed a high PCE of 6.58%.

By alternating BDT and TPD, Leclerc (Zou et al, 2010) presented polymer **P44**, with an E_g of 1.8eV and HOMO/LUMO energy levels -5.56eV/-3.75eV, respectively. The device ITO/PEDOT:PSS/**P53**:PC71BM/LiF/Al showed a high PCE of 5.5%. By incorporating BDT with DTBT, You (Price et al., 2010) synthesized **P45-46**. Compared with **P45**, **P46** exhibited higher molecular weight while lower hole mobility than other BDT-containing **P37-43**. **P45-46** device showed a PCE of 3.85% and 4.31%, respectively. By introducing thiopene moiety onto BDT, Yang (Huo et al., 2010) synthesized **P45** analogous polymer **P47**, with an E_g of 1.75eV. Spin-coated from DCB, **P47** device (ITO/PEDOT:PSS/**P47**:PC71BM(1:2 w/w)/Ca/Al) contributed the optimized PCE of 5.66% without annealing treatment.

By changing the alky substituent in TPD unit, Frechet (Piliego et al., 2010) synthesized **P44** analogue polymers **P48-50**, with E_g ranging from 1.70 to 1.75 eV and HOMO levels lowered down to -5.4~ -5.57 eV. Spin-coated from CB, **P48-50**:PCBM devices achieved a PCE of 2.8%, 3.9%, 6.4%, respectively. Clearly, the device performance increased as the branch length was decreased in **P48-50**. By spin-coating from CB solution with a small amount of DIO, **P48** and **P49** OSCs achieved improved PCE of 4.0% and 5.7%, respectively (Peet et al., 2008). However, for **P50**, the addition of DIO led to only slight improvements, which was attributed to a high level of order of **P50** in the blend without DIO. It was concluded that alkyl-substituted groups had much impact on the PCE of **P48-50**. The use of DIO is effective in promoting the packing of the polymer by preventing excessive crystallization of PCBM.

Dithieno[3,2-b:2',3'-d]pyrrole (DTP) has good planarity and stronger electron-donating ability of nitrogen atoms. Geng (Yue et al. 2009) synthesized BT-based P(DTP)s **P51-52** (Chart 5), with an E_g ~1.43eV. Owing to different alkyl chain, **P52** exhibited stronger absorption in the range of 600-900nm compared to **P51**. **P51-52** devices showed a PCE of 2.06% and 2.8%, respectively with the structure ITO/PEDOT:PSS/**P51-52**:PCBM(1:3 w/w)/LiF/Al, at the optimized blend film thickness of 100 nm for **P51** and 90 nm for **P52**. The device of **P51-52** delivered a V_{oc} in the range of 0.4-0.54V, leading to relatively low PCE.

By adopting DTP and DTBT units as the donor and acceptor segments, Hashimoto (Zhou et al., 2008) synthesized **P53** with an E_g of 1.46eV, which was the lowest compared with other D-A copolymers based on DTBT acceptor segment with other donor segments such as FL, silafluorene, carbazole, DTS, and CPDT (Slooff, 2007; Blouin, 2007; Moulle, 2008). **P53** device exhibited a PCE of 2.18% and a high J_{sc} of 9.47mA/cm^2, which was attributed to its broad absorption spectrum. However, **P53** device presented a V_{oc} of 0.52V, which is lower than that of the devices based on DTBT-containing D-A copolymers with FL (1V), silafluorene (0.9V) or carbazole (0.9V) as donating unit.

3,6-Dithien-2-yl-2,5-dialkylpyrrolo[3,4-c]pyrrole-1,4-dione (DTDPP) emerged as a promising electron acceptor in the design of photovoltaic polymers (Wienk et al., 2008). With DTDPP as accepting unit, Hashimoto (Zhou et al., 2010b) presented **P54**, with an E_g lowed down to 1.13eV. **P54** presented a hole mobility of 0.05cm^{2}V^{-1}s^{-1}, 1-2 orders of magnitude higher than that of other D-A polymers. Spin-coated with mixing solution of CF and DCB (4:1 v/v), **P54** device with structure of ITO/PEDOT:PSS/**P54**:PCBM(or PC71BM)/LiF/Al showed a PCE of 2.34% with PCBM and 2.71% with PC71BM as acceptor in the blend films. The PCE of **P54** was lower that of P3HT, which was mainly attributed to the lower V_{oc}, resulting from its high-lying HOMO level (-4.90eV). It was expected that the device performance would be improved by further reducing the HOMO and LUMO level.

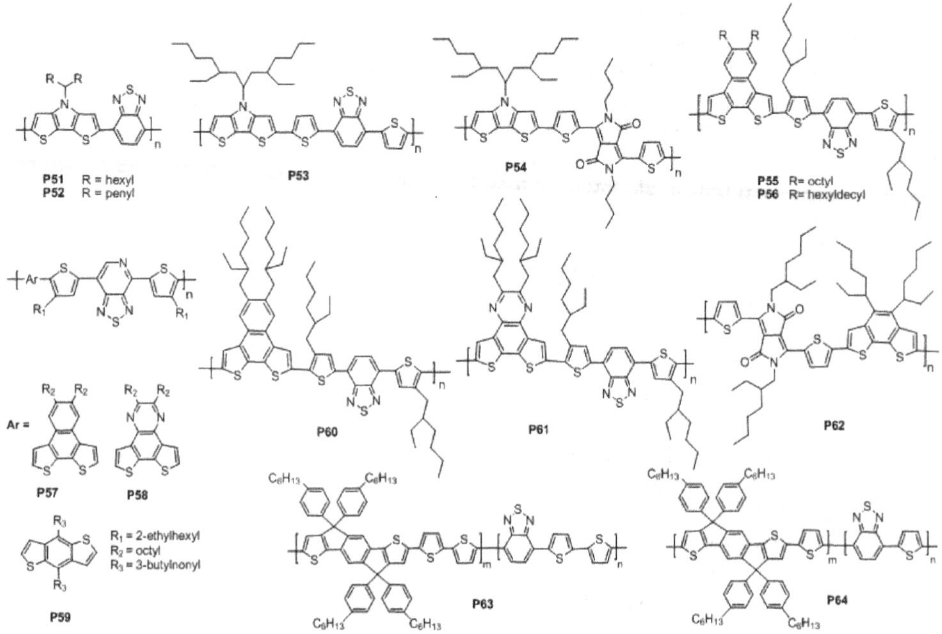

Chart 5. Fused Thiophene based narrow band gap polymers.

Recently, You (H. Zhou, 2010b) proposed a design strategy for narrow band gap polymers by alternating "weak donor" and "strong acceptor" to approach ideal polymers with both low HOMO energy level and small E_g in order to achieve both high V_{oc} and J_{sc} (H. Zhou, 2010e). By alternating DTBT or thiadiazolo[3,4-c]pyridine as strong acceptor, with naphtho[2,1-b:3,4-b']-dithiophene (NDT), dithieno[3,2-f:2',3'-h]quinoxaline (QDT) or BDT as weak donor, polymers P55-61 were developed (Yang, 2010; H. Zhou, 2009; H. Zhou, 2010b, 2010c). For NDT-containing P55-56, a 4-(2-ethylhexyl)thiophene unit in DTBT was used to reduce the steric hindrance of polymer backbone and hence to achieve near identical E_g and energy levels (H. Zhou, 2010a). It was believed that E_g and energy levels of a conjugated polymer were primarily determined by the molecular structure of conjugated backbone, while the solubilizing alkyl chains should have a negligible impact on these properties and hence a little impact on device's J_{sc} and V_{oc}. However, Yu's study showed variations of length and shape of alkyl chains in NDT had significant influence on device performance: long and branched side-chains would weaken the intermolecular interaction, benefiting for improvement of V_{oc}; while short and straight side-chains would promote the intermolecular interaction, rendering a large J_{sc}. The desirable balance between V_{oc} and J_{sc} could be tuned through adopting suitable short and branched side-chains. P55-56:PCBM (1:1,w/w) devices showed a best PCE of 3.0% and 3.36%, respectively.

Polymer	λmax nm	E_g eV	μ_h cm²V⁻¹s⁻¹	HOMO/ LUMO, eV	Polymer: PCBM	J_{sc} mA/cm²	V_{oc} V	FF	PCE
P51	764	1.42	-	-4.86/-3.07	1:3	11.1	0.43	0.43	2.06
P52	764	1.43	-	-4.81/-3.08	1:3	11.9	0.54	0.44	2.80
P53	697	1.46	-	-5.00/-3.43	1:1	9.47	0.52	0.44	2.18
P54	790	1.13	0.05	-4.90/-3.63	1:2ᵃ	14.87	0.38	0.48	2.71
P55	547	-	-	-	1:1	10.93	0.59	0.46	3.00
P56	548	-	-	-	1:1	10.67	0.69	0.46	3.36
P57	580	1.53	-	-5.36/-3.42	1:1	14.16	0.71	0.62	6.20
P58	650	1.56	-	-5.50/-3.44	1:1	13.49	0.75	0.55	5.57
P59	675	1.51	-	-5.47/-3.44	1:1	12.78	0.85	0.58	6.32
P60	681	1.61	1.7×10⁻⁵	-5.34/-3.29	1:0.8	14.20	0.67	0.54	5.10
P61	647	1.70	7.2×10⁻⁶	-5.46/-3.28	1:1.2	11.38	0.83	0.46	4.3
P62	928	1.34	-	-5.21/-3.63	1:2ᵃ	10.1	0.68	0.63	4.31
P63	520	1.76	7.0×10⁻⁴	-5.46/-3.56	1:3ᵃ	8.7	0.84	0.53	3.9
P64	590	1.70	3.4×10⁻³	-5.43/-3.66	1:3ᵃ	10.1	0.80	0.53	4.3

Definitions for all parameters are the same as those in Table 1.

Table 5. The optical, electrochemical, hole mobility, and PSC characteristics of **P51-64**

By replacing BT with a new stronger acceptor, thiadiazolo[3,4-c]pyridine, You (H. Zhou, 2010d) developed **P57-59** by adopting NDT, QDT or BnDT as donating unit. **P57-59** exhibit an E_g of 1.53eV, 1.56eV and 1.51eV, respectively, which is noticeably reduced (ca.0.09–0.19eV) compared with the band gaps of their BT counterparts (Price, 2010; H. Zhou, 2010c). The device with ITO/PEDOT:PSS/**P57-59**:PCBM/Ca/Al achieved a PCE of 6.20% for **P57**, 5.57% for **P58**, and 6.32%% for **P59**.By alternating DTBT acceptor with NDT or QDT weak donor, You (H. Zhou, 2010e) developed **P60-61**, with an E_g of **P60-61** of 1.61eV and 1.70eV, respectively. The devices based on **P60-61**:PCBM achieved a PCE of 5.1% and 4.3%, respectively. The donor unit with electron-withdrawing atoms was demonstrated to lead to a lower HOMO level while have little impact on the LUMO level. You thought that future research should be focused on employment of even weaker donors and stronger acceptors via innovative structural modification in order to concurrently achieve a higher V_{oc} and a higher J_{sc}. By alternating DPP and benzo[1,2-b:4,5-b']dithiophene, Yang (Huo et al., 2009) synthesized **P62**. Its device with the structure of ITO/PEDOT:PSS/**P62**:PC71BM(1:2 w/w)/Ca/Al achieved improved PCE from 4.31% to 4.45% after device annealing under 110ºC for 30 min. **P62** was expected to a very promising candidate for highly efficient OSCs owing to relatively low E_g and suitable HOMO level.

Ladder-type coplanar thiophene-phenylene-thiophene (TPT) structure tends to lead to strong molecular π-π interaction and thus exhibits remarkable hole mobility (Wong et al., 2006). Poly(TPT)s demonstrated an E_g ~2.1eV and hole mobility as high as 10⁻³cm²V⁻¹s⁻¹ (Chan et al., 2008). By alternating TPT donor and DTBT acceptor, Ting (Chen et al., 2008) developed **P63-64**, which exhibited an E_g of 1.76eV and 1.70eV, and absorption band in the range of 350-700nm and 350-730nm, respectively. **P64** displayed a hole mobility up to 3.4×10⁻³cm²V⁻¹s⁻¹. Both of them showed good solubility at room temperature in organic solvents. Using DCB as blending solvent, the devices ITO/PEDOT:PSS/**P63-64**:PCBM(1:3 w/w)/Ca/Al harvested a PCE of 2.0% and 2.5%, respectively, a little lower than that (3.9% PCE) of P3HT:PCBM(1:1 w/w) device under the same conditions. The optimized device with **P63-64**:PC71BM (1:3 w/w) showed a PCE of 3.9% for **P63** and 4.3% for **P64**, higher than that of the device based on P3HT:PC71BM (1:1 w/w) (3.4%) under the same conditions.

5. Conclusions

Narrow band gap polymers **P1-P64** developed by alternating donor (ca. fluorene, carbazole and thiophene) and acceptor (ca.benzothiadiazole, quinoxaline and diketopyrrolopyrrole) units in recent 4 years are summarized, with their fullerene blend-based BHJ OSCs contributing PCE over 3%. The design criteria for ideal polymer donors to achieve high efficiency OSCs is: (1) a narrow E_g (1.2-1.9eV) with broad absorption to match solar spectrum; (2) a HOMO energy level ranging from -5.2 to -5.8 eV and a LUMO level ranging from -3.7 to -4.0eV to ensure efficient charge separation while maximizing V_{oc}; and (3) good hole mobility to allow adequate charge transport. Besides, device structure and morphology optimizations of polymer:fullerene blend film have been extensively demonstrated to be crucial for PCE improvement in OSCs. The current endeavors boosted OSCs PCEs up to 7% would encourage further efforts toward a next target of efficiency in excess of 10%.

6. References

Allard, N., Aich, RB., Gendron, D., Boudreault, P-LT., Tessier, C., Alem, S., Tse, S-C., Tao, Y. & Leclerc, M. (2010). Germafluorenes: new heterocycles for plastic electronics. *Macromolecules*, Vol. 43, No. 5, (January 2010), pp. (2328-2333), ISSN: 1520-5835

Ameri, T., Dennler, G., Lungenschmied, C. & Brabec, CJ. (2009). Organic tandem solar cells: a review. *Energ. Environ. Sci.*, Vol. 2, No. 4, (February 2009), pp. 347-363, ISSN: 1754-5692

Beal, RM., Stavrinadis, A., Warner, JH., Smith, JM. & Beal, HE. (2010). The molecular structure of polymer-fullerene composite solar cells and its influence on device performance. *Macromolecules*, Vol. 43, No. 5, (February 2010), pp. 2343-2348, ISSN: 1520-5835

Blouin N, Michaud A. & Leclerc, M. (2007). A low-bandgap poly(2,7-Carbazole) derivative for use in high-performance solar cells. *Adv. Mater.*, Vol.19, No.17, (September 2007), pp. (2295–2300), ISSN: 1521-4095

Blouin, N., Michaud, A., Gendron, D., Wakim, S., Blair, E., Neagu-Plesu, R., Belletete, M., Durocher, G., Tao, Y. & Leclerc, M. (2008). Toward a rational design of poly(2,7-carbazole) derivatives for solar cells. *J. Am. Chem. Soc.*, Vol. 130, No. 2, (December 2007), pp. (732-742), ISSN: 0002-7863

Boudreault, P-LT., Michaud, A. & Leclerc, M. (2007). A new poly(2,7-dibenzosilole) derivative in polymer solar cells. *Macromol. Rapid Commun.*, Vol. 28, No. 22, (November 2007), pp. (2176–2179), ISSN: 1521-3927

Boudreault, P-LT., Wakim, S., Blouin, N., Simard, M., Tessier, C., Tao, Y. & Leclerc, M. (2007). Synthesis, characterization, and application of indolo[3,2-b]carbazole semiconductors. *J. Am. Chem. Soc.*, Vol. 129, No. 29, (June 2007), pp. (9125-9136), ISSN: 0002-7863

Chan, S-H., Chen, C., Chao, T., Ting, C. & Ko, B-T. (2008). Synthesis, characterization, and photovoltaic properties of novel semiconducting polymers with thiophene-phenylene-thiophene (TPT) as coplanar units. *Macromolecules*, Vol. 41, No.15, (June 2008), pp. (5519-5526), ISSN: 1520-5835

Chen, C., Chan, S-H., Chao, T., Ting, C. & Ko, BT. (2008). Low-bandgap poly(thiophene-phenylene-thiophene) derivatives with broaden absorption spectra for use in high-performance bulk-heterojunction polymer solar cells. *J. Am. Chem. Soc.*, Vol. 130, No. 38, (August 2008), pp. (12828–12833), ISSN: 0002-7863

Chen, H., Hou, J., Zhang, S., Liang, Y., Yang, G., Yang, Y., Yu, L., Wu, Y. & Li, G. (2009). Polymer solar cells with enhanced open-circuit voltage and efficiency. *Nat. Photonics*, Vol. 3, No. 11, (November 2009), pp. (649-653), ISSN: 1749-4893

Chen H-Y., Hou, J., Zhang, S., Liang, Y., Yang, G., Yang, Y., Yu, L., Wu, Y. & Li, G. (2009). Polymer solar cells with enhanced open-circuit voltage and efficiency. *Nat. Photonics*, Vol. 3, No.11, (November 2009), pp. (649-653), ISSN: 1749-4885

Cheng, YJ., Yang, SH. & Hsu, CS. (2009). Synthesis of conjugated polymers for organic solar cell applications. *Chem. Rev.*, Vol. 109, No. 11, (September 2009), pp. 5868-5923, ISSN: 1520-6890

Claire, HW., Pierre, MB., Thomas, WH., Olivia, PL. & Fréchet, JMJ. (2010). Incorporation of furan into low band-gap polymers for efficient solar cells. *J. Am. Chem. Soc.*, Vol. 132, No. 44, (October 2010), PP. (15547–15549) ISSN: 0002-7863

Dennler, G. Scharber, MC., Ameri, T., Denk, P., Forberich, K., Waldauf, C. & Brabec, CJ. (2008). Design rules for donors in bulk-heterojunction tandem solar cells - Towards 15% energy-conversion efficiency, *Adv. Mater.*, Vol. 20, No.3, (February 2008), pp. (579-583), ISSN: 1521-4095.

Dennler, G., Scharber, M. C. Brabec, C. J. (2009). Polymer-fullerene bulk-heterojunction solar cells. *Adv. Mater.*, Vol. 21, No. 13, (April 2009), pp. (1323-1338), ISSN: 1521-4095

Duan, C., Cai, W., Huang, F., Zhang, J., Wang, M., Yang, T., Zhong, C., Gong, X. & Cao, Y. (2010). Novel silafluorene-based conjugated polymers with pendant acceptor groups for high performance solar cells. *Macromolecules*, Vol. 43, No. 12, (January 2010), pp. (5262-5268), ISSN: 1520-5835

Duan, C., Chen, K., Huang, F., Yip, H-L., Liu, S., Zhang, J., Jen, AK-Y. & Cao, Y. (2010). Synthesis, characterization, and photovoltaic properties of carbazole-based two-dimensional conjugated polymers with donor-π-bridge-acceptor side chains. *Chem. Mater.*, Vol. 22, No. 23, (November 2010), pp. (6444-6452), ISSN: 1520-5002

Gadisa, A., Mammo ,W., Andersson, LM., Admassie, S., Zhang, F, Andersson, MR. & Inganäs, O. (2007). A new donor–acceptor–donor polyfluorene copolymer with balanced electron and hole mobility. *Adv. Funct. Mater.*, Vol. 17, No.18, (November 2007), pp. (3836–3842), ISSN: 1616-3028

Hou, J., Chen, H-Y., Zhang, S., Li, G. & Yang, Y. (2008). Synthesis, characterization, and photovoltaic properties of a low band gap polymer based on silole-containing polythiophenes and 2,1,3-benzothiadiazole. *J. Am. Chem. Soc.*, Vol. 130, No. 48, (November 2008), pp. (16144-16145), ISSN: 0002-7863

Huang, F., Chen, K-S., Yip, H-L., Hau, SK., Acton, O., Zhang, Y., Luo, J. & Jen, AK-Y. (2009). Developme -nt of new conjugated polymers with donor−π-bridge−acceptor side chains for high performance solar cells. *J. Am. Chem. Soc.*, Vol. 131, No. 39, (September 2009), pp. (13886–13887), ISSN: 0002-7863

Huo, L., Chen, H., Hou, J., Chen, T., Yang, Y. (2009). Low band gap dithieno[3,2-b:2,3-d]silole-con taining polymers, synthesis, characterization and photovoltaic application, *Chem. Commun.*, No. 37, (July 2009) ,PP. (5570–5572), ISSN: 1359-7345

Huo, L., Hou, J., Zhang, S., Chen, H-Y. & Yang, Y. (2010). A polybenzo[1,2-*b*:4,5-*b'*]dithiophene derivative with deep homo level and its application in high-performance polymer solar cells. *Angew. Chem. Int. Ed.*, Vol. 49, No. 8, (February 2010), pp. (1500–1503), ISSN: 1521-3773

Johan, CB., Arjan, PZ., Simon, GJM., Martijn, MW., Mathieu, T., Dago, ML. & Rene, AJJ. (2009). Poly(diketopyrrolopyrrole-terthiophene) for ambipolar logic and photovoltaics. *J. Am. Chem. Soc.*, Vol. 131, No. 46, (September 2009), pp. (16616-16617) ISSN: 0002-7863

Kitazawa, D., Watanabe, N., Yamamoto, S. & Tsukamoto, J. (2009). Quinoxaline-based π-conjugated donor polymer for highly efficient organic thin-film Solar Cells. *Appl. Phys. Lett.*, Vol. 95, No.5, (May 2009), pp. (053701-053703), ISSN: 1077-3188

Krebs, F.C. (2009). Fabrication and processing of polymer solar cells: a review of printing and coating techniques. *Solar Energ. Mater. Solar C.*, Vol. 93, No. 4, (April 2009), pp. 394-412, ISSN: 0927-0248

Li, G., Shrotriya, V., Huang, J. S.; Yao, V. & Moriarty, T. (2005). High-efficiency solution processable polymer photovoltaic cells by self-organization of polymer blends. *Nat. Mater.*, Vol. 4, No. 11, (October 2005), pp. 864-868, ISSN: 1476-4660

Li, K., Huang, J., Hsu, YC., Huang, P., Chu, CW., Lin, J., Ho, KC., Wei, K., Lin, H. (2009). Tunable novel cyclopentadithiophene-based copolymers containing various numbers of bithiazole and thienyl units for organic photovoltaic cell applications. *Macromolecules*, Vol. 42, No. 11, (April 2009), pp. (3681–3693), ISSN: 1520-5835

Li, R., Lv, X., Shi, D., Zhou, D., Cheng, Y., Zhang, G. & Wang, P. (2009). Dye-sensitized solar cells based on organic sensitizers with different conjugated linkers: furan, bifuran, thiophene, bithiophene, selenophene, and biselenophene. *J. Phys. Chem. C*, Vol. 113, No. 17, (April 2009), pp. (7469–7479), ISSN: 1932-7455

Li, Y., Wu, Y. & Ong, BS. (2006). Polyindolo[3,2-*b*]carbazoles: a new class of *p*-channel semiconductor polymers for organic thin-film transistors. *Macromolecules*, Vol. 39, No. 19, (August 2006), pp. (6521-6527), ISSN: 1520-5835

Liang, Y., Wu, Y., Feng, D., Tsai, S-T., Son, H-J., Li, G. & Yu, L. (2009). Development of new semiconducting polymers for high performance solar cells. *J. Am. Chem. Soc.*, Vol. 131, No. 1, (December 2008), pp. (56–57), ISSN: 0002-7863

Liang, Y., Feng, D., Wu, Y., Tsai, S-T., Li, G., Ray, C. & Yu, L. (2009). Highly efficient solar cell polymers developed via fine-tuning of structural and electronic properties. *J. Am. Chem. Soc.*, Vol. 131, No. 22, (May 2009), PP. (7792–7799), ISSN: 0002-7863

Liang, Y., Xu, Z., Xia, J B., Tsai, S-T., Wu, Y., Li, G., Ray, C. & Yu, L. (2010). For the bright future — bulk heterojunction polymer solar cells with power conversion efficiency of 7.4%. *Adv. Mater.*, Vol. 22, No. 20, (May 2010), pp. (E135–E138), ISSN: 1521-4095

Liao, L., Dai, L., Smith, A., Durstock, M., Lu, J., Ding, J. & Tao, Y. (2007). Photovoltaic-active dithienosilole-containing polymers. *Macromolecules*, Vol. 40, No. 26, (November 2007), pp. (9406-9412), ISSN: 1520-5835

Lu, G., Usta, H., Risko, C., Wang, L., Facchetti, A., Ratner, MA. & Marks, TJ. (2008), Synthesis, characterization, and transistor response of semiconducting silole polymers with substantial hole mobility and air stability: experiment and theory. *J. Am. Chem. Soc.*, Vol.130, No. 24, (May 2008), pp. (7670-7685), ISSN: 0002-7863

Lu, J., Liang, F., Drolet, N., Ding, J F., Tao, Y. & Movileanu, R. (2008). Crystalline low band-gap alternating indolocarbazole and benzothiadiazole-cored oligothiophene copolymer for organic solar cell applications. *Chem. Commun.*, Vol. 14, No. 42, (September 2008), pp. (5315–5317), ISSN: 1359-7345

Markov, DE., Amsterdam, E., Blom, PWM., Sieval, AB. & Hummelen, JC. (2005). Accurate measurement of the exciton diffusion length in a conjugated polymer using a heterostructure with a side-chain cross-linked fullerene layer. *J. Phys. Chem. A*, Vol.109, No. 24, (June 2005), pp. (5266–5274), ISSN: 1089-5639

Mcculloch, I., Heeney, M., Bailey, C., Genevicius, K., Macdonald, I., Shkuno,v M., Sparrowe, D., Tierney, S., Wagner, R., Zhang, W., Chabinyc, M., Kline R., McGehee, M., Toney, M. (2006). Liquid-crystalline semiconducting polymers with high charge-carrier mobility. *Nat. Mater.*, Vol. 5, (April 2006), pp. (328-333), ISSN: 1476-4660

Moulle, AJ. ,Tsami, A., Bunnagel, TW., Forster, M. Kronenberg, NM., Scharber, M., Koppe, M., Morana, M., Brabec, CJ., Meerholz, K. & Scherf, U. (2008). Two novel cyclopentadithiophene-based alternating copolymers as potential donor components for high-efficiency bulk-heterojunction-type solar cells. *Chem. Mater.*, Vol. 20, No. 12, (April 2008), pp. (4045-4050), ISSN: 1520-5002

Mühlbacher, D., Scharber, M., Morana, M., Zhu, Z., Waller, D., Gaudiana, R. & Brabec, CJ. (2006). High photovoltaic performance of a low-bandgap polymer. *Adv. Mater.*, Vol. 18, No. 21, (November 2006), pp. (2884–2889), ISSN: 1521-4095

Pan, H., Li, Y., Wu, Y., Liu, P., Ong, BS., Zhu, S. & Xu, G. (2007). Low-temperature, solution-processed, high-mobility polymer semiconductors for thin-film transistors. *J. Am. Chem. Soc.*, Vol. 129, No.14, (March 2007), pp. (4112-4113), ISSN: 0002-7863

Payne, MM., Parkin, SR., Anthony, JE., Kuo, C-C., Jackson, TN. (2005). Organic field-effect transistors from solution-deposited functionalized acenes with mobilities as high as 1cm²/V.s. *J. Am. Chem. Soc.*, Vol.127, No.14, (December 2004), pp. (4986-4987), ISSN: 0002-7863

Peet, J., Cho, NS., Lee, SK. & Bazan, GC. (2008). Transition from solution to the solid state in polymer solar cells cast from mixed solvents. *Macromolecules*, Vol. 41, No. 22, (October 2008), pp. (8655–8659), ISSN: 1520-5835

Peet, J., Heeger, AJ. & Bazan, GC. (2009). "Plastic" solar cells: self-assembly of bulk heterojunction nanomaterials by spontaneous phase separation. *Acc. Chem. Res.*, Vol. 42, No. 11, (July 2009), pp. 1700-1708, ISSN: 1520-4898

Piliego, C., Holcombe, TW., Douglas, JD., Woo, CH., Beaujuge, PM. & Frechet, JM. (2010). Synthetic control of structural order in n-alkylthieno[3,4-c]pyrrole-4,6-dione-based polymers for efficient solar cells. *J. Am. Chem. Soc.*, Vol. 132, No. 22, (May 2010), pp. (7595–7596), ISSN: 0002-7863

Price, SC., Stuart, AC. & You, W. (2010). Low band gap polymers based on benzo [1,2-b:4,5-b]dithiophene: rational design of polymers leads to high photovoltaic performance. *Macromolecules*, Vol. 43, No. 10, (March 2010), pp. (4609-4612), ISSN: 1520-5835

Qin, R., Li, W., Li, C., Du, C., Veit, C., Schleiermacher, H-F., Andersson, M., Bo, Z., Liu, Z., Inganäs, O., Wuerfel, U. & Zhang, FL. (2009). A planar copolymer for high efficiency polymer solar cells. *J. Am. Chem. Soc.*, Vol. 131, No. 41, (September 2009), pp. (14612–14613), ISSN: 0002-7863

Roncali, J. (1992). Conjugated poly(thiophenes): synthesis, functionalization, and applications. *Chem. Rev.*, Vol. 92, No. 4, (June 1992), pp. (711-738), ISSN: 1520-6890

Scharber, MC., Wuhlbacher, D., Koppe, M., Denk, P., Waldauf, C., Heeger, AJ. & Brabec, CJ. (2006). Design rules for donors in bulk-heterojunction solar cells - Towards 10% energy conversion efficiency, *Adv. Mater.*, Vol. 18, No. 6, (March 2006), pp. (789-794), ISSN: 1521-4095.

Scully, S R. & McGehee, MD. (2006). Effects of optical interference and energy transfer on exciton diffusion length measurements in organic semiconductors. *J. Appl. Phys.*, Vol. 100, (August 2006), pp. (034-907), ISSN: 1089-7550

Shaw, PE., Ruseckas, A. & Samuel, IDW. (2008). Exciton diffusion measurements in poly(3-hexyl thiophene). *Adv. Mater.* Vol. 20, No. 18, (September 2008), pp. (3516–3520), ISSN: 1521-4095

Slooff LH., Veenstra SC., Kroon JM., Moet DJD., Sweelssen J. & Koetse MM. (2007). Determining the internal quantum efficiency of highly efficient polymer solar cells through optical modeling. *Appl. Phys. Lett.*, Vol. 90, No.14, (April 2007), pp. (143506-143508), ISSN: 1077-3188

Sotzing, GA. & Lee, KH. (2002). Poly(thieno[3,4-b]thiophene): A *p*- and *n*-dopable polythiophene exhibiting high optical transparency in the semiconducting state. *Macromolecules*, Vol. 35, No. 19, (August 2002), pp. (7281-7286), ISSN: 1520-5835

Tang, W., Hai, J., Dai,Y., Huang, Z., Lu, B., Yuan, F., Tang, J., Zhang, F. (2010). Recent development of conjugated oligomers for high-efficiency bulk-heterojunction solar cells. Solar Energ. Mater. Solar C., Vol. 94, No. 12, (December 2010), pp. (1963-1979), ISSN: 0927-0248

Veldman, D., Ipek, O., Meskers, SCJ., Sweelssen, J., Koetse, MM., Veenstra, SC., Kroon, JM., Bavel, SS., Loos, J. & Janssen, RAJ. (2008). Compositional and electric field dependence of the dissociation of charge transfer excitons in alternating polyfluorene copolymer/fullerene blends. *J. Am. Chem. Soc.*, Vol. 130, No. 24, (May 2008), pp. (7721–7735), ISSN: 0002-7863

Wang, E., Wang, L., Lan, L., Luo, C., Zhuang, W., Peng, J. & Cao, Y. (2008). High-performance polymer heterojunction solar cells of a polysilafluorene derivative, *Appl. Phys. Lett.*, Vol. 92, No. 3, (January 2008), pp. (303-307), ISSN: 1077-3118

Wienk, MM., Turbiez, M., Gilot, J. & Janssen, RAJ. (2008). Narrow-bandgap diketo-pyrrolo-pyrrole polymer solar cells: the effect of processing on the performance. *Adv. Mater.*, Vol. 20, No. 13, (May 2008), pp. (2556–2560), ISSN: 1521-4095

Wong, KT., Chao, T-C., Chi, L-C., Chu, Y-Y., Balaiah, A., Chiu, S-F., Liu, Y-H., Wang, Y. (2006) Syntheses and structures of novel heteroarene-fused coplanar π-conjugated chromophores, *Org. Lett.*, Vol. 8, No. 22, (September 2006), pp. (5033-5036), ISSN: 1523-7052

Yang, L., Zhou, H. & You, W. (2010). Quantitatively analyzing the influence of side chains on photovoltaic properties of polymer-fullerene solar cells. *J. Phys. Chem. C*, Vol. 114, No. 39, (August 2010), pp. (16793–16800), ISSN: 1932-7455

Yu, G., Gao, J., Hummelen, JC., Wudl, F. & Heeger, AJ. (1995). Polymer photovoltaic cells: enhanced efficiencies via a network of internal donor-acceptor heterojunctions. *Science*, Vol. 270, No. 5243, (December 1995), pp. 1789-1791, ISSN: 1095-9203

Yuan, M-C., Chiu, M-Y., Liu, S-P., Chen, C-M. & Wei, K-H. (2010). A thieno[3,4-c]pyrrole-4,6-dione-based donor-acceptor polymer exhibiting high crystallinity for photovoltaic applications. *Macromolecules*, Vol. 43, No. 17, (Aug, 2010), pp. (6936-6938), ISSN: 1520-5835

Yue, W., Zhao, Y., Shao, S., Tian, H., Xie, Z., Geng, Y. &Wang, F. (2009). Novel NIR-absorbing conjugated polymers for efficient polymer solar cells: effect of alkyl chain length on device performance. *J. Mater. Chem.*, Vol. 19, No. 15, (January 2009), pp. (2199-2206), ISSN: 1364-5501

Zhang, L., He, C., Chen, J., Yuan, P., Huang, L., Zhang, C., Cai, W., Liu, Z. & Cao, Y. (2010). Bulk-heterojunction solar cells with benzotriazole-based copolymers as electron donors: largely improved photovoltaic parameters by using PFN/Al bilayer cathode. *Macromolecules*, Vol. 43, No. 23, (November 2010), pp. (9771-9778) ISSN 1520-5835

Zhang, M., Fan, H., Guo, X., He, Y., Zhang, Z., Min, J., Zhang, J., Zhao, G J., Zhan, X. & Li, Y. (2010). Synthesis and photovoltaic properties of bithiazole-based donor–acceptor copolymers. *Macromolecules*, Vol. 43, No 13, (June 2010), pp. (5706-5712), ISSN: 1520-5835

Zheng, Q., Jung, BJ., Sun, J. & Katz, HE. (2010). Ladder-type oligo-*p*-phenylene-containing copolymers with high open-circuit voltages and ambient photovoltaic activity. *J. Am. Chem. Soc.*, Vol. 132, No. 15, (October 2009), pp. (5394–5404), ISSN: 0002-7863

Zhou, E., Nakamura, M., Nishizawa, T., Zhang, Y., Wei, Q., Tajima, K., Yang, C. & Hashimoto, K. (2008). Synthesis and photovoltaic properties of a novel low band gap polymer based on n-substituted dithieno[3,2-*b*:2,3-*d*]pyrrole. *Macromolecules*, Vol. 41, No. 22, (October 2008), pp. (8302-8305), ISSN: 1520-5835

Zhou, E., Yamakawa, S., Zhang, Y., Tajima, K., Yang, C. & Hashimoto, K. (2009). Indolo[3,2-*b*] carbazole -based alternating donor–acceptor copolymers: synthesis, properties and photovoltaic application. *J. Mater. Chem.*, Vol. 19, No. 41, (August 2009), pp. (7730–7737), ISSN: 1364-5501

Zhou, E., Wei, Q., Yamakawa, S., Zhang, Y., Tajima, K., Yang, C. & Hashimoto, K. (2010). Diketopyrrolopyrrole-based semiconducting polymer for photovoltaic device with photocurrent response wavelengths up to 1.1 μm. *Macromolecules*, Vol. 43, No. 2, (December 2009), pp. (821-826), ISSN: 1520-5835

Zhou, E., Cong, J., Yamakawa, S., Wei, Q., Nakamura, M., Tajima, K., Yang, C. & Hashimoto, K. (2010). Synthesis of thieno[3,4-b]pyrazine-based and 2,1,3-benzothiadiazole-based donor–a cceptor copolymers and their application in photovoltaic devices. *Macromolecules*, Vol. 43, No. 6, (February 2010), pp. (2873 -2879), ISSN: 1520-5835

Zhou, H., Yang, L., Xiao, S., Liu, S. & You, W. (2010). Donor–acceptor polymers incorporating alkylated dithienylbenzothiadiazole for bulk heterojunction solar cells: pronounced effect of positioning alkyl chains, *Macromolecules*, Vol. 43, No. 2, (December 2009), pp. (811–820), ISSN: 1520-5835

Zhou, H., Yang, L., Stoneking, S. & You, W. (2010). A weak donor-strong acceptor strategy to design ideal polymers for organic solar cells. *ACS Appl. Mater. Interfaces*, Vol. 2, No. 5, (May 2010), pp. (1377-1383), ISSN: 1944-8252

Zhou, H., Yang, L. & You, W. (2010). Quantitatively analyzing the influence of side chains on photovoltaic properties of polymer–fullerene solar cell, *J. Phys. Chem. C*, Vol. 114, No. 39, (September 2010), pp. (16793–16800), ISSN: 1932-7455

Zhou, H., Yang, L., Price, SC., Knight, KJ. & You, W. (2010). Enhanced photovoltaic performance of low-bandgap polymers with deep LUMO levels. *Angew. Chem. Int. Ed.*, Vol. 49, No. 43, (October 2010), pp. (7992–7995), ISSN: 1521-3773

Zhou, H., Yang, L., Liu, S. & You, W. (2010). A tale of current and voltage: interplay of band gap and energy levels of conjugated polymers in bulk heterojunction solar cells. *Macromoleccules*, Vol. 43, No. 24, (November 2010), pp. (10390–10396), ISSN: 1520-5835

Zhou, Q., Hou, Q., Zheng, L., Deng, X., Yu, G. & Cao, Y. (2004). Fluorene-based low band-gap copolymers for high performance photovoltaic devices. *Appl. Phys. Lett.*, Vol. 84 , No. 10, (January 2004), pp. (1653-1655), ISSN: 1077-3188

Zou, Y., Najari, A., Berrouard, P., Beaupre, S., Aich, BR., Tao, Y. & Leclerc, M. (2010). A thieno[3,4-c]pyrrole-4,6-dione-based copolymer for efficient solar cells. *J. Am. Chem. Soc.*, Vol. 132, No. 15, (August 2010), pp. (5330-5331), ISSN: 0002-7863

11

Optical Absorption and Photocurrent Spectra of CdSe Quantum Dots Adsorbed on Nanocrystalline TiO₂ Electrode Together with Photovoltaic Properties

Taro Toyoda and Qing Shen
The University of Electro-Communications
Japan

1. Introduction

There is a great deal of interest in the technological applications of titanium dioxide (TiO₂) to dye-sensitized solar cells (DSCs) made from nanostructured TiO₂ electrodes because of their high photovoltaic conversion efficiency, which exceeds 10% (Chiba et al., 2006). Since the initial pioneering work on DSCs (O'Regan & Grätzel, 1991), they have often been proposed as a sustainable energy source. In DSCs, the applications of organic dye molecules as a photosensitizer, nanostructured TiO₂ as an electron transport layer, and an iodine redox couple for hole transport dramatically improve the light harvesting efficiency. With Ru-based organic dyes adsorbed on nanostructured TiO₂ electrodes, the large surface area enables more efficient absorption of the solar light energy. The main undertaking for those developing next-generation solar cells is to improve the photovoltaic conversion efficiency, together with the long time stability. Nowadays, there exists an intense effort aimed at developing third-generation solar cells. One of a promising approach is to replace the organic dyes by inorganic substances with strong optical absorption characteristics and longer stability over time. Recently, as an alternative to organic dyes, semiconductor quantum dots (QDs) have been studied for their light harvesting capability (Niitsoo et al., 2006; Diguna et al., 2007; Mora-Seró, 2009). The enormous potential of science and technology on nanoscale to impact on industrial output has been recognized all over the world. One emerging area of nanoscience being at the interface of chemistry, physics, biology and materials science is the field of semiconductor QDs, whose unique properties have attracted great attention by researchers during the last two decades. Different strategies for the synthesis of semiconductor QDs have been developed, so that their composition, size, shape, and surface protection can be controlled nowadays with an exceptionally high degree. The surface chemistry of semiconductor QDs is another key parameter, in many respects determining their properties related to their assembly. Semiconductor QDs exhibit attractive characteristics as sensitizers due to their tunable bandgap (or HOMO-LUMO gap) by size control (Yu et al., 2003), which can be used to match the absorption spectrum to the spectral distribution of solar light. Moreover, semiconductor QDs possess higher extinction coefficients than conventional metal-organic dyes, and larger intrinsic dipole moments leading to rapid charge separation (Underwood et al., 2001). The demonstration of multiple exciton generation (MEG) by impact ionization has fostered an interest in colloidal semiconductor QDs (Schaller et al, 2006; Trinh et al., 2008). One of the most attractive configurations to exploit these fascinating properties of semiconductor QDs is the quantum dot-sensitized solar cell (QDSC) (Nozik, 2002; Klimov, 2006). The efficient formation of more than one photoinduced electron-hole pair (exciton) upon the absorption of a single photon is a process not only of a great current scientific interest but is potentially important for optoelectronic devices that directly convert solar radiant energy into electricity. The demonstration of MEG by impact ionization in colloidal semiconductor QDs could push the thermodynamic photovoltaic conversion efficiency limit of solar cells up to 44% (Klimov, 2006) from the current 31% of the Shockley-Queisser detailed balance limit (Shockley & Queisser, 1961). The optimization of QDSCs can benefit from the intensive effort carried out with DSC. Although the photovoltaic conversion efficiencies of QDSCs lag behind those of DSCs and the use of semiconductor QDs as light absorbers requires the development of new strategies in order to push the performance of QDSCs, QDSCs have attracted significant attention among researchers as promising third-generation photovoltaic devices.

In this chapter, we describe the performance of QDSCs based on CdSe QD sensitizer on nanostructured TiO_2 electrode with a pre-adsorbed layer of CdS QDs (termed combined CdS/CdSe QDs) proposed by Niitsoo et al (Niitsoo et al, 2006) and developed by other groups (Lee & Lo, 2009; Sudhagar et al., 2009). They showed that a pre-adsorbed layer of CdS prior to CdSe adsorption improved the QDSC's performance. Hence it is interesting and useful to investigate the detailed function of combined CdS/CdSe QDs sensitizer on performance of QDSCs, together with the basic studies of optical absorption and photocurrent characteristics. Information regarding the optical absorption properties is initially necessary in order to investigate the electronic states of combined CdS/CdSe QDs for future photovoltaic cell applications. However, few accurate studies of the optical absorption properties of combined CdS/CdSe QDs adsorbed on nanostructured TiO_2 electrodes have been carried out. The main reason for this is the difficulty in using the conventional transmission method because of strong light scattering by the highly porous structure of the nanostructured TiO_2 electrodes. However, scattering effects can be minimized by employing the photothermal (PT) technique. In general, an optically excited solid relaxes to thermal equilibrium by the emission of photons (radiative processes) or phonons (nonradiative processes). In the PT technique, the signal detected is directly proportional to the thermal energy (heat production) induced by the absorbed photons through nonradiative processes (emission of phonons). Heat production by nonradiative processes has been detected by several methods (Tam, 1986). The PT signal is less sensitive to light scattering effects than conventional spectroscopy signals, and the ability of the PT technique to produce optical absorption spectra from strongly scattering media has been demonstrated, in particular using photoacoustic (PA) method which is a PT technique (Inoue et al., 2006; Toyoda et al., 2009). Thus, the PT technique is a useful technique for studying the optical absorption spectra of the strongly scattering and/or opaque samples with which we are dealing in our investigations. PA method detects the acoustic energy produced by heat generation through nonradiative processes in materials (Rosencwaig & Gersho, 1977). The PA cell, which is a small gas-tight enclosure with a sensitive acoustic microphone built in one wall, monitors the temperature changes in the sample produced by absorbed photons through nonradiative processes. Periodic temperature changes in the sample surface by modulated light by PT effect cause periodic pressure changes in the enclosed gas, which creates acoustic waves. The intensities of the acoustic waves are converted to an electrical signal by the microphone. The PA method can be applied to evaluate the optical absorption properties of combined CdS/CdSe QDs adsorbed on nanostructured TiO_2 electrodes. In order to investigate the subsequent photosensitization of combined CdS/CdSe QDs adsorbed on nanostructured TiO_2 electrodes after optical absorption, photoelectrochemical current measurements were carried out and the incident photon-to-current conversion efficiency (IPCE, namely the quantum efficiency of photocurrent) was evaluated. Also, photovoltaic properties were characterized by solar simulator under the illumination of one sun intensity (AM 1.5: 100 mW/cm²).

2. Experimental procedure

Experimental sections are divided into three sections, 1) sample preparations on TiO_2 electrodes and CdS/CdSe quantum dots, 2) optical absorption measurements by photoacoustic technique, and 3) photoelectrochemical current (incident photon to current conversion efficiency) and photovoltaic measurements.

2.1 Sample preparation

The method for the preparation of nanostructured TiO_2 electrodes has been reported in a previous paper (Shen & Toyoda, 2003). A TiO_2 paste was prepared by mixing 15 nm TiO_2 nanocrystalline particles (Super Titanai, Showa Denko; anatase type structure) and polyethylene glycol (molecular weight: 500,000) in pure water. The resultant paste was then deposited onto transparent conducting substrates [F-doped SnO_2 (FTO), sheet resistance: 10 $\mu\Omega$/sq]. The TiO_2 electrodes were then sintered in air at 450 °C for 30 min to obtain good necking and to sublimate polyethylene glycol. The highly porous nanostructure of the films (the pore sizes were on the order of a few tens of nanometers) was confirmed from scanning electron microscopy (SEM) images. The thicknesses of the films were measured and found to be ~ 5 μm by examining the cross sectional SEM images.

At first, CdS QDs were adsorbed onto nanostructured TiO_2 electrodes (pre-adsorbed layer) from the common NH_3 bath with a solution composition of 20 mM $CdCl_2$, 66 mM NH_4Cl, 140 mM thiourea, and 0.23 M ammonia to obtain a final pH ~9.5 (Niitsoo et al., 2006; Jayakrishnan et al., 1996). The TiO_2 electrodes were immersed in a container filled with the final solution. The adsorption was carried out at room temperature in the dark for 40 min.

The CdSe QDs were prepared by using a chemical bath deposition (CBD) technique (Shen & Toyoda, 2004; Shen et al., 2004; Gorer & Hodes, 1994). First, for the Se source, an 80 mM sodium selenosulphate (Na_2SeSO_3) solution was prepared by dissolving elemental Se powder in a 200 mM Na_2SO_3 solution. Second, an 80 mM $CdSO_4$ and 120 mM of a trisodium salt of nitrilotriacetic acid [$N(CH_2COONa)_3$] were mixed with the 80 mM Na_2SeSO_3 solution in a volume ratio of 1: 1: 1. TiO_2 electrodes adsorbed with CdS QDs were placed in a glass container filled with the final solution at 10 °C in the dark for various times (from 2 to 24 h) to promote CdSe QDs adsorption. To investigate the role of pre-adsorbed layer of CdS QDs, the CdSe QDs only were adsorbed directly on nanostructured TiO_2 electrodes with the same sample preparation conditions as mentioned above.

After the adsorption of CdSe QDs, the samples were coated with ZnS for surface passivation of the QDs by successive ionic layer adsorption and reaction (SILAR) for three times in 0.1

M $Zn(CH_2COO)$ and 0.1 M Na_2S aqueous solution for 1 min for each dip (Yang et al., 2002; Shen et al., 2008). In reference (Shen et al., 2008), we showed that the short-circuit-current density (J_{sc}), open-circuit voltage (Voc), and photovoltaic conversion efficiency (η) were enhanced by the ZnS coating, except for fill factor (FF). Although we applied the scanning electron microscopy (SEM) observation for visual investigation, we could not observe the difference of the morphologies with and without the ZnS coating up to 50,000 magnifications, indicating that the ZnS is coated with several atomic layers. In the future, we are going to observe the morphology of the ZnS coating layer by applying the transmission electron microscopy (TEM). To our knowledge, there are few reports in which ZnS coating has been applied to CdSe QD-sensitized solar cells, although ZnS-capped CdSe QDs dispersed in solution have been used for strong photoluminescence applications (Hines & Sionnet, 1996).

2.2 Optical absorption measurements

The optical absorption properties of nanostructured TiO_2 electrodes adsorbed with combined CdS/CdSe QDs were investigated using PA spectroscopy. The scattering effects in the optical absorption of nanostructured TiO_2 electrodes adsorbed with combined CdS/CdSe QDs can be minimized by employing the PA method. Figure 1 shows the schematic diagram of a photoacoustic spectrometer. Typical gas-microphone method was applied in the PA spectroscopic investigation (Rosencwaig & Gersho, 1977). The PA cell was composed of an aluminum cylinder with a small channel at the periphery in which a microphone (electret condenser type) was inserted (Shen &Toyoda, 2004). The inside volume of the cell was approximately 0.5 cm^3. The cell was suspended by four rubber bands

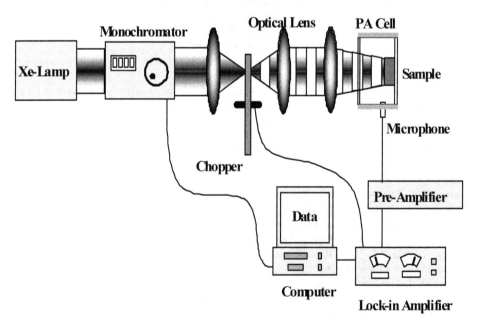

Fig. 1. Schematic diagram of a photoacoustic spectrometer, composed with light source, monochrometer, mechanical chopper, cell, preamplifier, lock-in amplifier, etc.

to prevent interference from external vibration. The cell window made of quartz was highly transparent throughout the observed wavelength range, and the sample holder could be easily removed from the cell to maintain the optical configuration. A 300 W xenon lamp was used as the light source. Monochromatic light through a monochromator was modulated at 33 Hz using a mechanical chopper and was focused within the PA cell. Light was focused on the sample over an impinging area of 0.20 cm^3. Modulation frequency of 33 Hz was determined to exclude the saturation effect of the spectrum. In this case (modulation frequency: 33 Hz), the optical absorption length is longer than the thermal diffusion length, indicating that the PA signal intensity is proportional to the optical absorption coefficient (no saturation effect) (Rosencwaig & Gersho, 1977). The PA signal was monitored by first passing the microphone output through a preamplifier and then into a lock-in amplifier. The data were averaged to improve the signal-to-noise ratio (S/N). The spectra were taken at room temperature in the wavelength range of 250 - 800 nm. The PA spectra were obtained by the normalization to the PA signal intensity of carbon black sheet that was proportional to the light intensity only. A UV cut filter was used for the measurements in the wavelength range of 600 - 800 nm to avoid the mixing of second harmonic light. The conditions for all the PA measurements (optical configuration, path-length, irradiation area, excitation light intensity etc.) were fixed as far as possible to compare each of the PA signals and spectra directly.

2.3 Photoelectrochemical current and photovoltaic measurements

Photocurrent measurements were performed in a sandwich structure cell (i.e., in the two-electrode configuration) with Cu_2S film on brass as the counter electrode (termed the Cu_2S counterelectrode). The applied electrolyte was polysulfide solution (1 M Na_2S + 1 M S). It is well known that the electrocatalytic activity of Pt with a polysulfide electrolyte is not satisfactory for photovoltaic cell applications and alternative counter electrode materials with higher activity such as Cu_2S and CoS have been reported (Hodes et al., 1980). The higher electro-catalytic activities of these materials are due to a reduction in the charge transfer resistance between the redox couple and the counterelectrode (Giménez et al. 2009). The Cu_2S counterelectrodes were prepared by immersing brass in HCl solution at 70°C for 5 min and subsequently dipping it into polysulfide solution for 10 min, resulting in a porous Cu_2S electrode (Hodes et al., 1980). The cells were prepared by sealing the Cu_2S counter-electrode and the nanostructured TiO₂ electrode adsorbed with combined CdS/CdSe QDs, using a silicone spacer (~ 50 μm) after the introduction of polysulfide electrolyte. The IPCE value was evaluated from the short-circuit photocurrent with a zero-shunt meter using the same apparatus and conditions as those used for the PA measurements. The incident light intensity was measured by an optical power-meter. The spectra were taken at room temperature in the wavelength of 250 - 800 nm. The conditions for all the measurements (optical configuration, path-length, irradiation area, excitation light intensity etc.) were fixed as far as possible to compare the IPCE values and spectra directly. Photovoltaic properties were characterized under a one sun illumination (AM 1.5: 100 mW/cm²) using a solar simulator by the measurements of photocurrent versus photovoltage to investigate J_{sc}, V_{oc}, FF, and η.

3. Results and discussion

Figure 2 shows the PA spectra of the nanostructured TiO₂ electrodes adsorbed with combined CdS/CdSe QDs for different adsorption times, together with that adsorbed with CdS QDs only. The pre-adsorption times for CdS QDs were fixed at 40 min (average diameter: ~ 4.2 nm).

The spectra were normalized to the photon energy of 4.0 eV. With increasing adsorption time, the red-shift of optical absorption at the shoulder point (indicated by arrows) can be clearly observed, implying the growth of CdSe QDs. Also, the comparison between the adsorption of CdSe QDs on the nanostructured TiO$_2$ electrodes with and without a pre-adsorbed CdS QD layer was carried out to evaluate the difference in PA spectra. For that, Figure 3 shows the PA spectra of the nanostructured TiO$_2$ electrodes adsorbed with CdSe QDs without a pre-adsorbed CdS QD layer for different adsorption times. The spectra are also normalized to the photon energy of 4.0 eV. As the PA spectra below the CdSe QDs adsorption time of 8 h agree with that of pure nanostructured TiO$_2$ electrode within the experimental accuracy, the CdSe QDs average size is very small less than 1 nm or no adsorption. Optical absorption in the visible light region due to the adsorbed CdSe QDs can be also observed both in Figs. 2 and 3. With increasing adsorption time, the red-shift of optical absorption at the shoulder point (indicated by arrows) can be clearly observed in Fig.3, also implying the growth of CdSe QDs. The exponential slopes at the fundamental absorption edges in combined CdS/CdSe QDs adsorbed on nanostructured TiO$_2$ electrodes in Fig. 2 are higher than those of CdSe QDs adsorbed on nanostructured TiO$_2$ electrodes without a pre-adsorbed CdS QD layer in Fig. 3, indicating that the uniformity of the average sizes or crystal quality of CdSe QDs in the former is better than that of the latter. Relative to the band-gap energy of 1.73 eV for bulk CdSe, the shoulder points in PA spectra of the nanostructured TiO$_2$ electrodes adsorbed with CdSe QDs shown in Figs. 2 and 3 exhibit blue-shifts, which is indicative of the quantum confinement effect. This fact implies that the radii of the CdSe QDs are smaller than the Bohr radius of bulk CdSe (~5.6 nm). We assume that the photon energy at the shoulder point corresponds to the lowest excitation energy of the CdSe QDs (Shen & Toyoda, 2004). The average diameter of the CdSe QDs for each adsorption time both in combined CdS/CdSe and CdSe without a preadsorbed CdS layer adsorbed on nanostructured TiO$_2$ electrodes can be estimated with the effective mass approximation (Shen et al., 2008; Bawendi et al., 1989). The dependence of the average diameter on the adsorption time is shown in Fig. 4, both of combined CdS/CdSe (●) and CdSe without apre-adsorbed CdS layer (○). It can be observed that the CdSe QDs on the nanostructured TiO$_2$ electrodes with a pre-adsorbed CdS QD layer grow rapidly during the initial adsorption process (less than 2 h adsorption). After a certain time, the crystal growth rate is slowed down, which is similar to the reference reported (Toyoda et al., 2007). When the adsorption time is sufficiently long (in this case ~ 24 h), the average diameter of CdSe QDs becomes constant at about 7.1 nm. On the other hand, it can be observed that the CdSe QDs adsorbed on the nanostructured TiO$_2$ electrodes without a pre-adsorbed CdS layer show very slow growth or no growth during the initial adsorption process (less than 8 hours) due to the fact that the PA spectra below the CdSe adsorption time below 8 hours agree with that of the pure nanostructured TiO$_2$ electrode within the experimental accuracy. Therefore, the crystal growth rate of CdSe QDs adsorbed on TiO$_2$ electrode with a pre-adsorbed CdS QD layer is faster than that of CdSe QDs adsorbed on the TiO$_2$ electrodes without a pre-adsorbed CdS QD layer. There are several possibilities for the faster crystal growth rate of CdSe QDs adsorbed on the nanostructured TiO$_2$ electrode with a pre-adsorbed CdS QD layer. First, it is possible that the active CdSe QDs form from the excess Cd remaining after CdS adsorption directly on the nanostructured TiO$_2$ electrode (Niitsoo et al., 2006). This is suggested that CdS QDs act as seed layer for subsequent CdSe growth (Sudhager et al., 2009). Second one is the passivation effect of CdS QDs on the nanostrctured TiO$_2$ surface to reduce defects or dislocations. Third, CdSe QDs are grown on the assembly of CdS QDs, indicating the growth on the similar crystal structure (or close lattice constant) different from the growth on TiO$_2$.

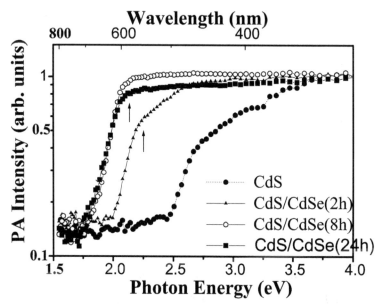

Fig. 2. Photoacoustic spectra of nanostructured TiO₂ electrodes adsorbed with combined
CdS/CdSe quantum dots for different adsorption times together with that adsorbed wth
CdS quantum dots only (modulation frequency: 33 Hz).

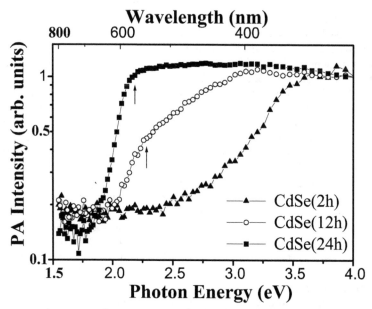

Fig. 3. Photoacoustic spectra of nanostructured TiO₂ electrodes adsorbed with CdSe
quantum dots without a preadsorbed CdS quantum dot layer for different adsorption times
(modulation frequency: 33 Hz).

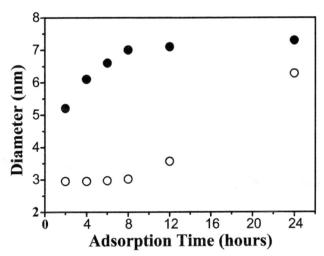

Fig. 4. Dependence of the average diameter on the adsorption time, both of combined CdS/CdSe (•) and CdSe without a preadsorbed CdS layer (○).

Figure 5 shows the IPCE spectra of the nanostructured TiO_2 electrodes adsorbed with combined CdS/CdSe QDs for different adsorption times, together with that adsorbed with CdS QDs only. The pre-adsorption time of CdS QD layer is fixed at 40 min. Photoelectrochemical current in the visible light region due to the adsorbed CdSe QDs can be observed, indicating the photosensitization by combined CdS/CdSe QDs. With increasing adsorption time, the red-shift of photoelectrochemical current can be clearly observed, implying the growth of CdSe QDs. The IPCE peak value increases with the increase of adsorption time up to 8 hours (~ 75%), then decreases until 24 h adsorption owing to the increase of recombination centers or interface states, together with the decrease of energy difference between LUMO in CdSe QDs and the bottom of conduction band of TiO_2. Also, the comparison between the adsorption of CdSe QDs on the TiO_2 electrodes with and without a pre-adsorbed CdS QD layer was carried out to evaluate the difference in IPCE spectra. For that, Figure 5 shows the IPCE spectra of the nanostructured TiO_2 electrodes adsorbed with CdSe QDs without a pre-adsorbed layer of CdS QD layer for different adsorption times. Photoelectrochemical current in the visible light region due to the adsorbed CdSe QDs can be observed in Fig. 6, also indicating the photosensitization by CdSe QDs. With increasing adsorption time, the red-shift of photoelectrochemical current can be clearly observed, implying the growth of CdSe QDs. However, the appearance of the spectrum in Fig. 6 is different from that of combined CdS/CdSe QDs, namely in the reduction of maximum IPCE value (~ 60%) and the adsorption time dependence of the spectrum shape. Also, the IPCE spectra below the CdSe QDs adsorption time of 8 h agree with that of pure nanostructured TiO_2 electrode within the experimental accuracy, indicating that the CdSe QDs adsorbed on the nanostructured TiO_2 electrode without a pre-adsorbed CdS layer show very slow growth or no growth similar to the results of PA characterization in Fig. 3. These results demonstrate that the spectral response of IPCE is enhanced upon combined CdS/CdSe sensitization rather than single CdSe QDs sensitization, indicating the possibility of the reduction in recombination centers and interface states owing to the possibilities of active CdSe QDs by the excess Cd remaining after CdS adsorption and passivation effect of CdS QDs on the nanostructured TiO_2 surface.

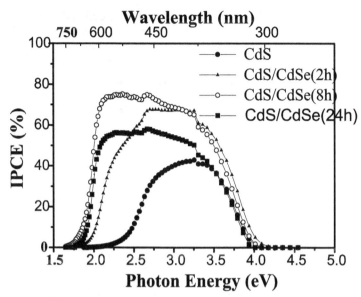

Fig. 5. IPCE spectra of nanostructured TiO₂ electrodes adsorbed with CdS/CdSe quantum dots for different adsorption times together with that adsorbed with CdS quantum dots only.

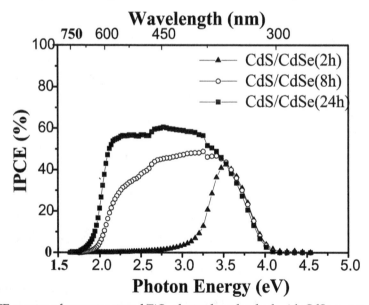

Fig. 6. IPCE spectra of nanostructured TiO₂ electrodes adsorbed with CdSe quantum dots without a preadsorbed CdS QD layer for different adsorption times.

The photocurrent-voltage cureves of (a) combined CdS/CdSe QD- and (b) CdSe QD-sensitized solar cells are shown in Fig. 7 (a) and (b), respectively, for different adsorption times, together with that obtained with cells adsorbed with CdS only. However, the appearance of the current-voltage curves of combined CdS/CdSe QD-sensitized solar cells is different from those of CdSe QD-sensitized solar cells. Figure 8 and 9 illustrates the photovoltaic parameters ((a) J_{sc}; (b) V_{oc}; (c) FF; (d) η) of combined CdS/CdSe QD-sensitized (●) and CdSe QD-sensitized (○) solar cells as a function of CdSe QDs adsorption times.

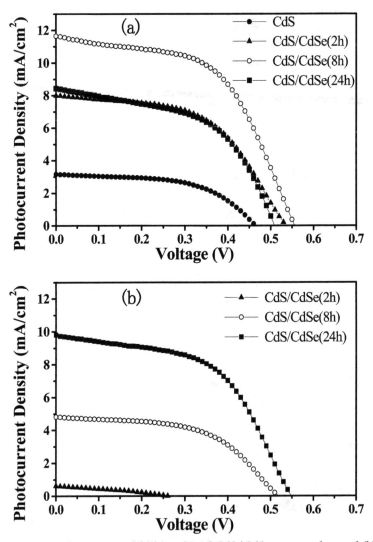

Fig. 7. Photocurrent-voltage curves of (a) combined CdS/CdSe quantm dot- and (b) CdSe quantum dot-sensitized solar cells for different adsorption times together with that adsorbed with CdS quantum dots only.

We observe that the parameter of J_{sc} in combined CdS/CdSe QD-sensitized solar cells increases with the increase of CdSe QDs adsorption times up to 8 h. On the other hand, V_{oc} and FF are independent of adsorption times. The performance of solar cells improved with an increase in adsorption time up to 8 h due, mainly, not only to the increase of the amount of CdSe QDs but the improvement in crystal quality and decrease of interface states. However, the increase in adsorption times after more than 8 h leads to deterioration in J_{sc} and V_{oc}. High adsorption time of CdSe QDs might cause an increase in recombination centers, poor penetration of CdSe QDs, and the decrease of energy difference between LUMO in CdSe QDs and the bottom of conduction band of TiO₂. Therefore, η of the combined CdS/CdSe QD-sensitized solar cell shows a maximum of 3.5% at 8 h adsorption times. On the other hand, J_{sc} and η below the CdSe QDs adsorption time of 8 h without a pre-adsorbed CdS layer show very small values close to zero, indicating the very small amount of CdSe QDs adsorption similar to the results of PA and IPCE characterization. We can observe that J_{sc}, V_{oc}, FF, and η in

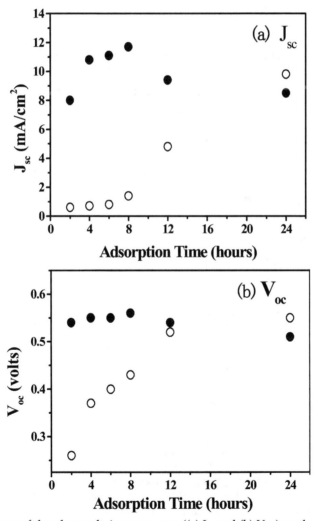

Fig. 8. Dependence of the photovoltaic parameters ((a) J_{sc} and (b) V_{oc}) on the adsorption time, both of combined CdS/CdSe (●) and CdSe without a preadsorbed CdS QD layer (○).

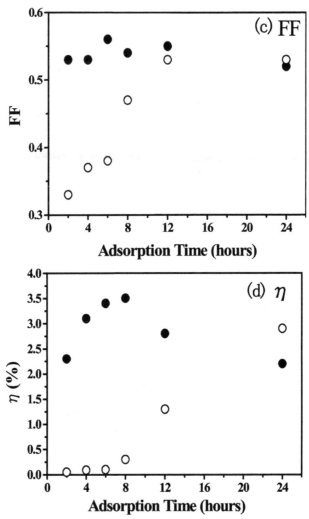

Fig. 9. Dependence of the photovoltaic parameters((c) FF and (d) η) on the adsorption time, both of combined CdS/CdSe (•) and CdSe without a preadsorbed CdS QD layer (○).

CdSe QD-sensitized solar cells without a pre-adsorbed CdS QD layer increase with the increase of adsorption times up to 24 h, indicating the difference of the crystal growth and the formation of recombination centers in combined CdS/CdSe and CdSe QDs.

Figure 9 shows the preliminary ultrafast photoexcited carrier dynamics characterization of combined CdS/CdSe and CdSe without a pre-adsorbed CdS layer (average diameters of both CdSe QDs are ~ 6 nm) using a improved transient grating (TG) technique (Katayama et al., 2003; Yamaguchi et al., 2003; Shen et al., 2010). TG signal is proportional to the change in the refractive index of the sample due to photoexcited carriers (electrons and holes). TG method is a powerful time-resolved optical technique for the measurements of various kinds of dynamics, such as carrier population dynamics, excited carrier diffusion, thermal diffusion, acoustic velocity and so on. Improved TG technique features very simple and compact optical setup, and is applicable for samples with rough surfaces. Comparing with transient absorption (TA) technique, improved TG method has higher sensitivity sue to its zero background in TG signals, which avoids the nonlinear effect and sample damage. Figure 9 shows that the hole and electron relaxation times of nanostructured TiO$_2$ electrodes adsorbed with combined CdS/CdSe QDs are faster about twice than those with CdSe QDs without a pre-adsorve CdS layer, indicating the decreases in recombination centers.

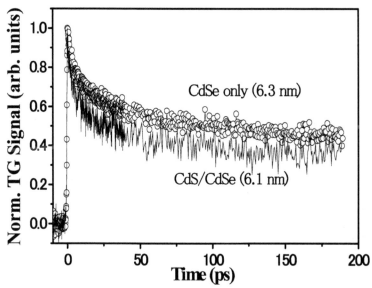

Fig. 10. Ultrafast carrier dynamics of combined CdS/CdSe and CdS without preadsorbed CdS quantum dots layer with a transient grating (TG) technique.

4. Conclusion

We have described the performance of quantum dot-sensitized solar cells (QDSCs) based on CdSe QD sensitizer on a pre-adsorbed CdS layer (combined CdS/CdSe QDs) together with the basic studies of optical absorption and photoelectrochemical current characteristics. It can be observed from optical absorption measurements using photoacoustic (PA) spectroscopy that the CdSe QDs on the nanostructured TiO$_2$ electrodes with a pre-adsorbed CdS layer grow more rapidly during the initial adsorption process than those without a pre-adsorbed CdS layer. Photoelectrochemical current in the visible light region due to the adsorbed CdSe QDs can be observed, indicating the photosensitization by combined CdS/CdSe QDs. The maximum IPCE value (~ 75%) of the CdSe QDs on the nanostructured TiO$_2$ electrodes with a pre-adsorbed CdS QD layer is 30% greater than that without a pre-adsorbed CdS layer. It indicates the possibilities of a decrease in recombination centers, interface states, and inverse transfer rate that is suggested by the preliminary ultrafast photoexcited carrier carrier dynamics characterization owing to the possibilities of active CdSe QDs by the excess Cd remaining after CdS adsorption and passivation effect of CdS QDs on the nanostructured TiO$_2$ surface. The short-circuit current (J$_{sc}$) in combined CdS/CdSe QD-sensitized solar cells shows maxima with the increase of CdSe QDs adsorption times between 2 h and 24 h, also indicating the decrease of recombination centers, interface states, and the increase in quasi Fermi level. The open-circuit voltage (V_{oc}) and fill factor (FF) are independent of adsorption times. The photovoltaic conversion efficiency (η) of the combined CdS/CdSe QD-sensitized solar cell shows a maximum value of 3.5%.

5. References

Bawendi, M. G.; Kortan, A. R.; Steigerwals, M.; & Brus, L. E. (1989), *J. Chem. Phys.*, Vol. 91, p. 7282.

Chiba, Y.; Islam, A.; Watanabe, Y.; Koyama, R.; Koide, N.; & Han, L. (2006), *Jpn. J. Appl. Phys.*Vol. 43, p. L638.

Diguna, L. J.; Shen, Q.; Kobayashi, J.; & Toyoda, T (2007), *Appl. Phys. Lett.*, Vol. 91, p. 023116.

Giménez, S; Mora-Seró, I; Macor, L.; Guijarro, N.; Lala-Villarreal, T.; Gómez, R.; Diguna, L.;Shen, Q.; Toyoda, T; & Bisquert, J (2009), *Nanotechnology*, Vol. 20, p. 295204.

Gorer, S.; & Hodes, G. (1994), *J. Phys. Chem.*, Vol. 98, p. 5338.

Hines, M. A.; & Sionnet, P. G. (1996), *J. Phys. Chem.*, Vol. 100, p. 468.

Hodes, G; Manassen, J; & Cahen, D. (1980), *J. Electrochem. Soc.*, Vol. 127, p. 544.

Inoue, Y.; Toyoda, T; & Morimoto, J (2006), *Jpn. J. Appl. Phys.*, Vol. 45, p. 4604.

Jayakrishnan, R.; Nair, J. P.; Kuruvilla, B. A.; & Pandy, R. K. (1996) *Semicond. Sci. Tech.*, Vol. 11, p. 116.

Katayama, K.; Yamaguchi, M.; & Sawada, T. (2003), *Appl. Phys. Lett.*, Vol. 82, p. 2775.

Klimov, V. I. (2006), *J. Phys. Chem. B*, Vol. 110, p. 16827.

Lee, Y-L.; & Lo, Y-S. (2009), *Adv. Func. Mater.*, Vol. 19, p. 604.

Mora-Seró, I; Giménez, S; Fabregat-Santiago, F.; Gómez, R.; Shen, Q.; Toyoda, T; & Bisquert, J. (2009), *Acc. Chem. Res.*, Vol. 42, 1848.

Niitsoo, O.; Sarkar, S. K.; Pejoux, P.; Rühle, S.; Cahen, D.; & Hodes, G., *J. Photochem. Photobiol. A*, Vol. 182, 306.

Nozik, A. J. (2002), *Physica E*, Vol. 14, p. 16827.

O'Regan, B.; & Grätzel (1991), *Nature*, Vol. 353, p. 737.

Rosencwaig, A. & Gersho, A. (1977), *J. Appl. Phys.*, Vol. 47, p. 64.

Schaller, R. D.; Sykora, M.; Pietryga, J. M.; & Klimov, V. I. (2006), *Nono Lett.*, Vol. 6. P. 424.

Shen, Q.; & Toyoda, T (2004), *Jpn. J. Appl. Phys.*, Vol. 43, p. 2946.

Shen, Q.; Arae, D.; & Toyoda, T. (2004), *J. Photochem. Photobiol. A*, Vol. 164, p. 75

Shen, Q.; Kobayashi, J.; Doguna, L. J.; & Toyoda, T. (2008), *J. Appl. Phys.*, Vol. 103, p. 084304.

Shen, Q.; Ayuzawa, Y.; Katayama, K.; Sawada, T.; & Toyoda, T. (2010), Appl. Phys. Lett., Vol. 97, p. 263113.

Shockley, W.; & Queisser, H. J. (1961), *J. Appl. Phys.*, Vol. 32, p. 510.

Sudhager, P.; Jung, J. H.; Park, S; Lee, Y-G.; Sathyamamoorthy, R.; Kang, Y. S.; & Ahn, H.(2009), *Electrochem. Commun.*, Vol. 11, p. 2220.

Tam, A. C. (1986), *Rev. Mod. Phys.*, Vol. 58, p. 381.

Toyoda, T.; Uehata, T.; Suganuma, R.; Tamura, S; Sato, A; Yamamoto, K.; Shen, Q. & Kobayashi, N. (2007), *Jpn. J. Appl. Phys.*, Vol. 46, p. 4616.

Toyoda, T; Tsugawa, S.; & Shen, Q. (2009), *J. Appl. Phys.*, Vol. 105, p. 034314.

Trinh, M. T.; Houtepen, A. J.; Schints, J. M; Hanrath, T.; Piris, J.; Knulst, W.; Goossens, P. L. M.; & Siebbeles, L. D. A. (2008), *Nano Lett*, Vol. 8, p. 1713

Yamaguchi, M.; Katayama, K.; & Sawada, T. (2003), *Chem. Phys. Lett.*, Vol. 377, p. 589.

Underwood, D. F.; Kippeny, T.; & Rosenthal, S. J. (2001), Eur. Phys, J. D, Vol. 16, p. 241.

Yang, S. M.; Huang, C. H.; Zhai, J.; Wang, Z. S.; & Liang, L. (2002), J. Mater. Chem., Vol. 12,p. 1459.

12

Conjugated Polymers for Organic Solar Cells

Qun Ye and Chunyan Chi
Department of Chemistry, National University of Singapore,
Singapore

1. Introduction

Energy shortage has become a worldwide issue in the 21st century (Lior, 2008). The urge to look for renewable energy to replace fossil fuel has driven substantial research effort into the energy sector (Hottel, 1989). The solar energy has enormous potential to take the place due to its vast energy stock and availability worldwide (Balzani et al., 2008). Conventional solar energy conversion device is based on silicon technology. However, wide use of silicon based solar cell technology is limited by its high power conversion cost (Wöhrle & Meissner, 1991). To address this issue, solution-processing based organic solar cell has been developed to replace Si-solar cell (Tang, 1986). Compared with conventional Si-based solar cell, conjugated polymer based solar cell (PSC) has several important advantages: 1) solution processability by spin-coating, ink-jet printing and roll-to-roll processing to reduce manufacturing cost; 2) tunable physical properties; and 3) mechanical flexibility for PSC application on curved surfaces (Sariciftci, 2004).

During the last decade, the power conversion efficiency (PCE) of organic based solar cell has increased from *ca.* 1% (Tang, 1986) to more than 7% (H. –Y. Chen et al., 2009) with the bulk heterojunction (BHJ) concept being developed and applied. During the pursuit of high efficiency, the importance of the structure-property relationship of the conjugated polymer used in the solar cell has been disclosed (J. Chen & Cao, 2009). It might be helpful to systematically summarize this structure-property relationship to guide polymer design and further improvement of the power conversion efficiency of PSCs in the future.

This chapter will be organized as follows. Firstly, we will discuss about the general criteria for a conjugated polymer to behave as an efficient sunlight absorbing agent. Secondly, we will summarize the properties of common monomer building blocks involved for construction of solar cell polymers. Only representative polymers based on the common building blocks will be discussed due to the space limit. More quality reviews and texts are directed to interested readers (C. Li, 2010; Günes et al., 2007; Sun & Sariciftci, 2005; Cheng et al., 2009).

2. Criteria for an efficient BHJ solar cell polymer

For a conjugated polymer to suit in organic photovoltaic bulk heterojunction solar cell, it should possess favorable physical and chemical properties in order to achieve reasonable device efficiency. Key words are: large absorption coefficient; low band gap; high charge mobility; favorable blend morphology; environmental stability; suitable HOMO/LUMO level and solubility.

2.1 Large absorption coefficient

For polymers used in solar cells, a large absorption coefficient in the film state is a prerequisite for a successful application since the preliminary physics regarding photovoltaic phenomenon is photon absorption. The acceptor component of the BHJ blend, usually $PC_{60}BM$ or $PC_{70}BM$, absorbs inefficiently longer than 400 nm (Kim et al., 2007). It is thus the responsibility for the polymer to capture the photons above 400 nm. Means to increase the solar absorption of the photoactive layer include: 1) increasing the thickness of the photoactive layer; 2) increasing the absorption coefficient; and 3) matching the polymer absorption with the solar spectrum. The first strategy is rather limited due to the fact that the charge-carrier mobilities for polymeric semiconductors can be as low as $10^{-4}cm^2/Vs$ (Sariciftci, 2004). Series resistance of the device increases significantly upon increasing the photoactive layer thickness and this makes devices with thick active layer hardly operational. The short-circuit current (J_{sc}) may drop as well because of the low mobility of charge carriers. With the limitation to further increase the thickness, large absorption coefficient (10^5 to 10^6) in the film state is preferred in order to achieve photocurrent >10 mA/cm^2 (Sariciftci, 2004). By lowering the band gap, absorption of the polymer can be broadened to longer wavelength and photons of $\lambda > 800nm$ can be captured as well.

2.2 Low band gap to absorb at long wavelength

The solar irradiation spectrum at sea level is shown in Fig 1 (Wenham & Watt, 1994). The photon energy spreads from 300 nm to > 1000 nm. However, for a typical conjugated polymer with energy gap $E_g \sim 2.0$ eV, it can only absorb photon with wavelength up to *ca.* 600 nm (blue line in Fig 1) and maximum 25% of the total solar energy. By increasing the absorption onset to 1000 nm ($E_g = \sim 1.2$ eV) (red line in Fig 1), approximately 70 to 80% of the solar energy will be covered and theoretically speaking an increase of efficiency by a factor of two or three can be achieved. A controversy regarding low band gap polymer is that once a polymer absorbs at longer wavelength, there will be one absorption hollow at the shorter wavelength range, leading to a decreased incident photon to electron conversion efficiency at that range. One approach to address this issue is to fabricate a tandem solar cell with both large band gap polymer and narrow band gap polymer utilized simultaneously for solar photon capture (Kim et al., 2007).

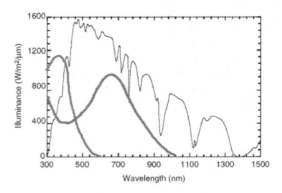

Fig. 1. Reference solar irradiation spectrum of AM1.5 illumination (black line). Blue line: typical absorption spectrum of a large band gap polymer; Red line: typical absorption spectrum of a narrow band gap polymer.

2.3 High charge carrier mobility

Charge transport properties are critical parameters for efficient photovoltaic cells. Higher charge carrier mobility of the polymer increases the diffusion length of electrons and holes generated during photovoltaic process and at the same time reduces the photocurrent loss by recombination in the active layer, hence improving the charge transfer efficiency from the polymer donor to the PCBM acceptor (G. Li et al., 2005). This charge transport property of the photoactive layer is reflected by charge transporting behavior of both the donor polymer and the PCBM acceptor. The electron transport property of pure PCBM thin film has been reported in details and is known to be satisfactory for high photovoltaic performance ($\sim 10^{-3}$ $cm^2V^{-1}s^{-1}$) (Mihailetchi et al., 2003). However, the mobility of the free charge carriers in thin polymer films is normally in the order of 10^{-3} to 10^{-11} $cm^2V^{-1}s^{-1}$, which limits the PCE of many reported devices (Mihailetchi et al., 2006). Therefore, it is promising to increase the efficiency by improving the charge carrier property of the polymer part, since there is huge space to improve if we compare this average value with the mobility value of some novel polymer organic field effect transistor materials (Ong et al., 2004; Fong et al., 2008).

2.4 Favorable blend morphology with fullerene derivatives

The idea that morphology of the photoactive layer can greatly influence the device performance has been widely accepted and verified by literature reports (Arias, 2002; Peet et al., 2007). However, it is still a 'state-of-art' to control the morphology of specific polymer/PCBM blend. Even though several techniques (Shaheen et al., 2001) have been reported to effectively optimize the morphology of the active layer, precise prediction on the morphology can hardly been done. It is more based on trial-and-error philosophy and theory to explain the structure-morphology relationship is still in infancy. Nevertheless, several reliable and efficient methods have been developed in laboratories to improve the morphology as well as the performance of the solar cell devices.

The first strategy is to control the solvent evaporation process by altering the choice of solvent, concentration of the solution and the spinning rate (Zhang et al., 2006). The slow evaporation process assists in self-organization of the polymer chains into a more ordered structure, which results in an enhanced conjugation length and a bathochromic shift of the absorption spectrum to longer wavelength region. It is reported (Peet. et al., 2007) that chlorobenzene is superior to toluene or xylene as the solvent to dissolve polymer/PCBM blend during the film casting process. The PCBM molecule has a better solubility in chlorobenzene and therefore the tendency of PCBM molecule to form clusters is suppressed in chlorobenzene. The undesired clustering of PCBM molecules will decrease the charge carrier mobility of electrons because of the large hopping boundary between segregated grains.

The second strategy is to apply thermal annealing after film casting process. This processing technique is also widely used for organic field effect transistor materials. The choice of annealing temperature and duration is essential to control the morphology. At controlled annealing condition, the polymer and PCBM in the blend network tend to diffuse and form better mixed network favorable for charge separation and diffusion in the photoactive layer (Hoppe & Sariciftci, 2006).

2.5 Stability

The air stability of the solar cell device, as it is important for the commercialization, has attracted more and more attention from many research groups worldwide. Even though industry pays more attention to the cost rather than the durability of the solar cell device, a shelf lifetime of several years as well as a reasonably long operation lifetime are requested to compete with Si-based solar cells. The air instability of solar cell devices is mainly caused by polymer degradation in air, oxidation on low work function electrode, and the degradation of the morphology of the photoactive layer.

For a conjugated polymer to achieve such a long lasting lifetime, it should have intrinsic stability towards oxygen oxidation which requires the HOMO energy level below the air oxidation threshold (*ca.* -5.27 eV) (de Leeuw et al., 1997). Device engineering can also provide the extrinsic stability by sophisticated protection of the conjugated polymer from air and humidity.

2.6 Desired HOMO/LUMO energy level

The Highest Occupied Molecular Orbital (HOMO) and Lowest Unoccupied Molecular Orbital (LUMO) of the polymer should be carefully tuned for several considerations. First of all, the HOMO energy level of a material, which describes the accessibility of the material molecule to be oxidized, reflects the air stability of the material. The oxidation threshold of air is -5.2 eV ~-5.3 eV against vacuum level. Therefore, the HOMO level cannot be more positive than this value to provide the air stability to the polymer. Secondly, the maximum open circuit voltage (V_{oc}) is correlated to the difference between the LUMO energy level of PCBM and the polymer's HOMO energy level based on experimental evidence (Brabec et al., 2001; Scharber et al., 2006). Therefore, in order to achieve high V_{oc} in the device, HOMO level should be reasonably low.

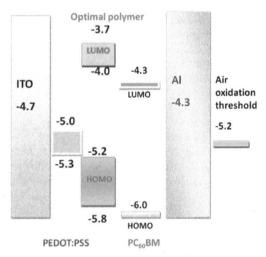

Fig. 2. Optimal HOMO/LUMO energy level of optical polymer used in BHJ solar cell with PC$_{60}$BM as acceptor (Blouin et al., 2008)

To ensure efficient electron transfer from the polymer donor to the PCBM acceptor in the BHJ blend, the LUMO energy level of the polymer material must be positioned above the LUMO energy level of the acceptor by at least 0.2-0.3 eV. Based on these factors, as shown in Fig 2, the ideal polymer HOMO level should range from -5.2 eV to -5.8 eV against vacuum level due to the compromise of stability, band gap and open circuit voltage. The ideal polymer LUMO level should range from -3.7 eV to -4.0 eV against vacuum level to assist electron injection from polymer to acceptor.

2.7 Solubility

Polymer prepared for solar cell application should possess reasonable solubility so that it can be analyzed by solution based characterization methods such as NMR spectroscopy. Meanwhile, polymer with poor solubility will be found inappropriate for solution processing and device performance is normally low due to unfavorable microscopic morphology of the thin film formed by spin coating. Aliphatic chains attached to the polymer backbone are essential to ensure solubility of the polymer. However, it should be noted that aliphatic chains, being electronically inactive, will dilute the conjugated part of the polymer and hence, reduce the effective mass of the polymer.

Some rules of thumb regarding the use of alkyl chains include that: 1) longer chain is better than shorter chain to solubilize polymer; 2) branched chain is better than linear chain to solubilize polymer; and 3) the more rigid or planar the polymer backbone is, the more or longer alkyl chains are needed.

3. Common building blocks for BHJ solar cell polymers

Common monomer building blocks to construct conjugated polymer for solar cells have been summarized in Chart 1. They are categorized by number of rings and way of linkage. Due to the space limit, we will only discuss monomers that are commonly encountered in the literature and the property of their representative polymers. Some important building blocks, even though not commonly used for PSC polymer, are also included for comparison.

3.1 Ethylene (double bond)

Ethylene (double bond) is a commonly adopted spacer or bridge in conjugated polymers. Common chemical methods to introduce double bond to the polymer include: Wittig-Horner reaction; Wessling sulfonium precursor method (Wessling, 1985); Gilch route (Gilch & Wheelwright, 1966) and palladium catalyzed coupling reactions.

By utilizing Wittig reaction, fully regioregular and regiorandom poly[(2-methoxy-5- ((3′,7′-dimethyloctyl)oxy)-1,4-phenylenevinylene] (MDMO-PPV, P1 and P2) were synthesized following the route shown in Scheme 1 (Tajima et al., 2008). The device study on these two polymers showed that the regioregular MDMO-PPV-based device had a PCE of 3.1%, which was much higher than 1.7% out of regiorandom MDMO-PPV. The higher crystallinity of the regioregular MDMO-PPV polymer and better mixing morphology with PCBM were ascribed to the improved PCE for regioregular MDMO-PPV. This study highlighted the importance of regioregularity of PPV-based polymer to achieve good solar cell performance.

3.2 Acetylene (triple bond)

Polyacetylene was the first discovered conducting conjugated polymer and inspired a lot of scientific interest in the research of conjugated polymers (Shirakawa et al., 1977). The synthetic chemistry of acetylene-containing polymers has been intensively reviewed by Liu et al.(Liu et al., 2009). In polymers designed for solar cell, acetylene is normally introduced to the polymer backbone via Sonogashira cross coupling reaction.

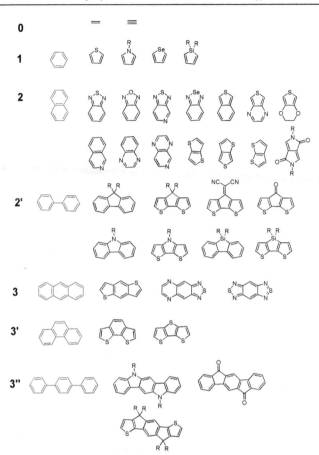

Chart 1. Common monomer building blocks used for construction of solar cell polymers

Scheme 1. synthesis of regioregular and regiorandom MDMO-PPV

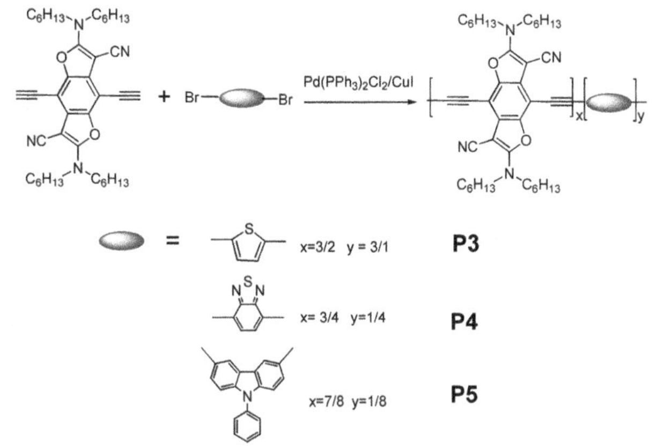

Scheme 2. Synthetic route of acetylene-containing polymers P3, P4 and P5

Benzodifuran moiety was copolymerized with thiophene, electron withdrawing benzothiadiazole and electron donating 9-phenylcarbazole, respectively, to form P3, P4 and P5 as shown in Scheme 2 (H. Li et al., 2010). The ratio of x/y is estimated from the integration of relevant peaks in their NMR spectra. The fraction of benzodifuran is more than 50% due to the self-coupling of diacetylene monomer. All three polymers absorbed beyond 600 nm in the film state and had a LUMO level above -4.0 eV. The high structural order of these three polymers was evidenced by power XRD study, as two reflection peaks, one at $2\theta = 4.95°–5.5°$ and the other at $2\theta = 19.95° – 21.75°$, were well observed. The highest PCE $\eta = 0.59\%$ was obtained based on P3/PCBM (1:4, w/w) blend.

Another category of acetylene containing polymer designed for PSC is π-conjugated organoplatinum polyyne polymers (Baek et al., 2008). The platinum-C_{sp} bond extends the conjugation of the polymer as a result of the fact that the d-orbital of the Pt can overlap with the p-orbital of the alkyne. This kind of Pt-C bond can be chemically accessible by a Sonogashira-type dehydrohalogenation between alkyne and platinum chloride precursor. Examples of this type of polymer and their synthetic routes are shown in Scheme 3 (Wong et al., 2007).

In order to tune the energy gap <2.0 eV, internal D-A function was introduced between electron rich Pt-ethyne groups. This effective band gap control strategy rendered P6 UV-vis absorption maximum at 548 nm and an optical band gap of 1.85 eV. This absorption behavior was proved to occur via the charge transfer excited state but not the triplet state of the polymer by photolumiscence lifetime study and PL temperature dependence study. The electrochemical HOMO and LUMO energy level were measured to be -5.37 eV and -3.14 eV, respectively. The best P6/PCBM (1:4, w/w) BHJ solar cell gave the open circuit voltage V_{oc}= 0.82 V, the short-circuit current density J_{sc}=15.43 mA, fill factor FF=0.39 and power conversion efficiency η =4.93%.

For polymers P7-P10 (Wong et al., 2007), the electron withdrawing moiety was replaced by bithiazole heterocycles. Nonyl chains were attached to the bithiazole to assist solvation of the polymer. By extending the conjugation (m: 0→3) along the polymer backbone, the band gaps of P7-P10 decreased from 2.40 eV to 2.06 eV. The power conversion efficiency (polymer/PCBM=1:4, w/w) was found significantly improved from ~0.2% to ~2.5% as the number of thiophene bridge increased from 0 to 3, most likely due to the improved charge carrier mobility of the active layer.

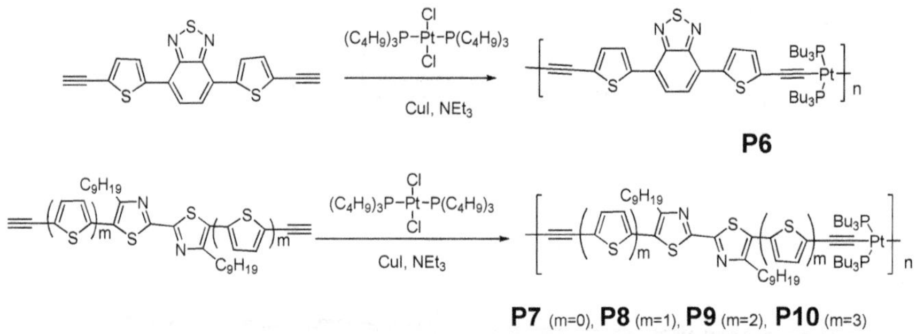

P6

P7 (m=0), **P8** (m=1), **P9** (m=2), **P10** (m=3)

Scheme 3. Synthesis of organoplatinum polyyne polymers: P6, P7-P10

3.3 Phenylene (benzene)

Benzene ring is the most fundamental building block for polymer solar cell materials. A lot of chemistry and reaction carried out in this research area are rooted back to the reactivity of benzene ring. Benzene can be polymerized by direct linkage at the 1,4-position to form poly(para-phenylene) (Chart 2). Poly(para-phenylene) without any substituents has a linear rod-like structure and poor solubility in common organic solvents which limits its application as organic electronics. One strategy to increase the solubility is to introduce alkyl or alkoxyl chain on the backbone. However, the planarity of the poly(para-phenylene) will be disturbed due to the steric effect of the R group attached (Chart 2, P12) and therefore the effective conjugation between adjacent benzene rings will be sacrificed. To address this issue, bridges can be introduced between benzene rings, e.g., double bond in poly(phenylvinylene) (PPV)(Chart 2, P13).

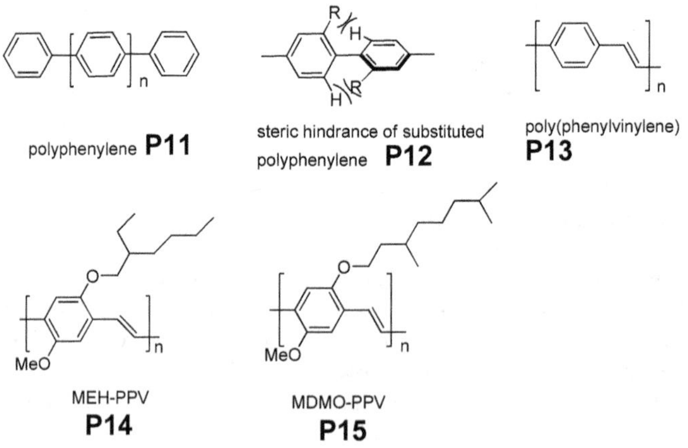

polyphenylene **P11**

steric hindrance of substituted
polyphenylene **P12**

poly(phenylvinylene)
P13

MEH-PPV
P14

MDMO-PPV
P15

Chart 2. Structures of polyphenylene and its derivatives

PPV and its derivatives have been intensively studied in organic electronics research for OLED and PSC materials due to their excellent conducting and photoluminescent properties (Burroughes et al., 1990). Poly[2-methoxy-5-((2'-ethylhexyl)oxy)-1,4- phenylenevinylene] (MEH-PPV, P14) was utilized to fabricate bilayer solar cell with C_{60} in the early days and it was reported that photoinduced electron transferred from electron donating MEH-PPV onto

Buckminsterfullerene, C_{60}, on a picosecond time scale (Sariciftci et al., 1992). This experiment explained one fundamental physical phenomenon present in organic photovoltaic cells and the concept developed by this study significantly inspired later research on organic solar cells.

Another derivative of PPV, poly[(2-methoxy-5-(3',7'-dimethyloctyl)oxy)-1,4-phenylene vinylene] (MDMO-PPV, Chart 2) is also widely studied for solar cells and still being used nowadays. The combination of MDMO-PPV and PCBM is used in BHJ solar cell and efficiency up to 3.1% has been reported (Tajima et al., 2008). However, the relatively low hole mobility of MDMO-PPV ($5 \times 10^{-11}cm^2V^{-1}s^{-1}$) (Blom et al., 1997) is reported to limit the charge transport inside the photoactive layer. Most PPV polymers have band gap greater than 2.0 eV and have maximum absorption around 500 nm. Furthermore, PPV materials are not stable in air and vulnerable to oxygen attack. Structural defects generated either during synthesis or by oxidation will substantially degrade the performance of the device. All these factors limit the application of PPV polymers in solar cells.

3.4 Thiophene

Thiophene has become one of the most commonly used building blocks in organic electronics due to its excellent optical and electrical properties as well as exceptional thermal and chemical stability (Fichou, 1999). Its homopolymer, polythiophene (PT), was first reported in 1980s as a 1D-linear conjugated system (Yamamoto et al., 1980; Lin & Dudek, 1980). Substitution by solubilizing moieties is adopted to increase the solubility of polythiophenes. The band gap of the polythiophene can also be tuned at the same time by inductive and/or mesomeric effect from the heteroatom containing substitution.

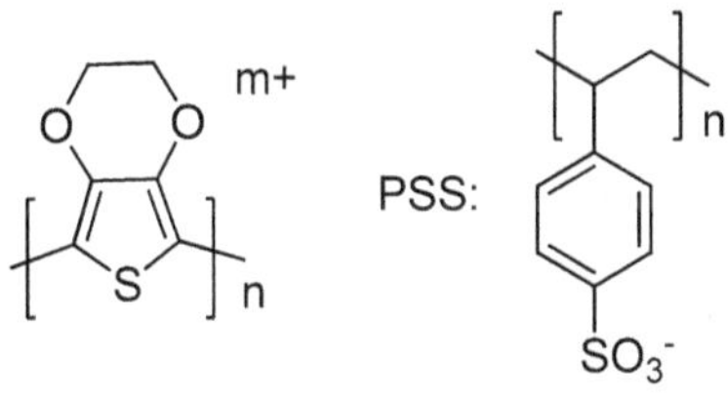

PEDOT:PSS **P16** 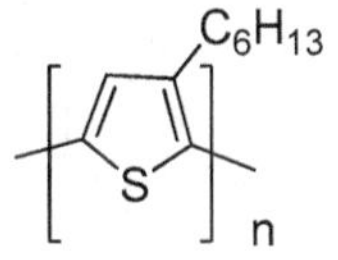   P3HT **P17**

Chart 3. Chemical structures of PEDOT:PSS and P3HT

Two frequently encountered thiophene-based conjugated polymers in literature are poly (3,4-ethylenedioxythiophene) poly(styrenesulfonate) (PEDOT-PSS, Chart 3) in conducting and hole transport layers for organic light emitting diodes (OLEDs) and PSCs and regioregular poly(3-hexylthiophene) (P3HT, Chart 3) as a hole transporting material in organic field effect transistors (OFETs) and PSCs.

As in PPV polymer, regioregularity is essential for the thiophene unit to conjugate effectively on the same plane since in regioregular form, steric consequence of the substitution is minimized, resulting in longer effective conjugation length and a lower band gap. As shown in Chart 4, three different regioisomers, head-to-head (HH), head-to-tail (HT) and tail-to-tail (TT) can be formed when two 3-alkyl thiophene units are linked via 2,5-position. Presence of HH and TT linkage in polythiophene will cause plane bending and generate structural disorder, which consequently weaken the intermolecular interaction.

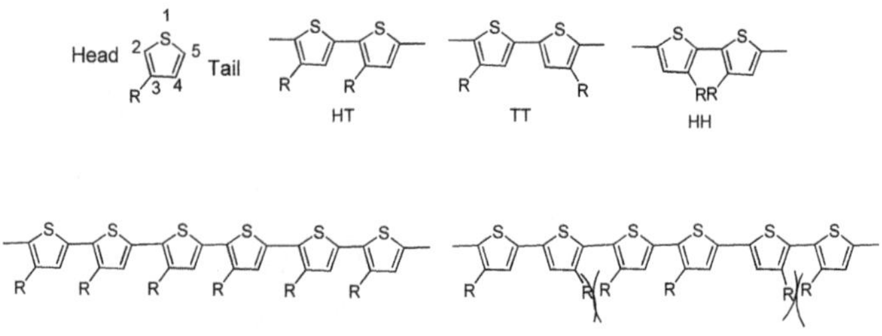

Chart 4. 3-substituted thiophene dimer isomers, regioregular P3HT and regioirregular P3HT

Regioregular P3HT was first synthesized by McCullough's group via a Grignard metathesis method (McCullough & Lowe, 1992, 1993). Polymerization with a Ni(0) catalyst yielded a highly regioregular (>99% HT) PT polymer (M_n=20000-35000, PDI=1.2-1.4). The mechanism of this Ni coupling reaction was proposed and justified to be a 'living' chain growth mechanism (Miyakoshi et al., 2005). Regioregular P3HT has been treated as a standard polymer for solar cell devices and commonly used for device testing and comparison. It represents the 'state of art' in the field of PSCs and efficiency up to ~5% has been reported based on P3HT/PCBM device (Ma et al., 2005).

3.5 Silole

Siloles or silacyclopentadienes, are a group of five-membered silacyclics with 4 accessible substitution positions on the butadiene and another 2 positions on the silicon atom. Compared with many other 5-membered heterocyclopentadiene, such as thiophene, furan or pyrrole, the silole (silacyclopentadiene) ring has the smallest HOMO-LUMO band gap and the lowest lying LUMO level due to the σ^* to π^* conjugation arising from interaction between the σ^* orbital of two exocyclic bonds on silicon and the π^* orbital of the butadiene moiety. The small band gap and lowest LUMO energy level render silole outstanding optoelectronic properties such as high PL efficiency and excellent electron mobility (Chen & Cao, 2007).

Random and alternating silole-containing copolymers P18 (Chart 5) (F. Wang et al., 2005) were synthesized via Suzuki coupling reaction from fluorene and 2,5-dithienyl-silole. The band gap of this series of polymer could be tuned from 2.95 eV to 2.08 eV by varying the silole content from 1% to 50% in the polyfluorene chain. The decrease of the band gap was found mainly due to the decrease of LUMO energy level while the HOMO of this series of polymer remained unchanged at ~ -5.7 eV. For the alternating copolymer, field effect charge mobility was measured to be 4.5×10^{-5} cm^{2}V^{-1}s^{-1} and the best PCE value was reported to be 2.01% using a P18(m=1)/PCBM (1:4, w/w) blend as active layer.

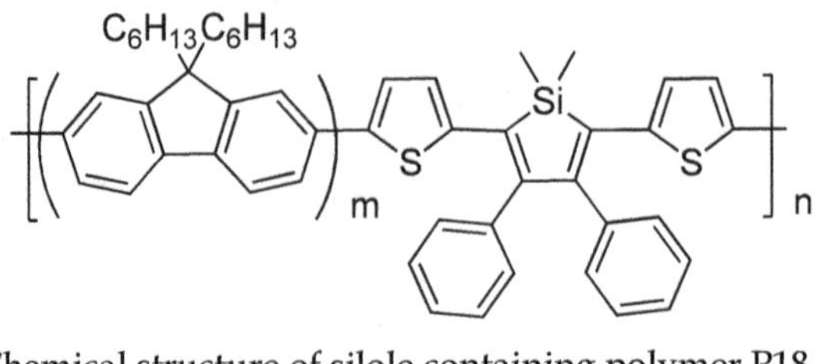

Chart 5. Chemical structure of silole containing polymer P18

3.6 Benzothiadiazole/ Aza-benzothiadiazole/ Se-benzothiadiazole

2,1,3-benzothiadiazole (BT, Chart 6) is an electron deficient heterocycle that has been incorporated with electron donating species to construct low band gap polymer donor for BHJ polymer solar cell. The electron withdrawing ability of BT can be further increased by replacing one carbon atom with sp^2-hybridized N atom (Chart 6). The sulfur atom in the BT unit can also be replaced to selenium, by doing so the band gap is further decreased.

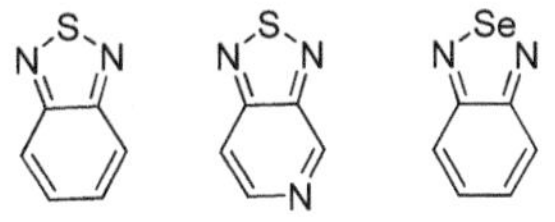

Chart 6. Structure of BT, aza-BT and Se-BT

A variety of low band gap polymers containing BT have been synthesized and tested for PSC performance (P19-P23) (Zhang et al., 2006; E. Zhou et al., 2008; Svensson et al., 2003; Slooff et al., 2007; Q. Zhou et al., 2004; Blouin et al., 2007). The electron donating moiety in the low band gap polymer varies from carbazole, fluorene, dibenzosilole, bridged thiophene-phenylene-thiophene, and dithienopyrrole. This type of polymer has a band gap <2.0 eV and gives moderate to good PCE value ranging from ~1% to ~5%, rendering a promising direction for the research of PSC donor material.

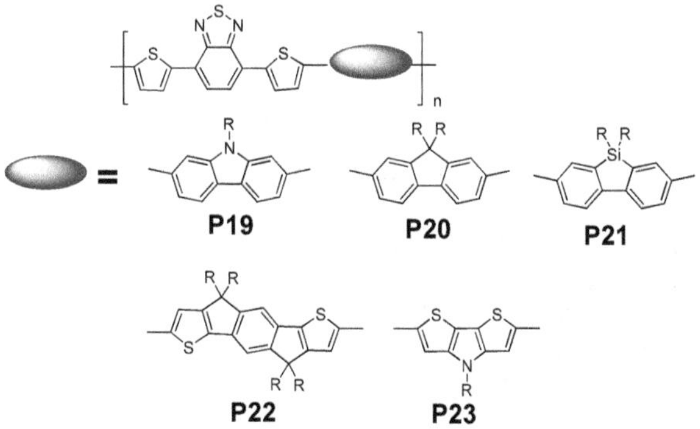

Chart 7. Benzothiadiazole containing low band gap polymers P19-P23

3.7 Isothianaphthene/ Thienopyrazine

The first example of poly(isothianaphthene) is reported by Wudl et al. in 1984 (Wudl et al., 1984) as shown in Scheme 4. Poly(isothianaphthene) has a greater tendency to adopt the quinoid structure due to the stabilization of the benzenoid ring formation (Chart 8). The quinoid structure adoption reduces the band gap of poly(isothianaphthene) to ca. 1.0 eV, which is about half that of polythiophene (~2.0 eV) (Kobayashi et al., 1984).

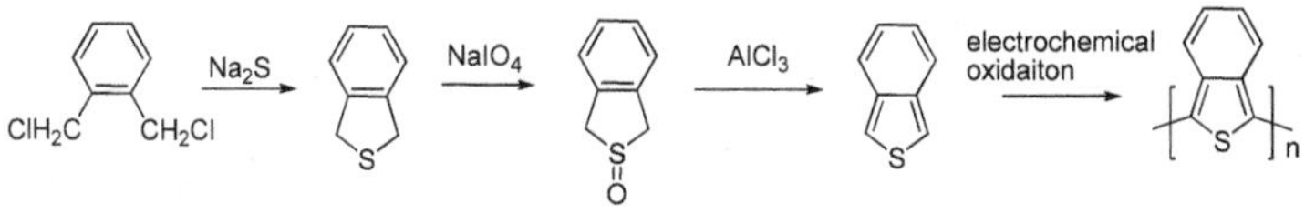

Scheme 4. Synthesis of poly (isothianaphthene)

Poly(thianaphthene) adopts a non-planar conjugation due to the steric hinderance present between benzo-H and thiophene-S atoms as shown in Chart 8. To increase the effective π-conjugation, one C-H can be replaced by sp^2-hybridized nitrogen to release the steric strain. As evidenced by X-ray structure analysis of 2,5-di(2-thienyl) pyridino[c]thiophene (Chart 8) (Ferraris et al., 1994), the torsional angle between the pyridinothiophene moiety and the thiophene unit on the N side is only 3.5º, while it is 39º on the other side. To further release the steric strain, the CH groups on both sides of the isothianaphthene can be replaced by N atom. Due to its effective conjugation and electron withdrawing nature, thieno[3,4-b]pyrazine is proposed as another type of electron withdrawing building block for the construction of low band gap polymers. In necessity of solubility, substituted thieno[3,4-b]pyrazines can be synthesized by condensation of 3,4-diaminothiophene with substituted 1,2-diones.

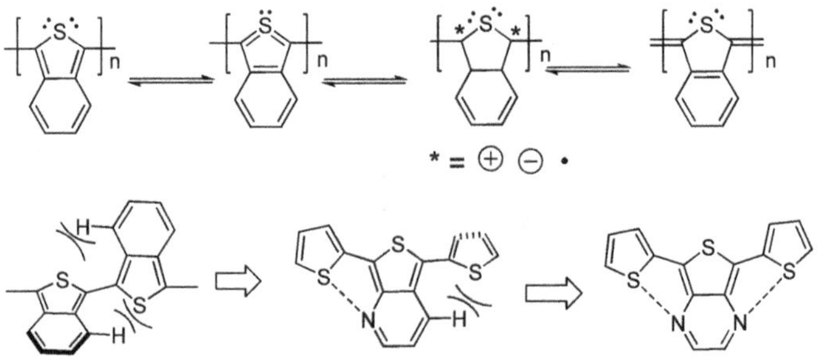

Chart 8. Resonance structure of poly(isothianaphthene) and demonstration of steric strains

The thienopyrazine unit is commonly linked to two thiophene rings at each side and coupled with electron donating moiety, such as fluorene to construct low band gap polymer (P24, P25, and P26) (Zhang et al., 2005, 2006; Mammo et al., 2007). P24 was reported to suffer from low solubility and low molecular weight. The polymerization yield was only 5% owing to the poor solubility in chloroform. The best PCE based on P24/PCBM(1:6, w/w) was obtained as η= 0.96%. To address the solubility issue, alkyl and alkoxyl chains were attached to the thienopyrazine moiety and another two low band gap polymers P25 and P26 were synthesized by copolymerization between thienopyrazine and fluorene. The addition of side chains did not change the band gap and HOMO/LUMO energy level as evidenced from absorption spectra and cyclic voltammetry measurement. These two polymers had almost identical absorption and HOMO/LUMO values. The best PCE value obtained for P25 was 1.4% while for P26 the optimal PCE value was 2.2%.

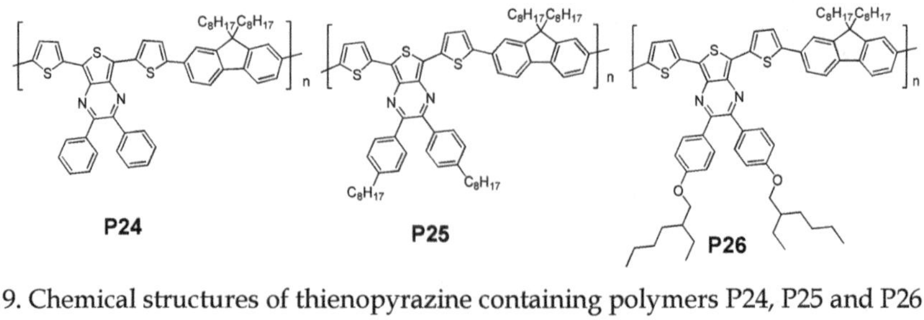

Chart 9. Chemical structures of thienopyrazine containing polymers P24, P25 and P26

3.8 Thieno[3,4-b]thiophene / Thieno[3,2-b]thiophene/ Thieno[2,3-b]thiophene

Annulations of two thiophene rings generates 4 isomers (Chart 10), namely, thieno[3,4-b] thiophene, thieno[3,2-b]thiophene, thieno[2,3-b] thiophene and thieno[3,4-c]thiophene. The first three isomers have been synthesized and isolated. Thieno[3,4-c]thiophene is predicted to be kinetically unstable and not isolated yet (Rutherford et al., 1992).

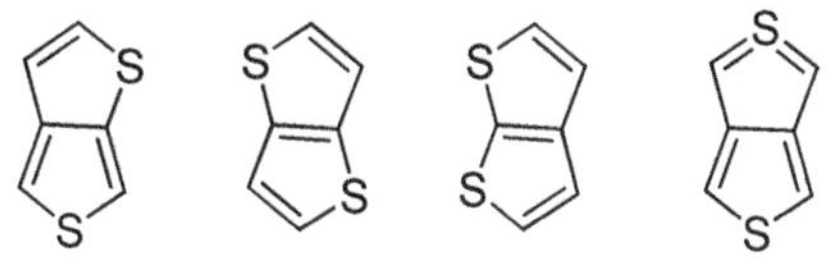

Chart 10. Isomer structure of thienothiophenes; from left to right: thieno[3,4-b]thiophene, thieno[3,2-b]thiophene, thieno[2,3-b]thiophene and thieno[3,4-c]thiophene

Thieno[2,3-b] thiophene, thieno[3,2-b]thiophene and thieno[3,4-b]thiophene are useful building blocks in preparing conjugated polymer for organic electronics applications due to their planarity and electron richness. Compared with thiophene, thienothiophene has a larger π-conjugated system. Therefore, it is introduced to the polymer backbone in the hope that it can facilitate interchain π-stacking to increase the structural order and improve the charge carrier mobility.

An efficient polymer donor P27 copolymerized by thieno[3,4-b]thiophene and benzodithiophene has been reported (Liang et al., 2009). Three dodecyl chains were used in each repeating unit to assist solvation of the polymer. BHJ solar cell fabricated using P27/PCBM(1:1 w/w) gave an excellent PCE of 4.76%, with V_{oc} = 0.58V, J_{sc} = 12.5 mA/cm^2 and FF = 0.654. A further improvement of the PCE to 5.3% was obtained by utilizing PC$_{70}$BM as the electron acceptor in the active layer. This high PCE value was ascribed to several factors including well tuned band gap (1.6 eV), proper HOMO/LUMO energy levels, balanced carrier mobility (P27 μ_h=4.5x10^{-4}cm^{2}V^{-1}s^{-1}), favorable morphology of the active layer and thieno-thiophene's ability of stabilizing the quinoid structure along the polymer backbone to enhance the planarity of the polymer.

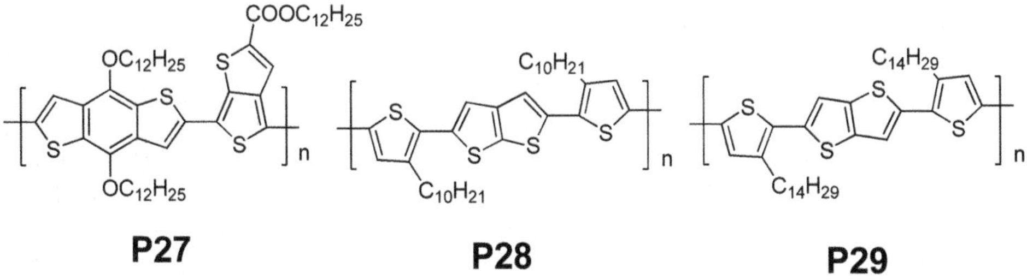

P27   **P28**   **P29**

Chart 11. Chemical structures of thienothiophene containing polymer P27, P28 and P29

Liquid-crystalline semiconductor polymers P28 and P29 were prepared by copolymerization of thienothiophene and thieno[2,3-b]thiophene, respectively, with 4,4'-dialkylbithiophene unit (McCulloch et al., 2006; Heeney et al., 2005). P28 had good field effect charge mobility of μ_h = 0.15 cm^{2}V^{-1}s^{-1}. However, its relatively large band gap (absorption maximum λ_{max} = 470 nm) limited its application as efficient solar cell material. For P29, the absorption maximum was red shifted to 547 nm and the field effect charge mobility was increased to

μ_h = ~0.7 cm^{2}V^{-1}s^{-1}. The improved mobility was suggested due to the improved control of crystallization. The PSC device fabricated from P29/PC$_{70}$BM(1:4 w/w) blend achieved an optimized PCE η= 2.3% in nitrogen atmosphere. The high lying HOMO energy level (-5.1eV) of P29, which is above the air oxidation threshold (-5.2 eV), makes the polymer relatively unstable in air.

3.9 Diketopyrrolopyrrole (DPP)

The DPP moiety has been utilized for construction of low band gap polymer for BHJ solar cells due to its electron deficient nature, planarity of the core and ability to accept H-bonding. D-A type low band gap polymers based on DDP have been synthesized by varying the electron donating part of the polymer (P30, P31, P32) (Wienk et al., 2008; Bronstein et al., 2011).

By combining electron rich quarterthiophene with electron deficient DDP unit, a low band gap polymer (1.4 eV in film state) P30 was obtained. P30 showed a good solubility in chloroform and tended to aggregate in dichlorobenzene (DCB). Device based on P30/PC$_{60}$BM (1:2, w/w) BHJ thin film prepared from solution in CHCl$_3$/DCB (4:1 v/v) gave a PCE of 3.2%. By utilizing PC$_{70}$BM as the acceptor in the active layer, an improved PCE of 4.0% was achieved under the same condition.

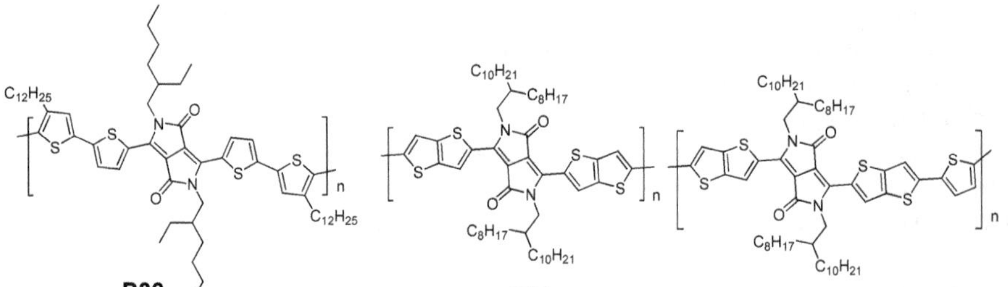

Chart 12. Chemical structures of DPP-containing polymers P30, P31 and P32

By replacing the thiophene unit with larger thieno[2,3-b] thiophene, P31 and P32 were prepared. Long branched chains have been incorporated at the DDP unit to assist solvation. Both polymers had band gap of ~1.4 eV and absorbed beyond 800 nm in the film state. Ambipolar charge transport behavior was found for both of the polymers. P31 had a high hole mobility of 0.04 cm^{2}V^{-1}s^{-1} and good PCE of 3.0% based on P31/PC$_{70}$BM (1:2 w/w) thin film prepared from CHCl$_3$/DCB (4:1 v/v) solution. By modifying the backbone with one more thiophene unit introduced to the repeating unit, P32 showed an even higher hole mobility of $ca.$ 2 cm^{2}V^{-1}s^{-1} and the BHJ solar cell device fabricated under the same condition as that of P31 showed an improved efficiency up to 5.4%.

3.10 Fluorene/ cylcopenta[2,1-b:3,4-b'] dithiophene/ silafluorene/ dithieno[3,2-b:2',3'-d] silole

Fluorene based polymers have been widely explored as organic electronic material in the field of OLED, OFET and PSC due to their high photoluminescence quantum yield, high thermal and chemical stability, good film-forming properties and good charge transport properties. Polyfluorene, however, has a band gap of ~3.0eV, which limits its application in solar cell. Therefore, fluorene is normally copolymerized with electron withdrawing moieties to construct polymers with band gap <2.0 eV so as to extend sunlight harvesting to longer wavelength. Although some solar cell polymers have been prepared by copolymerization of fluorene and electron-rich moieties, such as thienothiophene and pentacene, their absorption behaviors and wide band gaps are found to account for the moderate to poor performance (Schulz et al., 2009; Okamoto et al., 2008). Palladium catalyzed cross coupling reaction is normally adopted for the polymerization due to the ease of halogenation at the 2,7-position of fluorene unit. Alkynation at the 9-position of the fluorene assists solvation for the D-A type polymer, whereas necessary, alkynation on the electron deficient counterpart is also required. Fluorene copolymers prepared from electron deficient benzothiadiazole and thienopyrazine have been discussed previously.

By replacing two benzene rings of fluorene with thiophene, cylcopenta[2,1-b:3,4-b'] dithiophene can be obtained as another novel building block to construct D-A type low band gap polymer. Alkynation at the bridge sp^3-carbon renders solubility for the polymer. Cyclopenta[2,1-b:3,4-b']dithiophene based polymer P33 has been synthesized with a low band gap of ca. 1.4 eV (Mühlbacher et al., 2006). It was utilized by Kim et al. (Kim et al., 2007) to fabricate an efficient tandem solar cell. This brilliant and excellent work addressed the issue that while most low band gap polymers absorb at wavelength longer than 700 nm, there is a hollow at shorter wavelength and the lack of sufficient absorption at the hollow will drag the power conversion efficiency. In this case, P33 had an absorption maximum at ca. 800nm and a hollow at ca. 450nm. Kim et al. fabricated a tandem BHJ solar cell by utilizing P3HT (λ_{max} =~ 550nm) to absorb at the hollow of P33 and low band gap polymer P33 to absorb light at the NIR region. Tandem solar cell device (Al/TiOx/P3HT:PC$_{70}$BM/PEDOT:PSS/TiOx/P33:PCBM/PEDOT:PSS/ITO/glass) based on P3HT and P33 gave a typical performance parameter of J_{sc} = 7.8 mA/cm², V_{oc} = 1.24 V, FF =0.67 and PCE= 6.5%, which was among the highest values reported.

Silafluorene and dithieno[3,2-b:2',3'-d]silole are two interesting electron rich moieties that are structurally analogous to fluorene. Low band gap polymer P34 was synthesized by copolymerization of 2,7-silafluorene and dithienyl-benzothiadiazole (E. Wang et al., 2008). Field effect charge mobility of P34 was found to be ~1x10^{-3} cm²V^{-1}s^{-1}. High efficiency up to 5.4% with V_{oc} = 0.9 V, J_{sc} = 9.5 mA/cm², FF = 0.51 was obtained by using P34/PCBM(1:2 w/w) as active layer. Polymer P35 was synthesized by Stille coupling between dithieno[3,2-b:2',3'-d]silole and benzothiadiazole (Hou et al., 2008). The optical band gap of P35 was found to be 1.45 eV, which was similar to that of P33. Hole transport mobility of the polymer was determined to be 3 x 10^{-3} cm²V^{-1}s^{-1}, about 3 times higher than that of P33. The best device based on P35 gave a PCE of 5.1% with J_{sc} = 12.7 mA/cm², V_{oc}=0.68 V and FF = 0.55.

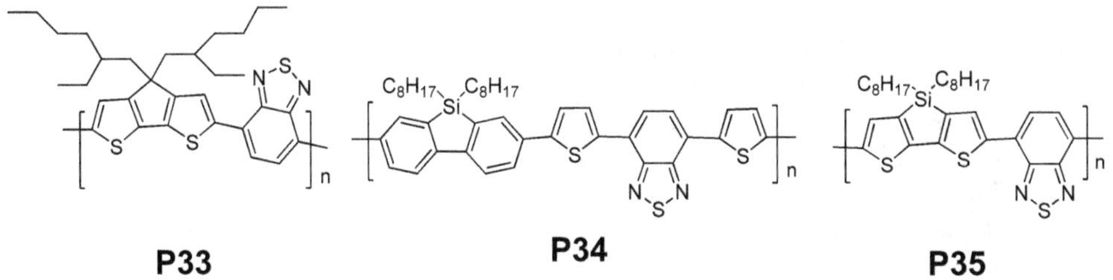

P33   **P34**   **P35**

Chart 13. Structures of low band gap polymers P33, P34 and P35

4. Conclusion

In this chapter, main effort has been directed to disclose the structure-property relationship for solar cell polymers. The requirements and criteria for an efficient polymer donor in BHJ solar cell have been discussed with representative examples. Key factors are: absorption efficiency, solubility, stability (thermal-, photo-), low band gap, HOMO/LUMO energy level, charge carrier mobility and morphology. In order to achieve high power conversion efficiency, a good balance among these factors should be met. On the other hand, choice of acceptor counterpart and device engineering for the BHJ device also play important roles for power conversion efficiency improvement. Nowadays choice of donor/acceptor combination and device fabrication is still 'a state of art' but more and more rules of thumb have been pointed out. Provided that if BHJ concept still prevails for the next 10 years or longer, newer device design is also urgently required. Tandem solar cell device reported is one example to address the efficiency issue from this point of view. But no matter what kind of new changes will be brought out, the photon flux capture material, which is conjugated polymer in PSC, will still be the core of the device.

5. Acknowledgment

This work was financially supported by National University of Singapore under MOE AcRF FRC Grant No. R-143-000-412-112 and R-143-000-444-112.

6. References

Arias, A. C., Corcoran, N., Banach, M., Friend, R. H., MacKenzie, J. D. & Huck, W. T. S. (2002). Vertically segregated polymer-blend photovoltaic thin-film structures through surface-mediated solution processing. *Applied Physics Letters*, Vol. 80, No. 10, (March 2002), pp. 1695-1697, ISSN 1077-3118

Baek, N. S., Hau, S. K., Yip, H. L., Acton, O., Chen, K. -S. & Jen, A. K. -Y. (2008). High performance amorphous metallated π-conjugated polymers for field-effect transistors and polymer solar cells. *Chemistry of Materials*, Vol.20, No.18, (September 2008), pp. 5734-5736, ISSN 1520-5002

Balzani, V., Credi, A. & Venturi, M. (2008). Photochemical conversion of solar energy. *ChemSusChem*, Vol.1, No.1-2, (February 2008), pp.26-58, ISSN 1864-5631

Blom, P. W. M., de Jong, M. J. M., & van Munster, M. G. (1997). Electric-field and temperature dependence of the hole mobility in poly(p-phenylene vinylene). *Physical Review B*, Vol.55, No.2, (January 1997), pp. R656-R659 ISSN 1550-235X

Blouin, N., Michaud, A., & Leclerc, M. (2007). A low-bandgap poly(2,7-carbazole) derivative for use in high-performance solar cells. *Advanced Materials*, Vol.19, No.17, (September 2007), pp. 2295-2300, ISSN 1521-4095

Blouin, N., Michaud, A., Gendron, D., Wakim, S., Blair, E., Plesu, R. N., Neagu-Plesu, R., Belletête, M., Durocher, G., Tao, Y. & Leclerc, M. (2008). Towards a rational design of poly(2,7-carbazole) derivatives for solar cells. *Journal of the American Chemical Society*, Vol.130, No.2, (January 2008), pp. 732-742, ISSN 1520-5126

Brabec, C. J., Cravino, A., Meissner, D., Sariciftci, N. S., Fromherz, T., Rispens, M. T., Sanchez, L. & Hummelen, J. C. (2001). Origin of the open circuit voltage of plastic solar cells. *Advanced Functional Materials*, Vol. 11, No.5, (October 2001), pp. 374-380, ISSN 1616-3028

Bronstein, H., Chen, Z., Ashraf, R. S., Zhang, W., Du, J., Durrant, J. R., Tuladhar, P. S., Song, K., Watkins, S. E., Geerts, Y., Wienk, M. M., Janssen, R. A. J., Anthopoulos, T., Sirringhaus, H., Heeney, M. & McCulloch, I. (2011). Thieno[3,2-b]thiophene-diketopyrrolopyrrole-Containing polymers for high-performance organic field-effect transistors and organic photovoltaic devices. *Journal of the American Chemical Society*, Vol.133, No.10, (March 2011), pp. 3272-3275, ISSN 1520-5126

Burroughes, J. H., Bradley, D. D. C., Brown, A. R., Marks, R. N., Mackay, K., Friend, R. H., Burns, P. L. & Holmes, A. B. (1990). Light-emitting diodes based on conjugated polymers. *Nature*, Vol.347, No.6293, (October 1990), pp. 539-541, ISSN 1476-4687

Chen, H. -Y., Hou, J., Zhang, S., Liang, Y., Yang, G., Yang, Y., Yu, L., Wu, Y. & Li, G. (2009). Polymer solar cells with enhanced open-circuit voltage and efficiency. *Nature Photonics*, Vol.3, No.11, (November 2009), pp. 649-653, ISSN 1749-4885

Chen, J. & Cao, Y. (2007). Silole-containing polymers: chemistry and optoelectronic properties. *Macromolecular Rapid Communications*, Vol.28, No.17, (September 2007), pp. 1714-1742, ISSN 1521-3927

Chen J. & Cao Y. (2009). Development of novel conjugated donor polymers for high-efficiency bulk-heterojunction photovoltaic devices. *Accounts of Chemical Research*, Vol.42, No.11, (November 2009), pp. 1709-1718, ISSN 1520-4898

Cheng, Y. J., Yang, S. H. & Hsu, C. S. (2009). Synthesis of conjugated polymers for organic solar cell applications. *Chemical Reviews*, Vol.109, No.11, (November 2009), pp. 5868-5923, ISSN 1520-6890

de Leeuw, D. M., Simenon, M. M. J., Brown, A. R. & Einerhard, R. E. F. (1997). Stability of n-type doped conducting polymers and consequences for polymeric microelectronic devices. *Synthetic Metals*, Vol.87, No.1, (February 1997), pp. 53-59, ISSN 0379-6779

Ferraris, J. P., Bravo, A., Kim, W. & Hrncir, D. C. (1994). Reduction of steric interactions in thiophene-pyridino[c]thiophene copolymers. *Journal of Chemical Society Chemical Communications*, Vol.1994, No.8, (April 1994), pp. 991-992, ISSN 0022-4936

Fichou, D. (Ed.). (1999). *Handbook of Oligo- and Polythiophenes*, Wiley-VCH Verlag GmbH, ISBN 3527294457, Weinheim

Fong, H. H., Pozdin, V. A., Amassian, A., Malliaras, G. G., Smilgies, D. -M., He, M., Gasper, S., Zhang, F. & Sorensen, M. (2008). Tetrathienoacene copolymers as high mobility, soluble organic semiconductors. *Journal of the American Chemical Society*, Vol.130, No.40, (October 2008), pp.13202-13203, ISSN 1520-5126

Gilch, H. & Wheelwright, W. (1966). Polymerization of α-halogenated *p*-xylenes with base. *Journal of Polymer Science Part A: Polymer Chemistry*, Vol.4, No.6, (March 2003), pp.1337-1349, ISSN 1099-0518

Günes, S., Neugebauer, H. & Sariciftci, N. S. (2007). Conjugated polymer-based organic solar cells. *Chemical Reviews*, Vol.107, No.4, (April 2007), pp. 1324-1338, ISSN 1520-6890

Heeney, M., Bailey, C., Genevicius, K., Shkunov, M., Sparrowe, D., Tierney, S. & McCulloch, I. (2005). Stable polythiophene semiconductors incorporating thieno[2,3-b]thiophene. *Journal of the American Chemical Society*, Vol.127, No.4, (February 2005), pp. 1078-1079, ISSN 1520-5126

Hoppe, H. & Sariciftci, N. S. (2006). Morphology of polymer/fullerene bulk heterojunction solar cells. *Journal of Materials Chemistry*, Vol.16, No.1, (January 2006), pp. 45-61, ISSN 1364-5501

Hottel, H. (1989). Fifty years of solar energy research supported by the Cabot Fund. *Solar Energy*, Vol.43, No.2, (1989), pp. 107-128, ISSN 0038-092X

Hou, J., Chen, H.-Y., Zhang, S., Li, G. & Yang, Y. (2008). Synthesis, characterization, and photovoltaic properties of a low band gap polymer based on silole-containing polythiophenes and 2,1,3-benzothiadiazole. *Journal of the American Chemical Society*, Vol. 130, No.48, (December 2008), pp. 16144-16145, ISSN 1520-5126

Kim, J. Y., Lee, K., Coates, N. E., Moses, D., Nguyen, T.-Q., Dante, M. & Heeger, A. J. (2007). Efficient tandem polymer solar cells fabricated by all-solution processing. *Science*, Vol.317, No.5835, (July 2007), pp. 222-225, ISSN 1095-9203

Kobayashi, M., Chen, J., Chung, T.-C., Moraes, F., Heeger, A. J. & Wudl, F. (1984). Synthesis and properties of chemically coupled poly(thiophene). *Synthetic Metals*, Vol.9, No.1, (January 1984), pp. 77-86, ISSN 0379-6779

Liang, Y., Wu, Y., Feng, D., Tsai, S.-T., Son, H, -J., Li, G. & Yu, L. (2009). Development of new semiconducting polymers for high performance solar cells. *Journal of the American Chemical Society*, Vol.131, No.1, (January 2009), pp. 56-57, ISSN 1520-5126

Li, C., Liu, M., Pschirer, N. G., Baumgarten, M. & Müllen, K. (2010). Polyphenylene-based materials for organic photovoltaics. *Chemical Reviews*, Vol.110, No.11, (November 2010), pp. 6817-6855, ISSN 1520-6890

Li, G., Shrotriya, V., Huang, J., Yao, Y., Mariarty, T., Emery, K. & Yang, Y. (2005). High-efficiency solution processable polymer photovoltaic cells by self- organization of polymer blends. *Nature Materials*, Vol.4, No.11, (November 2005), pp. 864-868, ISSN 1476-4660

Li, H., Jiang, P., Yi, C., Li, C., Liu, S., Tan, S., Zhao, B., Braun, J., Meier, W., Wandlowski, T. & Decurtins, S. (2010). Benzodifuran-based π-conjugated copolymers for bulk heterojunction solar cells. *Macromolecules*, Vol.43, No.19, (October 2010), pp. 8058-8062, ISSN 1520-5835

Lin, J. W. P. & Dudek, L. P. (1980). Synthesis and properties of poly(2,5-thienylene). *Journal of Polymer Science Polymer Chemistry Edition*, Vol.18, No.9, (September 1980), pp.2869-2873, ISSN 1099-0518

Lior, N. (2008). Energy resources and use: The present situation and possible paths to the future. *Energy*, Vol.33, No.6, (June 2008), pp. 842-857, ISSN 0360-5442

Liu, J., Lam, J. W. Y. & Tang, B. (2009). Acetylenic polymers: syntheses, structures, and functions. *Chemical Reviews*, Vol.109, No.11, (November 2009), pp.5799-5867, ISSN 1520-6890

Ma, W., Yang, C., Gong, X., Lee, K. & Heeger, A. J. (2005) Thermally stable, efficient polymer solar cells with nanoscale control of the interpenetrating network morphology.

Advanced Functional Materials, Vol.15, No.10, (October 2005), pp. 1617-1622, ISSN 1616-3028

McCulloch, I., Heeney, M., Bailey, C., Genevicius, K., Macdonald, I., Shkunov, M., Sparrowe, D., Tierney, S., Wagner, R., Zhang, W., Chabinyc, M. L., Kline, R. J., Mcgehee, M. D. & Toney, M. F. (2006). Liquid-crystalline semiconducting polymers with high charge-carrier mobility. *Nature Materials*, Vol.6, No.4, (April 2006), pp. 328-333, ISSN 1476-4660

McCullough, R. D. & Lowe, R. D. (1992). Enhanced electrical conductivity in regioselectively synthesized poly(3-alkylthiophenes). *Journal of Chemical Society Chemical Communications*, Vol.70, No.1, (January 1992), pp. 70-72, ISSN 0022-4936

McCullough, R. D. & Lowe, R. D. (1993). Design, synthesis, and control of conducting polymer architectures: structurally homogeneous poly(3-alkylthiophenes). *The Journal of Organic Chemistry*, Vol.58, No.4, (February 1993), pp. 904-912, ISSN 1520-6904

Mihailetchi, V. D., Duren, J. K. J., Blom, P. W. M., Hummelen, J. C., Janssen, R. A. J., Kroon, J. M., Rispens, M. T., Verhees, W. J. H. & Wienk, M. M. (2003). Electron transport in a methallofullerene. *Advanced Functional Materials*, Vol. 13, No.1, (January 2003), pp. 43-46, ISSN 1616-3028

Mihailetchi, V. D., Xie, H., Boer, B., Popescu, L. M., Hummelen, J. C. & Blom, P. W. M. (2006). Origin of the enhanced performance in poly(3-hexylthiophene): [6,6]-phenyl C_{61}-butyric acid methyl ester solar cells upon slow drying of the active layer. *Applied Physics Letters*, Vol.89, No.1, (July 2006), pp. 012107, ISSN 1077-3118

Mammo W., Admassie, S., Gadisa, A., Zhang, F., Inganäs, O. & Andersson, M. R. (2007). New low band gap alternating polyfluorene copolymer-based photovoltaic cells. *Solar Energy Materials and Solar Cells*, Vol.91, No.11, (July 2007), pp. 1010-1018, ISSN 0927-0248

Miyakoshi, R., Yokoyama, A. & Yokozawa, T. (2005). Catalyst-transfer polycondensation. Mechanism of Ni-Catalyzed chain-growth polymerization leading to well-defined poly(3-hexylthiophene). *Journal of the American Chemical Society*, Vol.127, No.49, (December 2005), pp. 17542-17547, ISSN 1520-5126

Mühlbacher, D., Scharber, M., Morana, M., Zhu, Z., Waller, D., Gaudiana, R. & Brabec, C. (2006). High photovoltaic performance of a low-bandgap polymer. *Advanced Materials*, Vol.18, No.21, (November 2006), pp. 2884-2889, ISSN 1521-4095

Okamoto, T., Jiang, Y., Qu, F., Mayer, A. C., Parmer, J. E., McGehee, M. D. & Bao, Z. (2008). Synthesis and characterization of pentacene- and anthradithiophene-fluorene conjugated copolymers synthesized by Suzuki reactions. *Macromolecules*, Vol.41, No.19, (October 2008), pp. 6977-6980, ISSN 1520-5835

Ong, B. S., Wu, Y., Liu, P. & Gardner, S. (2004). High performance semiconducting polythiophenes for organic thin-film transistors. *Journal of the American Chemical Society*, Vol.126, No.11, (March 2004), pp. 3378-3379, ISSN 1520-5126

Peet, J., Kim, J. Y., Coates, N. E., Ma, W. L., Moses, D., Heeger, A. J. & Bazan, G. C. (2007). Efficiency enhancement in low-bandgap polymer solar cells by processing with alkane dithiols. *Nature Materials*, Vol.6, No.7, (July 2007), pp. 497-500, ISSN 1476-4660

Rutherford, D. R., Stille, J. K., Elliott, C. M. & Reichert, V. R. (1992). Poly(2,5-ethynylene thiophenediylethynylenes), related heteroaromatic analogs, and poly(thieno[3,2-b]thiophenes): synthesis and thermal and electrical properties. *Macromolecules*, Vol.25, No.9, (April 1992), pp. 2294-2306, ISSN 1520-5835

Sariciftci, N. S. (2004). Plastic photovoltaic devices. *Materials Today*, Vol.7, No.9, (September 2004), pp. 36-40, ISSN 1369-7021

Sariciftci, N. S., Smilowitz, L., Heeger, A. J. & Wudl, F. (1992). Photoinduced electron transfer from a conducting polymer to buckminsterfullerene. *Science*, Vol.258, No.5087, (November 1992), pp. 1474-1476, ISSN 1095-9203

Scharber, M. C., Mühlbacher, D., Koppe, M., Denk, P., Waldauf, C., Heeger, A. J. & Brabec, C. J. (2006). Design rules for donors in bulk-heterojunction solar cells- towards 10% energy-conversion efficiency. *Advanced Materials*, Vol. 18, No.6, (March 2006), pp. 789-794, ISSN 1521-4095

Schulz, G. L., Chen, X. & Holdcroft, S. (2009). High band gap poly(9,9-dihexylfluorene- alt-bithiophene) blended with [6,6]-phenyl C_{61} butyric acid methyl ester for use in efficient photovoltaic devices. *Applied Physics Letters*, Vol.94, No.2, (January 2009), pp. 023302, ISSN 1077-3118

Shaheen, S. E., Brabec, C. J. & Sariciftci, N. S. (2001). 2.5% efficient organic plastic solar cells. *Applied Physics Letters*, Vol.78, No.6, (February 2001), pp. 841-843, ISSN 1077-3118

Shirakawa, H., Louis, E. J., MacDiarmid, A. G., Chiang, C. K., & Heeger, A. J. (1977). Synthesis of electrically conducting organic polymers: halogen derivatives of polyacetylene, $(CH)_x$. *Journal of Chemical Society Chemical Communications.*, Vol.1977, No.16, (August 1977), pp. 578-580, ISSN 0022-4936

Slooff, L. H., Veenstra, S. C., Kroon, J. M., Moet, D. J. D., Sweelssen, J., Koetse, M. M. (2007). Determining the internal quantum efficiency of highly efficient polymer solar cells through optical modeling. *Applied Physics Letters*, Vol.90, No.14, (April 2007), pp. 143506, ISSN 1077-3118

Sun, S. S. & Sariciftci, N. S. (Eds.). (March 29 2005). *Organic Photovoltaics: Mechanisms, Materials, and Devices*, CRC press, ISBN 9780824759636, Florida, USA

Svensson, M., Zhang, F., Veenstra, S. C., Verhees, W. J. H., Hummelen, J. C., Kroon, J. M., Inganäs, O., & Andersson, M. R. (2003). High-performance polymer solar cells of an alternating polyfuorene copolymer and a fullerene derivative. *Advanced Materials*, Vol.15, No.12, (June 2003), pp. 988-991, ISSN 1521-4095

Tajima, K., Suzuki, Y. & Hashimoto, K. (2008). Polymer Photovoltaic Devices Using Fully Regioregular Poly[(2-methoxy-5- (3',7'-dimethyloctyloxy))-1,4- phenylenevinylene]. *Journal of Physiccal Chemistry C*, Vol.112, No.23, (June 2008), pp. 8507-8510, ISSN 1932-7455

Tang, C.W. (1986). Two-layer organic photovoltaic cell. *Applied Physics Letters*, Vol.48, No.2, (January 1986), pp. 183-185, ISSN 1077-3118

Wang, E., Wang, L., Lan, L., Luo, C., Zhuang, W., Peng, J. & Cao, Y. (2008). High performance polymer heterojunction solar cells of a polysilafluorene derivative. *Applied Physics Letters*, Vol.92, No.3, (January 2008), pp. 033307, ISSN 1077-3118

Wang, F., Luo, J., Yang, K., Chen, J., Huang, F. & Cao, Y. (2005). Conjugated fluorene and silole copolymers: synthesis, characterization, electronic transition, light emission, photovoltaic cell, and field effect hole mobility. *Macromolecules*, Vol.38, No.6, (March 2005), pp. 2253-2260, ISSN 1520-5835

Wenham, S. R., & Watt, M. E. (1994). *Applied Photovoltaics,* Bridge Printery, ISBN 0867589094, Syndney

Wessling, R. A. (1985). The polymerization of xylylene bisdialkyl sulfonium salts. *Journal of Polymer Science: Polymer Symposia*, Vol.72, No.1, (March 2007), pp. 55-66, ISSN 1936-0959

Wienk, M. M., Turbiez, M., Gilot, J. & Janssen, R. A. J. (2008). Narrow-bandgap diketo-pyrrolo-pyrrole polymer solar cells: The effect of processing on the performance. *Advanced Materials*, Vol.20, No.13, (July 2008), pp. 2556-2560, ISSN 1521-4095

Wöhrle, D. & Meissner, D. (1991). Organic solar cells. *Advanced Materials*, Vol.3, No.3, (March 1991), pp. 129-138, ISSN 1521-4095

Wong, W. -Y., Wang, X. -Z., He Z., Chan, K. -K., Djurišić, A. B., Cheung, K. -Y., Yip, C. -T., Ng, A. M. -C., Xi, Y. Y., Mak, C. S. K. & Chan, W. -K. (2007). Tuning the absorption, charge transport properties, and solar cell efficiency with the number of thienyl rings in platinum-containing poly(aryleneethynylene)s. *Journal of the American Chemical Society*, Vol.129, No.46, (November 2007), pp. 14372-14380, ISSN 1520-5126

Wong, W. -Y., Wang, X. -Z., He, Z., Djurišić, A. B., Yip, C.-T, Cheung, K. -Y., Wang, H., Mak, C. S. K. & Chan, W. -K. (2007). Metallated conjugated polymers as a new avenue towards high-efficiency polymer solar cells. *Nature Materials*, Vol.6, No.7, (July 2007), pp. 521-527, ISSN 1476-4660

Wudl, F., Kobayashi, M. & Heeger, A. J. (1984). Poly(isothianaphthene). *The Journal of Organic Chemistry*, Vol.49, No.18, (September 1984), pp. 3382-3384, ISSN 1520-6904

Yamamoto, T., Sanechika, K. & Yamamoto, A. (1980) Preparation of thermostable and electric-conducting poly(2,5-thienylene). *Journal of Polymer Science: Polymer Letters Edition*, Vol.18, No.1, (January 1980), pp. 9-12, ISSN 1543-0472

Zhang, F., Mammo, W., Andersson, L. M., Admassie, S., Andersson, M. R. & Inganäs, O. (2006). Low-bandgap alternating fluorene copolymer/methanofullerene heterojunctions in efficient near-infrared polymer solar cells. *Advanced Materials*, Vol.18, No.16, (August 2006), pp. 2169-2173, ISSN 1521-4095

Zhang, F., Jespersen, K. G., Björström, C., Svensson, M., Adderson, M. R., Sundström, V., Magnusson, K., Moons, E., Yartsev, A. & Inganäs, O. (2006). Influence of solvent mixing on the morphology and performance of solar cells based on polyfluorene copolymer/fullerene blends. *Advanced Functional Materials*, Vol.16, No.5, (March 2006), pp. 667-674, ISSN 1616-3028

Zhang, F., Perzon, E., Wang, X., Mammo, W., Andersson, M. R. & Inganäs, O. (2005). Polymer solar cells based on a low-bandgap fluorene copolymer and a fullerene derivative with photocurrent extended to 850nm. *Advanced Functional Materials*, Vol.15, No.5, (May 2005), pp. 745-750, ISSN 1616-3028

Zhou, E., Nakamura, M., Nishizawa, T., Zhang, Y., Wei, Q., Tajima, K., Yang, C. & Hashimoto, K. (2008). Synthesis and photovoltaic properties of a novel low band gap polymer based on N-substituted dithieno[3,2-b:2',3'-d]pyrrole. *Macromolecules*, Vol.41, No.22, (November 2008), pp. 8302-8305, ISSN 1520-5835

Zhou, Q., Hou, Q., Zheng, L., Deng, X., Yu, G. & Cao, Y. (2004). Fluorene-based low band-gap copolymers for high performance photovoltaic devices. *Applied Physics Letters*, Vol.84, No.10, (March 2004), pp. 1653, ISSN 1077-3118

13

Investigation of Lattice Defects in GaAsN Grown by Chemical Beam Epitaxy Using Deep Level Transient Spectroscopy

Boussairi Bouzazi[1], Hidetoshi Suzuki[2], Nobuaki Kijima[1],
Yoshio Ohshita[1] and Masafumi Yamaguchi[1]
[1]*Toyota Technological Institute*
[2]*Myazaki University*
Japan

1. Introduction

With only 3 % of N and 9 % of In, InGaAsN with a band gap of 1.04 eV was obtained and could be lattice matched to GaAs and Ge. This dilute nitride semiconductor has been selected as a promising candidate for high efficiency multijunction tandem solar cells (Geisz and Friedman, 2002). However, the diffusion length of minority carriers and the mobility are still lower than of that in GaAs or InGaAs and showed a considerable degradation with increasing the N concentration. These electrical properties are insufficient to insure the current matching in the multijunction solar cell structure AlInGaP/GaAs/InGaAsN/Ge (Friedman et al., 1998). An obvious reason of such degradation is the high density of N-related lattice defects that can be formed during growth to compensate for the tensile strain caused by the small atomic size of N compared with that of arsenic (As) and to the large miscibility of the gap between GaAs and GaN. These defect centers are expected to act as active recombination and/or scattering centers in the forbidden gap of the alloy (Zhang & Wei, 2001). However, no experimental evidence has yet been reported. On the other hand, the conductivity of undoped p-type InGaAsN or GaAsN and their high background doping (Friedman et al., 1998; Kurtz et al., 1999; Moto et al., 2000; Krispin et al., 2000) prevent the design of wide depletion region single junction solar cell and the fabrication of intrinsic layer to overcome the short minority carrier lifetime. This serious problem was expected in the first stage to the density of unintentional carbon in the film (Friedman et al., 1998; Kurtz et al., 1999; Moto et al., 2000). However, the carrier density in some InGaAsN semiconductors was found to be higher than that of carbon (Kurtz et al., 2002). Furthermore, the high density of hydrogen (up to 10^{20} cm^{-3}) and the strong interaction between N and H in InGaAsN to form N-H related complex were confirmed to be the main cause of high background doping in InGaAsN films (Li et al., 1999; Janotti et al., 2002, 2003; Kurtz et al., 2001, 2003; Nishimura et al., 2007). In addition, N-H complex was found theoretically to bind strongly to gallium vacancies (V_{Ga}) to form N-H-V_{Ga} with a formation energy of 2 eV less than that of isolated V_{Ga} (Janotti et al., 2003). These predictions were supported experimentally using positron annihilation spectroscopy results (Toivonen et al., 2003).

On the other hand, similar electrical properties were obtained in InGaAsN grown by metal-organic chemical vapor deposition (MOCVD) and molecular beam epitaxy (MBE) despite the large difference in the density of residual impurities, which excludes them as a main cause of low mobilities and short minority carrier lifetimes. For that, lattice defects, essentially related to the N atom, were expected to be the main reason of such degradation. Several theoretical and experimental studies have investigated carrier traps in InGaAsN films. Theoretically, using the first principles pseudo-potential method in local density approximation, four N-related defects were proposed: $(As_{Ga}-N_{As})_{nn}$, $(V_{Ga}-N_{As})_{nn}$, $(N-N)_{As}$, and $(N-As)_{As}$ (Zhang & Wei, 2001). While the two first structures were supposed to have lower formation probabilities, the two split interstitials $(N-N)_{As}$, and $(N-As)_{As}$ were suggested to compensate the tensile strain in the film and to create two electron traps at around 0.42 and 0.66 eV below the conduction band minimum (CBM) of InGaAsN with a band gap of 1.04 eV, respectively (Zhang & Wei, 2001). Experimentally, the ion beam analysis provided a quantitative evidence of existence of N-related interstitial defects in GaAsN (Spruytte et al., 2001; Ahlgren et al., 2002; Jock, 2009). Furthermore, several carrier traps were observed in GaAsN and InGaAsN using deep level transient spectroscopy (DLTS). A deep level (E2/H1), acting as both an electron and a hole trap at 0.36 eV below the CBM, was observed (Krispin et al., 2001). Other electron traps in GaAsN grown by MBE were recorded: A2 at 0.29 eV and B1 at 0.27 eV below the CBM of the alloy (Krispin et al., 2003). In addition, a well known electron trap at 0.2 ~ 0.3 eV and 0.3 ~ 0.4 eV below the CBM of p-type and n-type GaAsN grown by MOCVD were observed, respectively (Johnston et al., 2006). Although the importance of these results as a basic knowledge about lattice defects in GaAsN and InGaAsN, no recombination center was yet experimentally proved and characterized. Furthermore, the main cause of high background doping in p-type films was not completely revealed.

Chemical beam epitaxy (CBE) has been deployed (Yamaguchi et al., 1994; Lee et al., 2005) to grow (In)GaAsN in order to overcome the disadvantages of MOCVD and MBE. It combines the use of metal-organic gas sources and the beam nature of MBE. (In)GaAsN films were grown under low pressure and low temperature to reduce the density of residual impurities and to avoid the compositional fluctuation of N, respectively. Furthermore, a chemical N compound source was used to avoid the damage of N species from N_2 plasma source in MBE. Although we obtained high quality GaAsN films gown by CBE, the diffusion length of minority carriers is still short (Bouzazi et al., 2010). This indicates that the electrical properties of GaAsN and InGaAsN films are independent of growth method and the problem may be caused by the lattice defects caused by N. Therefore, it is necessary to investigate these defects and their impact on the electrical properties of the film. For that, this chapter summarizes our recent results concerning lattice defects in GaAsN grown by CBE. Three defect centers were newly obtained and characterized. The first one is an active non-radiative N-related recombination center which expected to be the main cause of short minority carrier lifetime. The second lattice defect is a N-related acceptor like-state which greatly contributes in the background doping of p-type films. The last one is a shallow radiative recombination center acceptor-like state.

2. Deep level transient spectroscopy

To characterize lattice defects in a semiconductor, several techniques were used during the second half of the last century. Between these methods, we cite the thermally stimulated current (TSC) (Leonard & Grossweiner, 1958; Bube, 1960), the admittance spectroscopy (Losee, 1974), the increase or decay curves of photoconductivity (Rose, 1951; Devore, 1959), the optically stimulated conductivity (Lambe, 1955; Bube, 1956), and the analysis of space-charge-limited currents as function of applied voltage (Smith & Rose, 1955; Rose, 1955; Lampert, 1956). The basic concept of using the change of capacitance under bias conditions by the filling and emptying of deep levels was already anticipated fifty years ago (Williams, 1966). The thermally stimulated capacitance (TSCAP), which gives the temperature dependence of junction capacitance (Sah et al., 1978; Sah & Walker, 1973), was used. By dressing the properties of all these techniques, D. V. Lang found that they lacked the sensitivity, the speed, the depth range of recorded trap, and the spectroscopic nature to make them practical for doing spectroscopy on non-radiative centers. For that, in 1974, he proposed DLTS as a characterization method of lattice defects that can overcome the disadvantages of the other methods (Lang, 1974). DLTS is based on the analysis of the change of capacitance due to a change in bias condition at different temperatures. It can be applied to Schottky contacts and p-n junctions. DLTS has advantages over TSC due to its better immunity to noise and surface channel leakage currents. It can distinguish between majority and minority carrier traps, unlike TSC, and has a strong advantage over admittance spectroscopy, which is limited to majority-carrier traps. Comparing with TSCAP, DLTS has much greater range of observable trap depths and improved sensitivity. Despite the success of optical techniques such as photoluminescence to characterize superficial levels, they are rarely used in the study of non-radiative deep levels. Furthermore, such experiences must be done in the infrared domain. However, sensors are less sensitive than in the visible domain. Thus, we need a technique, which can separate between minority and majority traps and evaluate their concentrations, their energies, and their capture cross sections.

2.1 Fundamental concept of DLTS
2.1.1 Capacitance transient
To fully understand DLTS, it is worth to have a basic knowledge of capacitance transients arising from the SCR of Schottky contacts or p^+-n/n^+-p asymmetric junctions. If a pulse voltage is applied to one of these device structures that is originally reverse-biased, the SCR width decreases and the trap centers are filled with carriers (majority or minority depending on the structure). When the junction is returned to reverse bias condition, the traps that remains occupied with carriers are emptied by thermal emission and results in a transient decay. The capacitance transients provide information about these defect centers. Here, we restrain our description to a p^+-n junction where the p-side is more much heavily doped than the n-side, which gives the SCR almost in the low doped side.

The causes of change in capacitance depend on the nature of applied voltage. In case of reverse biased voltage, the junction capacitance, due to the change in SCR width, is dominant. However, when the applied voltage is forward biased, the diffusion capacitance, due to the contribution of minority carrier density, is dominant. The basic equation governing the capacitance transient in the p^+-n junction is expressed by

$$C(t) = A\sqrt{\frac{\varepsilon\varepsilon_0 e N_D}{2(V_R + V_b)}}\left[1 - \frac{N_T}{2N_D}\exp\left(-\frac{t}{\tau}\right)\right] = C_0\left[1 - \frac{N_T}{2N_D}\exp\left(-\frac{t}{\tau}\right)\right] \qquad (1)$$

where A is the contact area, V_b is the built-in potential, $\varepsilon\varepsilon_0$ is the permittivity of the semiconductor material, and e is the elementary charge of an electron. C_0, N_T, N_D, and τ

denote the junction capacitance at reverse bias, the density of filled traps under steady state conditions, the ionized donor concentration, and the time constant that gives the emission rate, respectively. The change in capacitance after the recharging of traps is given by

$$\Delta C = C_0 \sqrt{1 - \frac{N_T}{N_D}}$$ (2)

In most cases of using transient capacitance, the trap centers form only a small fraction of the SCR impurity density, i.e., $N_T << N_D$. Hence, using a first-order expansion of Eq. (2) gives

$$|\Delta C| = C_0 |1 - N_T/2N_D| \cong C_0 N_T/2N_D$$ (3)

Thus, the trap concentration calculates from the capacitance change ΔC is expressed by

$$N_T = 2\frac{\Delta C}{C_0} N_D$$ (4)

Note that Eq. (3) assumes that $N_T << N_D$ and the traps are filled throughout the total depletion width. To be more accurate, N_T should be adjusted to N_{Tadj} according to [30]

$$N_{Tadj} = 2\frac{\Delta C}{C_0} N_D \frac{W_R^2}{L_1^2 - L_2^2}$$ (5)

where W_R is the total SCR at reverse bias voltage V_R, $L_1 = W_R - \lambda$, $L_2 = W_p - \lambda$, and

$$\lambda = (\frac{2\varepsilon\varepsilon_0}{e^2 N_D}(E_F - E_T))^{1/2}$$ (6)

where W_p, E_F, and E_T denote the SCR at V_p, the Fermi level, and the trap energy level.

2.1.2 Thermal emission of carriers from deep levels
The emission rates for electrons and holes are given, respectively by

$$e_n = \sigma_n v_{thn} N_c \exp\left(-\frac{E_{CBM} - E_T}{kT}\right)$$ (7)

$$e_p = \sigma_p v_{thp} N_v \exp\left(-\frac{E_T - E_{VBM}}{KT}\right)$$ (8)

where σ_n, N_c, and v_{thn} are the thermal capture cross section, the density of states, and the thermal velocity of holes, respectively. σ_p, N_v, and v_{thp} are the same parameters for holes. E_{CBM}, E_{VBM}, and E_T are the energy levels of the conduction band minimum, the valence band maximum, and the trap, respectively.

2.2 Other DLTS related techniques
The isothermal capacitance transient spectroscopy (ICTS) and the double carrier pulse DLTS (DC-DLTS) are two DLTS related methods. They are used to obtain the density profiling of lattice defects and to check whether they act as recombination centers or not, respectively.

2.2.1 Isothermal capacitance transient spectroscopy

ICTS is used to analyze the profiling of lattice defects in the SCR of the semiconductor. It can be done through three different methods. The first one is obtained by fixing V_R and varying V_p to build difference of transients among the SCR of the device. The second method is evaluated by measuring at constant V_p and varying V_R. The last option is obtained by varying V_R and V_p, where the profiling analysis is also possible without building difference of transients. Using the first method, the medium trap density $\overline{N_T}(x_{ij})$ at a point x_{ij} is given by

$$\overline{N_T}(x_{ij}) = \frac{2\,N_D(\varepsilon\varepsilon_0)^2\,A^2}{C_R^3}\,\frac{\Delta C_{ij}}{(L_{P_i}^2 - L_{P_j}^2)} \tag{9}$$

where ΔC_{ij} is the amplitude difference of the two capacitance transients.

2.2.2 Double carrier pulse DLTS

DC-DLTS is used in asymmetric n^+-p or p^+-n junctions (Khan et al., 2005). It aims to check whether a trap is a recombination center or not. As shown in Fig. 1, two pulsed biases are applied to the sample, in turn, to inject majority and minority carriers to an electron trap. At the initial state, the junction is under reverse bias, and the energy level E_T of the trap is higher than the Fermi level (E_{Fn}).When the first pulse voltage is applied to the sample, E_{Fn} is higher than E_T, which allows the trap to capture electrons. During the second reverse biased pulse, with a duration t_{ip}, holes are injected to the SCR from the p–side of the junction. After the junction pulse is turned off, electrons and holes are thermally emitted. The amount of trapped carriers can be observed as a change in the DLTS peak height of the trap. If the trap captures both electrons and holes, the DLTS maximum of the corresponding level decreases compared with that in conventional DLTS. Such a decrease is explained by the electron–hole (e–h) recombination process, which indicates that the level is a recombination center.

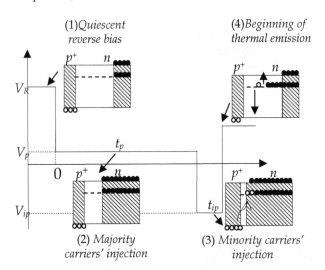

Fig. 1. Basic concept of capture and thermal emission processes from an electron trap located at an energy level E_T in p^+-n junction. A saturating injection pulse is applied to the reverse biased junction to fill the trap with holes.

To formulate the recombination process, we consider the same notation in § 2.1.2, with assuming that $n \approx N_D$. The relationship between the total density of recombination centers and that only occupied by electrons in the n-side of the junction can be expressed by

$$\frac{dn_T}{dt} = -e_p n_T(t) + \langle p \rangle c_p [N_T - n_T(t)] - N_D c_n n_T(t) + e_n [N_T - n_T(t)] \tag{10}$$

where $\langle p \rangle$ is the average of injected holes. As a solution of Eq. (10), we have

$$\frac{dn_T}{dt} = n_T(t_{ip}) = n_T(\infty) + [n_T(0) - n_T(\infty)] \exp(-\frac{t_{ip}}{\tau}) \tag{11}$$

where $n_T(\infty) = (\langle p \rangle c_p + e_n)/(\tau^{-1} N_T)$, $\tau^{-1} = \langle p \rangle c_p + N_D c_n + e_n + e_p$, and t_{ip} is the width of the injected pulse. Considering the I_{DLTS} and $I_{DC\text{-}DLTS}$ the peak heights of the recombination center in conventional and DC-DLTS, respectively. Equation (11) can be rewritten properly as

$$\frac{n_T(t_{ip})}{N_T} = \frac{(I_{DLTS} - I_{DCDLTS})}{I_{DCDLTS}} = \langle p \rangle \sigma_p v_{thp} t_{ip} \tag{12}$$

Similarly for hole trap in the p-side of a n^+-p junction, which acts as recombination center, we obtain

$$\frac{n_T(t_{ip})}{N_T} = \frac{(I_{DLTS} - I_{DCDLTS})}{I_{DCDLTS}} = \langle n \rangle \sigma_n v_{thn} t_{ip} \tag{13}$$

where $\langle n \rangle$ is the average of injected electrons.

3. Experimental procedure

All GaAsN films were grown by CBE on high conductive n- or p-type GaAs 2° off toward [010] substrate using Triethylgallium (($C_2H_5)_3$Ga, TEGa), Trisdimethylaminoarsenic ([($CH_3)_2N]_3$As, TDMAAs), and Monomethylhydrazine ($CH_3N_2H_3$, MMHy) as Ga, As, and N sources, respectively. The flow rates TEGa = 0.1 sccm and TDMAAs = 1.0 sccm were considered as conventional values. The growth temperatures of 420 °C and 460 °C were used for p-type and n-type GaAsN, respectively. Concerning the doping, p-type GaAsN films are unintentionally doped. The n-type alloys were obtained using a silane (SiH_4) source or by growing the films under lower MMHy and high growth temperature.

Three different device structures are used in this study: (i) n- and p-type GaAsN schottky contacts, (ii) n^+-GaAs/p-GaAsN/p-GaAs, and (iii) n-GaAsN/p^+-GaAs hetro-junctions. The N concentration in all GaAsN layers was evaluated using XRD method. Aluminum (Al) dots with a diameter of 0.5/1 mm were evaporated under vacuum on the surface of each sample. Alloys of Au-Ge (88:12 %) and Au-Zn (95:05 %) were deposited at the bottom of n-type and p-type GaAs substrates for each device, respectively. Some samples were treated by post-thermal annealing under N_2 liquid gas and using GaAs cap layers to avoid As evaporation from the surface. The temperature and the time of annealing will be announced depending on the purpose of making annealing. The background doping and the doping profile in the extended depletion region under reverse bias condition were evaluated using the capacitance-voltage (C-V) method. The leakage current in all used samples ranged from 0.3

nA to 10 μA for a maximum reverse bias voltage of -4 V. A digital DLTS system *Bio-Rad DL8000* was used for DLTS and C-V measurements. The activation energy E_t and the capture cross section $\sigma_{n,p}$ were determined from the slope and the intercept values of the Arrhenius plot, respectively.

4. Lattice defects in GaAsN grown by CBE

In this section, the distribution of electron and hole traps in the depletion region of GaAsN grown by CBE will be dressed using DLTS and related methods.

4.1 Electron traps in GaAsN grown by CBE

4.1.1 DLTS spectra and properties of a N-related electron trap

The DLTS spectrum of Fig. 2(a) shows an electron trap (*E*2) at 0.69 eV below the CBM of GaAs. After rapid thermal annealing at 720°C for 2 min, *E*2 disappears completely whereas a new electron trap (*E*3) appears at 0.34 eV below the CBM. From the Arrhenius plots of Fig. 2(d), the capture cross sections of *E*2 and *E*3 are calculated to be σ_{E2} = 8.1 × 10^{-15} cm^2 and σ_{E3} = 7.5 × 10^{-18} cm^2, respectively. Based on previous results about native defects in n-type GaAs, *E*2 and *E*3 are independent of *N* and considered to be identical to *EL*2 and *EL*3, respectively (Reddy et al., 1996). In order to focus only on *N*-related lattice defects, these two energy levels will be excluded from the DLTS spectra of *N*–containing n-type GaAsN. The addition

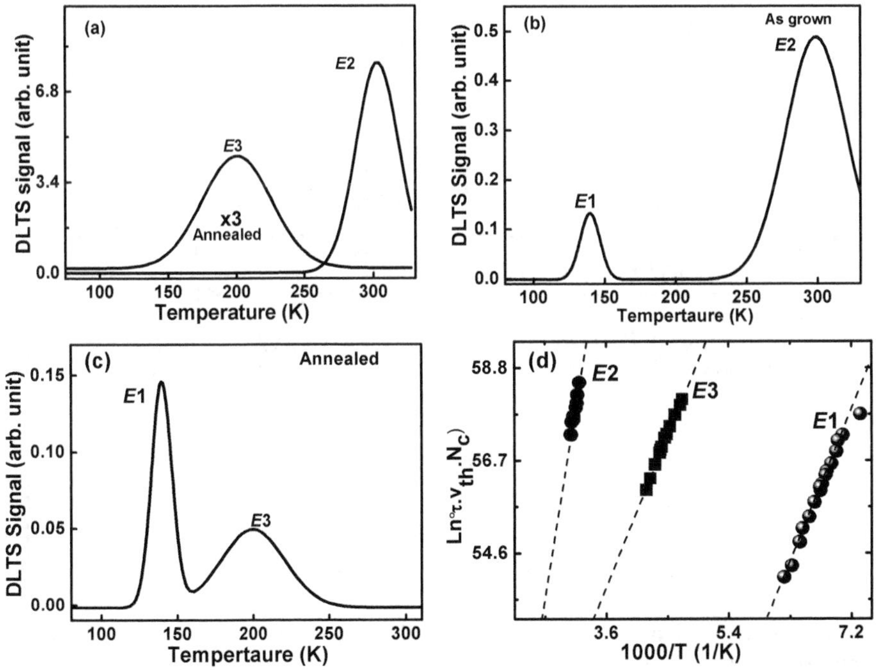

Fig. 2. DLTS spectra of (a) *N* free as grown and annealed GaAs, (b) as grown n-type $GaAs_{0.998}N_{0.002}$, (c) annealed n-type $GaAs_{0.998}N_{0.002}$, and (d) Arrhenius plots of DLTS spectra.

of a small atomic fraction of N to GaAs leads to the record of a new electron trap ($E1$), at an average activation energy 0.3 eV below the CBM of GaAsN. The DLTS spectra of as grown and annealed n-type $GaAs_{0.998}N_{0.002}$ are given in Figs. 2. (b) and (c), respectively. The activation energies (E_{E1}) and the capture cross sections (σ_{E1}) of $E1$ for N varying GaAsN samples are given Fig. 3 (a) and (b), respectively. The fluctuation of E_{E1} from one sample to another can be explained by the effect of Poole–Frenkel emission, where the thermal emission from $E1$ is affected by the electric field (Johnston and Kurtz, 2006). As illustrated in Fig. 3(c), with increasing the filling pulse duration, the DLTS peak height of $E1$ saturates

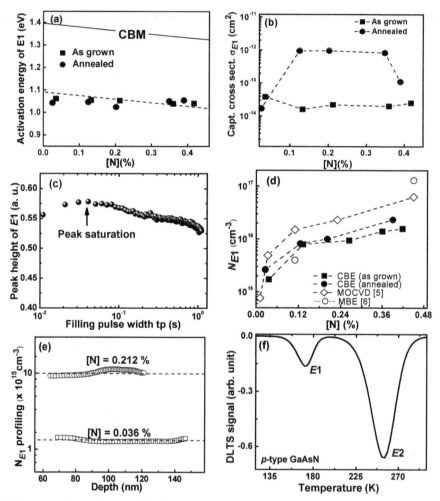

Fig. 3. Nitrogen dependence of (a) thermal activation energy, (b) capture cross section, and (d) adjusted density of $E1$ in as grown and annealed GaAsN samples. The large capture cross section is confirmed with (c) the filling pulse width dependence of the DLTS peak height of $E1$. (e) Density profiling of $E1$ in the bulk of GaAsN films, and (f) DLTS spectrum of undoped p-type GaAsN grown by CBE.

rapidly. This behavior is explained by the large value of σ_{E1}, compared with that of $E2$, $E3$, and other native defects in GaAs. The adjusted densities of $E1$ (N_{E1}) in as grown and annealed samples are plotted in Fig. 3(d). N_{E1} increases considerably with increasing [N] in the film and persists to post thermal annealing. This indicates that $E1$ is a N-related and a stable electron trap.

The defect center $E1$ was not observed previously in N free GaAs grown by CBE despite the existence of N species in the chemical composition of the As source. This can be explained through three possible scenarios. First, the absence of tensile strain in GaAs prevents the formation of $E1$. Second, the N atom in the atomic structure of $E1$ comes from the N source. Finally, the N atom comes from the N and As compound sources and in presence of tensile strain $E1$ can be formed. The tensile strain was reported in most theoretical and experimental studies. As given in Fig. 3(e) and using the ICTS, this idea is supported by the uniform distribution of N_{E1} in the bulk of GaAsN. This indicates that $E1$ is formed during growth to compensate for the tensile strain in the GaAsN films caused by the small atomic size of N compared with that of As.

Furthermore, the properties of $E1$ are identical to that of the famous electron traps reported by Johnston et Kurtz (Johnston and Kurtz, 2006) and Krispin et al. (Krispin et al., 2003) in MOCVD and MBE grown n–type GaAsN, respectively. As illustrated in Fig. 3(d), the densities of these traps are approximately similar to N_{E1} despite the large difference in the density of residual impurities between the three growth methods. Therefore, the atomic structure of $E1$ may be free from impurities. Furthermore, by carrying out DLTS measurements for minority carriers in undoped p-type film, $E1$ was also observed. This indicates clearly that $E1$ is independent of doping atoms (see Fig. 3(f)).

4.1.2 Nature of the electron trap E1

Two methods are used to verify whether $E1$ is a recombination center or not. The first method is indirect, in which the activation energy of deep levels is correlated with that of the reverse bias current in the depletion region of n-type GaAsN Schottky junction and n^+-GaAs/p-GaAsN heterojunction. These two device structures are selected because the current is due mainly to electrons. The second method is direct, in which DC-DLTS is used to show the behavior of the electron traps in simultaneous injection of majority and minority carriers in the depletion region.

4.1.2.1 Origin of reverse bias current in GaAsN

The temperature dependence of the reverse bias current in the depletion region of n-type GaAsN Schottky junction and n^+-GaAs/p-GaAsN is shown in Fig. 4(a) for reverse bias voltages of 0.5 and -0.5 V, respectively. At lower temperature, the dark current changes slowly in the two structures, then fellows an Arrhenius type behavior. As shown in Fig. 4(b), the same result is obtained by applying reverse bias voltages of 1 and -1 V. Under these conditions, the reverse bias current $I_d(T)$ can be expressed by

$$I_d(T) = I_\infty \exp(-\frac{\Delta E}{kT})$$ (14)

where I_∞, ΔE, k, and T denote the limit of the high-temperature current, the thermal activation energy of the reverse bias current, the Boltzmann constant, and the temperature, respectively. The I-V characteristics deviate in the two samples from the thermionic emission. This is explained by the fact that supplying the *p-n* junction under reverse bias conditions decreases the product of excess carriers to less than the square of intrinsic carriers. Hence, the Shockley-Read-Hall (SRH) generation mechanism is activated to increase the product of excess carriers to assure the balance of charge. The generated carriers are swept to the transition regions by the electric field in the depletion region. Therefore, an SRH center, with a thermal activation energy arround 0.3 eV, is the origin of the dark current in the SCR of GaAsN. The thermal activation energies are measured with respect to majority and minority carriers in *n*-type GaAsN schottky junction and *n*⁺-GaAs/*p*-GaAsN heterojunction, respectively. This corresponds to the conduction band in the two structures. By correlating the conduction mechanism and DLTS measurements, the thermal activation energy of the reverse bias current and the activation energy of the *N*-related electron trap *E1* are typically identical. Therefore, *E1* is responsible for the generation/recombination current in the depletion region of GaAsN grown by CBE.

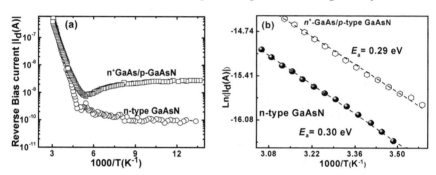

Fig. 4. Temperature dependence of dark current under reverse bias voltages of (a) 0.5 and -0.5V and (b) 1 and -1V in *n*-type GaAsN schottky junction and the *n*⁺-GaAs/*p*-GaAsN heterojunction, respectively.

4.1.2.2 DC-DLTS measurements

DC-DLTS is used to confirm the recombination nature of *E1* and to characterize the recombination process via this defect center. An unintentionally doped *n*–type GaAsN layer (~ 1 μm) was grown on a *p*–type GaAs by CBE. This structure is not commonly used for DLTS measurements. However, the absence of a *p*-type doping source prevented us to obtaining a *p*⁺-*n* junction. Here, the *p*-type substrate is used as source of minority carriers. As shown in Fig. 5(a), the DC-DLTS spectrum is compared with that of the conventional DLTS. A decrease in the peak height of *E1* is observed by varying the voltage of the second injected pulse and also confirmed by varying its duration. The obvious reason for such reduction is the mechanism of *e-h* recombination at the energy level *E1* in the forbidden gap of GaAsN. Hence, *E1* is reconfirmed to act as a *N*-related recombination center. To verify the non-radiative recombination process, the temperature dependence of σ_{E1} is obtained by varying the emission rate window e_{rw} from 0.5 - 50 s⁻¹. The value of σ_{E1} is obtained from the fitting of the Arrhenius plots for each e_{rw}. As shown in Fig. 5(b), the natural logarithmic of σ_{E1} shows a linear increase with the reciprocal of the temperature. It can be expressed as

$$\ln(\sigma_{E1}) = -E_{cap,e}/kT + \ln(\sigma_{\infty}) \qquad (15)$$

where $E_{cap,e}$ = 0.13 ± 0.02 eV, k, T, and σ_{∞} = 1.38 x 10⁻⁹ cm² denote the barrier height for the capture of electron, the Boltzmann constant, the temperature, and the capture cross section of electrons at an infinite temperature, respectively. At room temperature, $\sigma_{E1}(300\ K)$ is evaluated to ~ $8.89 \times 10^{-12}\ cm^2$. Such a value is large enough to shorten the lifetime of electrons in p-type GaAsN. This indicates that $E1$ is a strongly active recombination center at room temperature and the e-h recombination process is non-radiative. In addition, from the temperature dependence of σ_{E1}, the true energy depth of $E1$ can be obtained by subtracting the barrier height for electron capture from the thermal activation energy obtained from the Arrhenius plot. The recombination center $E1$ is localized at $E_a\ (E1) = 0.20 \pm 0.02$ eV from the CBM of GaAsN. Furthermore, the average capture cross section of holes σ_p, at a temperature of $T = 175$ K, is estimated using Eq. 12 to be $\sigma_p(175\ K)$ ~ $5.01 \times 10^{-18}\ cm^2$. The physical parameters of $E1$ can be summarized in a configuration coordinate diagram (CCD), in which the energy state of $E1$ is described as a function of lattice configuration (Q). As shown in Fig. 5(c), the CCD of $E1$ can be presented in three different branches: (i)[0, f.e + f.h]: the charge state of $E1$ is neutral, with a free electron and a free hole, (ii)[-, t.e + f.h]: the electron is trapped and the hole remains free, (iii)[0]: the free hole is captured at the crossed point B and recombined with the already-trapped electron. $E1$ losses its charge and becomes neutral. As the recombination process is non-radiative, the lattice relaxation occurs with the emission of multi-phonon. The energy of multi-phonon emission can be evaluated as function of N concentration according to

$$E_{phonon}(N) = E_g(N) - (E_a(E1) - E_{cap,e}) \tag{16}$$

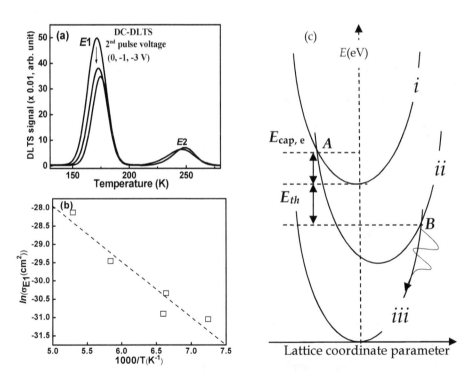

Fig. 5. (a) Reduction of peak height of $E1$ under minority carrier injection spectra, (b) temperature dependence of σ_{E1} for electrons, and (c) Configuration-coordinate-diagram showing the different charge states of $E1$ as function of lattice coordinate parameter.

4.1.3 Possible origin of the N-related recombination center E1

It is worth remembering that the atomic structure of $E1$ may be free from impurities and doping atoms owing to the difference in the density of residual impurities in GaAsN grown with MOCVD, MBE, and CBE. Furthermore, the uniform distribution of N_{E1} in the bulk of GaAsN indicates that $E1$ is formed during growth to compensate for the tensile strain caused by the small atomic size of N compared with that of As. Therefore, the origin of $E1$ has high probability to depend only on the atoms of the alloy (N, As, Ga). To confirm these expectations, the origin of $E1$ is investigated qualitatively by considering the results of two different experiments: (*i*) the dependence of N_{E1} to the As source flow rate and (*ii*) the effect of H implantation on the distribution of lattice defects in *n*-type GaAsN.

4.1.3.1 Dependence of N_{E1} to As source flow rates

The objective of this experiment is to clarify whether the density of $E1$ is sensitive to the As atom or not. The MMHy was supplied to 9.0 sccm and the TDMAAs was varied between and 0.7 and 1.5 sccm. As shown in Fig. 6. (a), increasing TDMMAs drops the N concentration in the film and tends to saturate for a flow rate higher than 1 sccm. For two emission rate $e_{rw} = \{100, 10\}$ s^{-1} and filling pulse $t_p = \{0.1, 5\}$ ms values, the DLTS spectra are normalized on the junction capacitance to exclude the effect of various carrier densities in the samples. The same N-related recombination center $E1$ was observed in all samples. The TDMAAs flow rate dependence of the DLTS peak height of $E1$ for two settings of measurement parameters is given in Fig. 6(b). A peaking behavior at approximately TDMAAs = 0.9 sccm was obtained. As N_{E1} is uniformly distributed in GaAsN films, the incorporation of N atom at the growth surface affects both the incorporation of N_{As} and the formation of $E1$. If $E1$ depends only on N atom, the decrease of [N] with increasing TDMMAs flow rate results in monotonically dropping of N_{E1}. However, the peaking behavior of N_{E1} indicates the sensitivity of $E1$ to As atom, either than N.

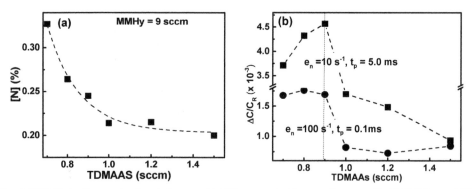

Fig. 6. TDMAAs flow rate dependence of (a) N concentration under supplied MMHy and (b) normalized DLTS peak height of E1 for two settings of measurement parameters.

4.1.3.2 Effect of H implantation on lattice defects in GaAsN

GaAsN films were treated by H implantation. This experiment was used because H bounds strongly to N in GaAsN films to form N-H complexes (Suzuki et al., 2008; Amore & Filippone, 2005). H ions with multi-energy from 10 to 48 keV were implanted into GaAsN layers with peaks concentration of 5×10^{18} (GaAsN$_{HD1}$) and 1×10^{19} atom/cm^3 (GaAsN$_{HD2}$), respectively. The depth of implantation was thought to be distributed between 110 and 410 nm from the surface of GaAsN by calculating the *SRIM* 2003 simulation code (Ziegler, 1985). After implantation, the samples were treated by post thermal annealing at 500 °C for 10 min under N_2 gas and GaAs cap layers. As plotted in Fig. 7(a) and (b), the crystal quality of GaAsN films after implantation was controlled using XRD curves and C-V measurements. DLTS spectra of implanted samples are shown in Fig. 7(c). After implantation, $E1$ was not observed; however, two new lattice defects appeared. The signature and the density of these traps are summarized in Table 1. The thermal emission from them is plotted as an Arrhenius plot in Fig. 7(d).

GaAsN	Traps	$E_a(eV)$	$\sigma(cm^2)$	$N_{t\text{-adj}}(cm^{-3})$	Possible origin
As grown	E1	E_{CM}-0.331	5.18×10^{-15}	3.37×10^{17}	$(N\text{-As})_{As}$
Implanted	EP1	E_{CM}-0.414	8.20×10^{-13}	5.88×10^{17}	EL5 in GaAs
	HP1	E_{VM}-0.105	5.42×10^{-18}	1.84×10^{16}	$N\text{-H-}V_{Ga}$

Table 1. Summary of E_a, σ, adjusted $N_{t\text{-adj}}$, and possible origin of defects in as grown and implanted GaAsN samples.

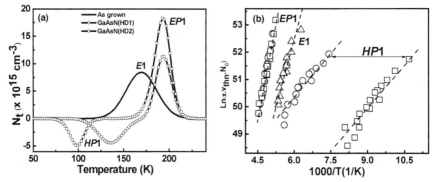

Fig. 7. (a)DLTS spectra of as grown and implanted GaAsN and (b) their Arrhenius plots.

The new electron trap ($EP1$) is located approximately 0.41 eV below the CBM of GaAsN. Its properties are identical to that of the native defect $EL5$ in GaAs (Reddy, 1996). Its atomic structure was discussed in many publications, where the common result indicated that $EL5$ is a complex defect free from impurities and dominated by As interstitials, such V_{Ga}-As_i or As_{Ga}-V_{Ga} (Deenapanray et al., 2000; Yakimova et al., 1993). The second new defect is a hole trap ($HP1$) at approximately an average activation energy 0.11 eV above the VBM of GaAsN. Compared with majority carrier traps in GaAs grown with various techniques, no similar hole trap to $HP1$ was reported. However, in p-type GaAsN grown by CBE with around the same N concentration, E_{HP1} and σ_{HP1} are identical to that of the hole trap $HC2$ in p-type GaAsN grown by CBE (bouzazi et al., 2011). This defect was confirmed recently to be an acceptor state in GaAsN films (see § 4.2.3) and to be related to the N-H bond. While the H impurity was provided by implantation, the N atom can originate through two possibilities: First; examining XRD results, the N atom can originate from its ideal site. However, [N] in as grown is much higher than that in implanted samples (~10^{19} cm^{-3}). It is also much higher than N_{HP1}. Second, N atom can be originated from the complete dissociation of $E1$, since the ratio N_{E1}/N_{HP1} is less than 2. If $E1$ is the split interstitial $(N-N)_{As}$, N_{HP1} must be at least equals to that of $E1$ and one N atom remains free. This expectation is disapproved by DLTS measurements, where N_{HP1} is largely less than N_{E1}. This means that $E1$ contains only one N atom in its atomic structure. Considering the results of last sub-section, $E1$ may be the split interstitial $(N-As)_{As}$ formed from one N and one As in a single As site. This result is supported by the theoretical calculation (Zhang et al., 2001).

4.1.4 Effect of E1 on minority carrier lifetime in GaAsN

The effect of $E1$ on the electrical properties of GaAsN can be evaluated through the calculation of minority carrier lifetime using the SRH model for generation–recombination (Hall, 1952; Shockley & Read, 1952). Such parameter has been estimated to be less than 0.2 ns as a result of the calculation according to

$$\tau_{E1} = \left(v_{thn}\sigma_{E1}N_{E1}\right)^{-1} < 0.2\ ns \tag{17}$$

Therefore, E1 is considered to be the main cause of short minority carrier in GaAsN. It is required to investigate the formation mechanism of this defect in order to decrease its density and to recover the minority carrier lifetime in GaAs.

4.2 Hole traps in GaAsN grown by CBE
4.2.1 DLTS spectra and properties of hole traps in GaAsN

Here, we only focus on the hole traps that coexist in all p-type GaAsN based Schottky junctions and n^+-GaAs/p-GaAsN heterojunction. The difference between these two structures is the temperature range in which the DLTS measurements can be carried out due to the freeze-out of carriers. The DLTS spectrum of p-type GaAsN in the heterojunction is shown in Fig. 8(a). Three hole traps $H0$, $H2$, and $H5$ are observed at 0.052, 0.185, and 0.662 eV above the VBM of GaAsN. Their peak temperatures are 35, 130, and 300 K, respectively. The thermal dependence of emission from the hole traps is plotted as an Arrhenius plot in Fig. 8(b). The activation energy, capture cross section, and density are given in Table 2.

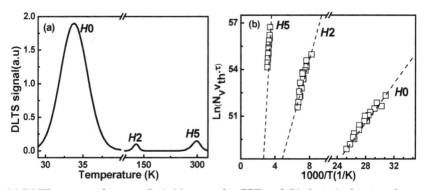

Fig. 8. (a) DLTS spectra of p-type GaAsN grown by CBE and (b) their Arrhenius plots.

The hole traps $H2$ and $H5$ were also observed in Schottky junctions and coexist in all samples. $H2$ is a N-related acceptor-like state and was proved to be in good relationship with high background doping in GaAsN films. These properties will be discussed later. The hole trap $H5$ presents the same properties as $H3$ and $HA5$ (Li et al., 2001; Katsuhata et al., 1978). It is proposed to be the double donor state $(+/^{++})$ of $EL2$ (Bouzazi et al., 2010). The hole trap $H0$ cannot be observed in Schottky junctions owing to the freeze-out effect.

Traps	E_t (eV)	$\sigma_p(cm^2)$	$N_{t\text{-adj}}(cm^{-3})$
H0	0.052	2.16×10^{-14}	4.64×10^{16}
H2	0.185	3.87×10^{-17}	4.52×10^{15}
H5	0.662	6.16×10^{-14}	6.24×10^{15}

Table 2. Summary of E_t, σ_p, and adjusted $N_{t\text{-adj}}$ of $H0$, $H2$, and $H5$.

4.2.2 Radiative shallow recombination center H0

DC-DLTS measurements were carried out to confirm whether there is a recombination center among the hole traps or not. As shown in Figs. 9(a) and (b), the DC-DLTS signal is compared to that of conventional DLTS. A decrease in the DLTS peak height of $H0$ is observed and confirmed by varying the voltage and the duration of the injected pulse.

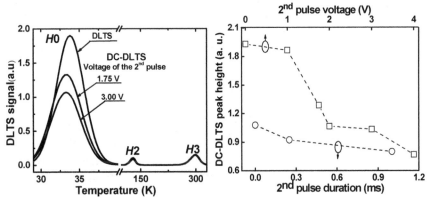

Fig. 9. (a)DC-DLTS spectra of p-type GaAsN for various second pulse voltage and (b) $H0$ DC-DLTS peak height dependence of second pulse voltage and duration.

The shallow hole trap H0 is observed and reported for the first time owing to the temperature range it was recorded, which cannot be reached with standard DLTS systems. Second, its capture cross section is large enough to capture majority carriers and minority carriers. The reduction in the peak height of H0 is explained by the electron hole-recombination. This implies that H0 is a shallow recombination center in p-type GaAsN grown by CBE and can also play the role of an acceptor state. To verify whether the recombination process via H0 is radiative or not, the temperature dependence of the capture cross section of electrons is obtained by varying the emission rate e_{nw} from 1 to 50 s^{-1}. As shown in Fig. 10(a), the peak temperature of H0 shifts to high temperatures with increasing e_{nw}. The value of σ_{H0} is obtained from the fitting of the Arrhenius plots for each e_{nw}. It is important to note that the fitting errors of activation energy and capture cross section are relatively large owing to the instability of temperature in the range of measurements. As shown in Fig. 10(b), the capture cross section of H0 does not exhibit an Arrhenius behavior, which excludes the non-radiative recombination process. Its shallow energy level suggests that $H0$ plays the role of an intermediate center in the recombination process, with the exception that the recombination is quite often radiative.

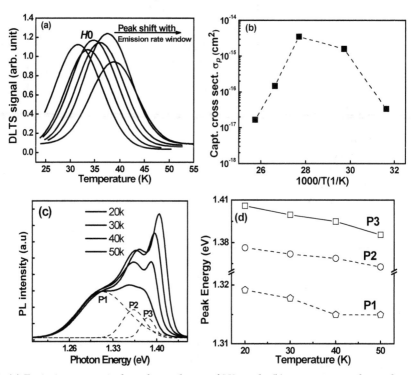

Fig. 10. (a) Emission rate window dependence of $H0$ peak, (b) temperature dependence of capture cross section and activation energy of $H0$, (c) PL spectra of p-type GaAsN at 20 to 50 K, and (d) temperature dependence of the peaks $P1$, $P2$, and $P3$ obtained from the fitting of PL spectra.

Furthermore, the capture cross section of electrons can be estimated using Eq. 13 and by the reduction of the peak height of $H0$, which follows from the injection of minority carriers. By varying the injected pulse voltage at fixed duration, the average capture cross section of electrons σ_n, at a temperature T = 35 K, is estimated to be $\sigma_n \sim 3.64 \times 10^{-16}$ cm^2. However, by varying the width of the injected pulse at fixed pulse voltage, it is estimated, at the same temperature, to be $\sigma_n \sim 3.05 \times 10^{-16}$ cm^2. These two values are nearly identical and indicate that the capture cross section of electrons and holes of $H0$ are approximately the same. To identify this radiative recombination, photoluminescence (PL) measurements were carried out at low temperature on the same GaAsN sample. The PL spectra at 20, 30, 40, and 50K are shown in Fig. 10 (c). Three different peaks $P1$, $P2$, and $P3$ can be distinguished from PL spectra fitting. The temperature dependence of their energies is plotted in Fig. 10 (d). The three peaks were sufficiently discussed in many N-varying GaAsN samples and they are proposed to be: (*i*) $P1$ is the result of band-to-band transition, (*ii*) $P2$ is caused by free exciton or related to shallow energy level, and (*iii*) $P3$ originates from the transition between a neutral donor and a neutral acceptor pair (DAP) (Inagaki et al., 2011). Therefore, the peak $P2$

is suggested to be in relation with $H0$. The band diagram of such recombination is shown in Fig. 11, where the transition occurs between $H0$ and the CBM and/or a donor-like defect (E_t). The transition between free electrons in the CBM and charged $H0$ is called free-to-charged bound transition.

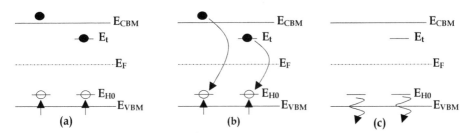

Fig. 11. Band diagram of radiative recombination through $H0$, where the transition occurs between $H0$ and the CBM and/or a donor-like defect (E_t).

Concerning the origin of the acceptor–like hole trap H0, more experiments are required to discuss it. However, considering its density and the distribution of shallow acceptors in GaAs, it can be suggested that $H0$ is a carbon-related acceptor, where the reported ionization energy from PL and Hall Effect measurements is 0.026 eV and its density in the 10^{-15} cm^{-3} range (Baldereschi & Lipari, 1974).

4.2.3 Deep N-H related acceptor state H2

The ionized acceptor density (N_A) is found to be in good linear dependence with N concentration in p-type GaAsN samples (see Fig. 12 (a)). As given in Figs. 12(b) and (c), the junction capacitance (C_j) showed a N-related sigmoid behavior with temperature in the range 70 to 100 K. This behavior has not yet been observed in GaAs and n-type GaAsN grown by CBE. It was recorded at 20 K in silicon p-n junction and explained by the ionization of a shallow energy level (Katsuhata, 1978; 1983). Hence, the N dependence of N_A and C_j is explained by the thermal ionization of a N-related acceptor-like state. The thermal ionization energy of this energy level was estimated in the temperature range 70 to 100 K to be between 0.1 and 0.2 eV. It is in conformity with the theoretical calculations, which suggested the existence of a N-related hole trap acceptor-like defect with an activation energy within 0.03 and 0.18eV above the VBM of GaAsN (Janotti et al., 2003; Suzuki et al., 2008). Experimentally, a deep acceptor level, A2, was confirmed in CBE grown undoped GaAsN with ionization energies of E_{A1} = 130 ± 20 meV (Suzuki et al., 2008). On the other hand, the properties of $H2$ in N-varying GaAsN schottky junctions are cited below: The peak temperature of $H2$ is within the temperature range of increase of C_j. This means that the electric field at this temperature range fellows the same behavior of C_j and depends on N concentration. Hence, the emission of carriers from the charged traps is affect by the Poole-Frenkel emission (Johnston and Kurtz, 2006). This is confirmed by the fluctuation of E_{H2} from one sample to another depending on N concentration (see Table 3). However, the average of E_{H2} is within the energy range of the acceptor level obtained from theoretical prediction and identical to E_{A2} (Suzuki et al., 2008; Janotti et al., 2003). Furthermore, as given in Fig. 12 (d), N_{H2-adj} is in linear dependence with N concentration. Therefore, $H2$ is proved to be the N-related hole trap acceptor-like state, which thermal ionization increased C_j and drops the depletion region width. The contribution of $H2$ in the background doping of p-type GaAsN films grown by CBE can be evaluated from the ratio between the real N_{H2-Cj} calculated from the change of C_j and N_A at room temperature (Bouzazi et al., 2010). As shown in Fig. 12 (e) and (f), This ratio comes closer to the unit for a N concentration greater than 0.2%. Thus, $H2$ is the main cause of the high background doping in p-type GaAsN.

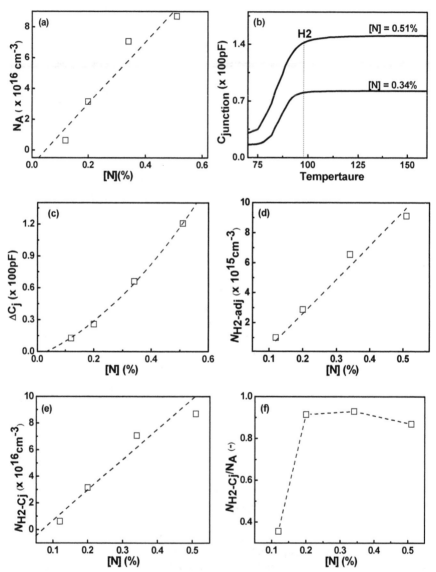

Fig. 12. N dependence of (a) N_A, (c) amplitude of C_j after thermal ionization of $H2$, and (d) N_{H2-adj}. (b) Sigmoid increase of C_j between 70 and 100 K for two different GaAsN samples. N dependence of (e) N_{H2-Cj} and (f) N_{H2-Cj}/N_A.

To investigate the origin of $H2$, it is worth remembering some previous results about carrier concentration and the density of residual impurities in undoped GaAsN grown by CBE, obtained using Hall Effect, Fourier transform infra-red (FTIR), and second ion mass spectroscopy (SIMS) measurements. On one hand, under lower Ga flow rates (TEGa), N_A and N_{H2} sowed a rapid saturation with $[N]$, despite the increase of $[N]$ (see Fig. 13 (a)). This means that the atomic structure of $H2$ depends on other atoms, either than N. In addition, the densities of C and O was found to be less than free hole concentration, which excludes these two atoms from the origin of $H2$. On the other hand, using SIMS measurements, the ratio $[H]_{TEGa = 0.02}/[H]_{TEGa = 0.1}$ was evaluated to be ~0.6 (Sato et al., 2008). Furthermore, the free hole concentration at room temperature showed a linear increase with the density of N-H bonds (Nishimura et al., 2006). This means that N_A depends strongly on $[H]$ and the saturation of N_A under lower TEGa can be explained by the desorption of H from the growth surface, since the growth rate in our films was found to be in linear dependence with TEGa. Hence, the structure of $H2$ is related to the N-H bond. However, the N-H bond may not be the exact structure of $H2$ because the slope of the linear relationship between N_A and $[N-H]$ increased with increasing growth temperature ($T_G \in [400, 430]$ °C). This indicates that N_A is determined by both the number of N–H and another unknown defect, which concentration increased with increasing T_G. The binding energy of this unknown defect can be determined from Arrhenius plot. Furthermore, the formation energy of $(N-H-V_{Ga})^{-2}$ was found to be lower than $(N-V_{Ga})^{-3}$, $(H-V_{Ga})^{-2}$, and isolated V_{Ga}^{-3} (Janotti et al., 2003). This means that the unknown defect may be V_{Ga}. These predictions were experimentally supported using positron annihilation spectroscopy results (Toivonen et al., 2003). Hence, $H2$ may be related to the N-H-V_{Ga} structure.

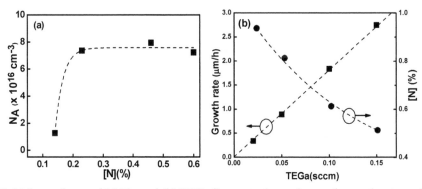

Fig. 13. N dependence of (a) N_A and (b) TEGa flow rate dependence of growth rate and N concentration at a growth temperature of 420 °C.

$[N](\%)$	E_{H2} (eV)	$\sigma_{H2}(cm^2)$	$N_{H2\text{-}adj}(cm^{-3})$	E_{max} (V/cm)	$N_{H2\text{-}Cj}(cm^{-3})$	$N_{H2\text{-}Cj}/N_A$
0.12	0.210	2.8×10^{-14}	2.64×10^{15}	6.2×10^4	2.23×10^{15}	0.36
0.20	0.150	6.3×10^{-16}	3.08×10^{15}	1.4×10^5	2.88×10^{16}	0.91
0.34	0.138	6.3×10^{-16}	5.20×10^{15}	2.1×10^5	6.56×10^{16}	0.93
0.51	0.103	1.3×10^{-17}	9.12×10^{15}	2.3×10^5	7.55×10^{16}	0.87

Table 3. Summary of E_{H2}, σ_{H2}, $N_{H2\text{-}adj}$, E_{max}, $N_{H2\text{-}est}$, and the ratio $N_{H2\text{-}Cj}/N_A$ for CBE grown undoped p-type GaAsN Schottky junctions.

5. Conclusion

Three defect centers, related to the optoelectronic properties of GaAsN, were identified and characterized using DLTS and some related methods:

- The first defect is a N-related non-radiative recombination center (E1), located approximately 0.33 eV below the CBM of GaAsN grown by CBE. E1 is a stable defect and exhibits a large capture cross section at room temperature. Its activation energy comes closer to the midgap by increasing the N concentration in the film, which increases the recombination rates of carriers. In addition, the density profiling of E1 was found to be uniformly distributed, which may indicates that this defect was formed during growth to compensate for the tensile strain caused by N. Using the SRH model for generation-recombination, the lifetime of electrons to E1 was evaluated to be ~0.2 ns. Therefore, E1 is suggested to be the main cause of poor minority carrier lifetime in GaAsN films. Its origin was suggested to be the split interstitial formed from one N and one As in a single arsenic site.
- The second defect is a radiative shallow hole trap acceptor-like state (H0), at 0.052 eV above the VBM of GaAsN. it was observed and reported for the first time in GaAs and GaAsN films. The radiative recombination process through H0 is confirmed to be a free-to-charged bound transition. This defect may be in relationship with carrier density and the optical properties of the film.
- The last defect is a N-related hole trap acceptor-like state (H2) located approximately 0.15 eV above the VBM of GaAsN. It was newly observed in GaAsN films. H2 was proved to have a good relationship with the carrier density and to be the main origin of high background doping in GaAsN films, essentially for relatively high N concentration. H2 is strongly suggested to be related with N-H bond. Its formation limits the junction depth and minimizes the contribution of the depletion region in the quantum efficiency of GaAsN based solar cells.

In conclusion, the results obtained in this study are very useful for scientific understanding of defects in III-V-N materials and to improve GaAsN and InGaAsN qualities for realizing high efficiency multi-junction solar cells.

6. Acknowledgment

Part of this work was supported by the New Energy Development Organization (NEDO) under the Ministry of Economy, Trade and Industry, Japan.

7. References

Ahlgren, T.; Vainonen-Ahlgren, E.; Likonen, J.; Li, W. & Pessa, M. (2002). Concentration of interstitial and substitutional nitrogen in GaN_xAs_{1-x}. Applied Physics Letters, Vol. 80, pp. 2314-2316.

Amore, B. A. & Filippone, F. (2005). Theory of Nitrogen-Hydrogen Complexes in N-containing III-V Alloys, chapter 13, Dilute Nitride Semiconductors, Henidi, M. Elsevier Ltd, UK.

Baldereschi, A. & Lipari, N. O. (1974). Cubic contributions to the spherical model of shallow acceptor states. Physical Review B, Vol.9, pp. 1525-1539.

Bouzazi, B.; Nishimura, K.; Suzuki H.; Kojima, N.; Ohshita, Y. & Yamaguchi M. (2010). Properties of Chemical Beam Epitaxy grown GaAs$_{0.995}$N$_{0.005}$ Homo-junction Solar Cell, Current Applied Physics, Vol.10, pp. 188-190.

Bouzazi, B.; Suzuki, H.; Kojima, N.; Ohshita, Y. & Yamaguchi, M. (2011). Japanese Journal of Applied Physics, Vol.49, pp. 121001-121006.

Bube, R. H. (1956). Comparison of Surface-Excited and Volume-Excited Photoconduction in Cadmium Sulfide Crystals. Physical Review, Vol.101, pp. 1668-1676.

Bube, R. H. (1960). Photoconductivity of Solids. John Wiley & Sons, Inc., New York, pp. 292-299.

Deenapanray, P. N. K.; Tan, H. H. & Jagadish, C. (2000). Investigation of deep levels in rapid thermally annealed SiO$_2$-capped n-GaAs grown by metal-organic chemical vapor deposition. Applied Physics Letters, Vol.77, pp. 696-698.

DeVore, H. B. (1959). Gains, response times, and trap distributions in powder photoconductors. RCA Review, Vol. 20, pp. 79-91.

Friedman, D. J.; Geisz, J. F.; Kurtz, S. R.; & Olson, J. M. (1998). 1-eV Solar Cells with GaInNAs Active Layer. Journal of Crystal Growth, Vol.195, pp. 409-415.

Geisz, J. F.; Friedman, D. J.; Olson, J. M.; Kurtz, S. & Keyes, B. M. (1998). Photocurrent of 1 eV GaInNAs lattice-matched to GaAs. Journal of Crystal Growth, Vol.195, pp. 401-408.

Geisz, J. F. & Friedman, D. J. (2002). III–N–V Semiconductors for Solar Photovoltaic Applications. Semiconductor Science and Technology, Vol.17, No.8, pp. 769-777.

Hall, R. N. (1952). Electron-Hole Recombination in Germanium. Physical Review, Vol. 87, pp. 387-387.

Inagaki, M.; Suzuki, H.; Suzuki, A.; Mutaguchi, K.; Fukuyama, A.; Kojima, N.; Ohshita, Y. & Yamagichi, M. (2011). Shallow Carrier Trap Levels in GaAsN Investigated by Photoluminescene. Japanese Journal of Applied Physics, Vol.50, 4, pp. 04DP141-144.

Janotti, A.; Zhang, S. B.; Wei, S. H. & Van de Walle, C. G. (2002). Effects of hydrogen on the electronic properties of dilute GaAsN alloys. Physical Review Letters, Vol.89, pp. 6403-6406.

Janotti, A.; Wei, S. H.; Zhang, S. B. & Kurtz, S. (2003). Interactions between nitrogen, hydrogen, and gallium vacancies in GaAs$_{1-x}$N$_x$ alloys. Physical Review B, Vol.67, pp. 161201-161204.

Jock, M. R. (2009). Effect of N Interstitials on the Electronic Properties of GaAsN Alloy Films. Thesis, the University of Michigan, 2009.

Johnston, S. W. & Kurtz, S. R. (2006). Comparison of a dominant electron trap in n-type and p-type GaNAs using deep-level transient spectroscopy. Journal of Vacuum Science & Technology, A24 (4), pp. 1252-1257.

Katsuhata, M.; Koura, K. & Yoshida S. (1978). Temperature Dependence of Capacitance of Silicon p-n Step Junctions. Japanese Journal of applied physics, Vol.17, pp. 2063-2064.

Katsuhata, M.; Yamagata, S., Miyayama, Y.; Hariu, T. & Shibata, Y. (1983). P-n Junction Capacitance Thermometers. Japanese Journal of Applied Physics, Vol.22, pp. 878-881.

Krispin, P.; Spruytte, S. G.; Harris, J. S. & Ploog, K. H. (2000). Electrical depth profile of p-type GaAs/Ga(As, N)/GaAs hetero-structures determined by capacitance–voltage measurements. Journal of Applied Physics, Vol.88, pp. 4153- 4158.

Krispin P.; Gambin V.; Harris J. S. & Ploog K. H. (2003). Nitrogen-related electron traps in Ga(As,N) layers (< 3% N). Journal of Applied Physics, Vol.93, pp. 6095-6099.

Krispin, P.; Spruytte, S. G.; Harris, J. S. & Ploog, K. H. (2001). Origin and annealing of deep-level defects in p-type GaAs/Ga(As,N)/GaAs hetero-structures grown by molecular beam epitaxy. Journal of Applied Physics, Vol.89, pp. 6294-6301.

Kurtz, S.; Allerman, A. A.; Jones, E. D.; Gee, J. M.; Banas, J. J. & Hammons, B. E. (1999). InGaAsN solar cells with 1.0 eV band gap, lattice matched to GaAs. Applied Physics Letters Vol.74, pp. 729-731.

Kurtz, S.; Webb, J.; Gedvilas, L.; Friedman, D.; Geisz, J.; Olson, J.; King, R., Joslin, D. & Karam, N. (2001). Structural changes during annealing of GaInAsN. Applied Physics Letters, Vol.78, pp. 748-750.

Kurtz, S.; Reedy, R.; Barber, G. D.; Geisz, J. F.; Friedman, D. J.; McMahon, W. E. & Olson, J. M. (2002) Incorporation of nitrogen into GaAsN grown by MOCVD using different precursors. Journal of Crystal Growth, Vol.234, pp. 318-322.

Kurtz, S. R.; Geisz, J. F.; Keyes, B. M.; Metzger, W. K.; Friedman, D. J.; Olson, J. M.; King, R. R. & Karam, N. H. (2003). Effect of growth rate and gallium source on GaAsN. Applied Physics Letters, Vol.82, pp. 2634-2636.

Lambe, J. (1955). Recombination Processes in Cadmium Sulfide. Physical Review, Vol.98, pp. 985-992.

Lampert, M. A. (1956). Simplified Theory of Space-Charge-Limited Currents in an Insulator with Traps, Physical Review, Vol.103, pp. 1648-1656.

Lang, D. V. (1974). Deep-level transient spectroscopy: A new method to characterize traps in semiconductors. Journal of applied Physics, Vol.45, pp.3023-3032.

Lee, H. S.; Nishimura, K.; Yagi, Y.; Tachibana, M.; Ekins-Daukes, N. J.; Ohshita, Y.; Kojima, N.; Yamaguchi, M. (2005). Chemical beam epitaxy of InGaAsN films for multi-junction tandem solar cells. Journal of Crystal Growth, Vol.275, pp. 1127-1130.

Leonard, I. & Grossweiner, A. (1953). A Note on the Analysis of First-Order Glow Curves. Journal of Applied Physics, Vol.24, pp. 1306-1307.

Li, J. Z.; Lin, J. Y.; Jiang, H. X.; Geisz, J. F. & Kurtz, S. R. (1999). Persistent photoconductivity in $Ga_{1-x}In_xN_yAs_{1-y}$. Applied Physics Letters, Vol.75, pp. 1899-1901.

Li, W.; Pessa, M.; Ahlgren, T.; & Dekker, J. (2001). Origin of improved luminescence efficiency after annealing of Ga(In)NAs materials grown by molecular-beam epitaxy. Applied Physics Letters, Vol.79, pp. 1094-1096.

Losee, D. L. (1974). Admittance spectroscopy of deep impurity levels: ZnTe Schottky barriers. Applied Physics Letters, Vol.21, pp. 54-56.

Moto, A.; Takahashi, M. & Takagishi, S. (2000). Hydrogen and carbon incorporation in GaInNAs. Journal of Crystal Growth, Vol.221, pp. 485-490.

Nishimura, K.; Suzuki, H.; Saito, K.; Ohshita, Y.; Kojima, N. & Yamaguchi, M. (2007). Electrical properties of GaAsN film grown by chemical beam epitaxy. Physica B, Vol.401, pp. 343-346.

Reddy, C. V.; Fung, S. & Beling, C. D. (1996). Nature of the bulk defects in GaAs through high-temperature quenching studies. Physical Review B, 54, pp. 11290-11297.

Rose, A. (1951). An outline of some photoconductive processes. RCA Review., Vol.12, pp. 362-414.

Rose, A. (1955). Space-Charge-Limited Currents in Solids. Physical Review, Vol.97, pp. 1538-1544.

Saito K.; Nishimura K.; Suzuki H.; Kojima N.; Ohshita Y. & Yamaguchi, M. (2008). Hydrogen reduction in GaAsN thin films by flow rate modulated chemical beam epitaxy. Thin Solid Films, Vol.516, pp. 3517-3520.

Sah, C. T.; Chan, W. W.; Fu, H. S. & Walker, J. W. (1972). Thermally Stimulated Capacitance (TSCAP) in p-n Junctions. Applied Physics Letters, Vol.20, pp. 193-195

Sah, C. T. & Walker, J. W. (1973). Thermally stimulated capacitance for shallow majority-carrier traps in the edge region of semiconductor junctions. Applied Physics Letters, Vol.22, pp. 384-385.

Shockley, W. & Read. W.T. (1952). Statistics of the Recombinations of Holes and Electrons. Physical Review, Vol. 87, pp. 835-842.

Smith, R. W. & Rose, A. (1955). Space-Charge-Limited Currents in Single Crystals of Cadmium Sulfide. Physical Review, Vol.97, pp. 1531-1537.

Spruytte, S. G.; Coldren, C. W.; Harris, J. S.; Wampler, W.; Ploog, K. & Larson, M. C. (2001). Incorporation of nitrogen in nitride-arsenides: Origin of improved luminescence efficiency after anneal. Journal of Applied Physics, Vol.89, pp. 4401-4406.

Suzuki, H.; Nishimura, K.; Saito, K.; Hashiguchi, T.; Ohshita, Y.; Kojima, N. & Yamaguchi, M. (2008). Effects of Residual Carbon and Hydrogen Atoms on Electrical Property of GaAsN Films Grown by Chemical Beam Epitaxy. Japanese Journal of Applied Physics, Vol.47, pp. 6910-6913.

Toivonen, J.; Hakkarainen, T.; Sopanen, M.; Lipsanen, H.; Oila, J. & Saarinen, K. (2003). Observation of defect complexes containing Ga vacancies in GaAsN. Applied Physics Letters, Vol.82, pp. 40-43.

Williams, R. (1966). Determination of Deep Centers in Conducting Gallium Arsenide. Journal of Applied Physics, Vol.37, pp. 3411-3416.

Yakimova, R.; Paskova, T. & Hardalov, Ch. (1993). Behavior of an EL5-like defect in metalorganic vapor-phase epitaxial GaAs:Sb. Journal of Applied Physics, Vol.74, pp. 6170-6173.

Yamaguchi, M.; Warabisako, T.; Sugiura, H. (1994). Chemical beam epitaxy as a breakthrough technology for photovoltaic solar energy applications. Journal of Crystal Growth, Vol.136, pp.29-36.

Zhang, S. B. & Wei, S. H. (2001). Nitrogen Solubility and Induced Defect Complexes in Epitaxial GaAs:N. Physical Review Letters, Vol.86, pp. 1789-1792.

Ziegler, J. F.; Biersack J. P. & U. Littmark. (1985). The Stopping and Ion Range of Ions in Solids. Pergamon, New York, Vol. 1; SRIM program for PCs available from J. F. Zieglerand: www.srim.org.

Permissions

The contributors of this book come from diverse backgrounds, making this book a truly international effort. This book will bring forth new frontiers with its revolutionizing research information and detailed analysis of the nascent developments around the world.

We would like to thank Professor, Doctor of Sciences, Leonid A. Kosyachenko, for lending his expertise to make the book truly unique. He has played a crucial role in the development of this book. Without his invaluable contribution this book wouldn't have been possible. He has made vital efforts to compile up to date information on the varied aspects of this subject to make this book a valuable addition to the collection of many professionals and students.

This book was conceptualized with the vision of imparting up-to-date information and advanced data in this field. To ensure the same, a matchless editorial board was set up. Every individual on the board went through rigorous rounds of assessment to prove their worth. After which they invested a large part of their time researching and compiling the most relevant data for our readers. Conferences and sessions were held from time to time between the editorial board and the contributing authors to present the data in the most comprehensible form. The editorial team has worked tirelessly to provide valuable and valid information to help people across the globe.

Every chapter published in this book has been scrutinized by our experts. Their significance has been extensively debated. The topics covered herein carry significant findings which will fuel the growth of the discipline. They may even be implemented as practical applications or may be referred to as a beginning point for another development. Chapters in this book were first published by InTech; hereby published with permission under the Creative Commons Attribution License or equivalent.

The editorial board has been involved in producing this book since its inception. They have spent rigorous hours researching and exploring the diverse topics which have resulted in the successful publishing of this book. They have passed on their knowledge of decades through this book. To expedite this challenging task, the publisher supported the team at every step. A small team of assistant editors was also appointed to further simplify the editing procedure and attain best results for the readers.

Our editorial team has been hand-picked from every corner of the world. Their multi-ethnicity adds dynamic inputs to the discussions which result in innovative outcomes. These outcomes are then further discussed with the researchers and contributors who give their valuable feedback and opinion regarding the same. The feedback is then collaborated with the researches and they are edited in a comprehensive manner to aid the understanding of the subject.

Apart from the editorial board, the designing team has also invested a significant amount of their time in understanding the subject and creating the most relevant covers. They scrutinized every image to scout for the most suitable representation of the subject and create an appropriate cover for the book.

The publishing team has been involved in this book since its early stages. They were actively engaged in every process, be it collecting the data, connecting with the contributors or procuring relevant information. The team has been an ardent support to the editorial, designing and production team. Their endless efforts to recruit the best for this project, has resulted in the accomplishment of this book. They are a veteran in the field of academics and their pool of knowledge is as vast as their experience in printing. Their expertise and guidance has proved useful at every step. Their uncompromising quality standards have made this book an exceptional effort. Their encouragement from time to time has been an inspiration for everyone.

The publisher and the editorial board hope that this book will prove to be a valuable piece of knowledge for researchers, students, practitioners and scholars across the globe.

List of Contributors

Shinji Yae
University of Hyogo, Japan

Qing Shen
PRESTO, Japan Science and Technology Agency (JST), Japan

Taro Toyoda
The University of Electro-Communications, Japan

Ayse Bedeloglu
Dokuz Eylül University, Turkey

Yoshihiro Irokawa, Mickael Lozac'h and Masatomo Sumiya
National Institute for Materials Science, Namiki, Tsukuba, Ibarak, Japan

Nobuyuki Matsuki
National Institute for Materials Science, Namiki, Tsukuba, Ibarak, Japan Institute of Science and Technology Research, Chubu University, Matsumoto, Kasugai, Aichi, Japan Department of Electrical and Electronic Engineering, Faculty of Engineering, Gifu University, Yanagido, Gifu, Japan

Yoshitaka Nakano
Institute of Science and Technology Research, Chubu University, Matsumoto, Kasugai, Aichi, Japan

AC Varonides
Physics & Electrical Engineering Dept., University of Scranton, Scranton, PA, USA

V.I. Laptev
Russian New University, Russian Federation

H. Khlyap
Kaiserslautern University, Germany

Hossein Movla, Foozieh Sohrabi and Arash Nikniazi
Faculty of Physics, University of Tabriz, Iran

Mohammad Soltanpour
Faculty of Humanities and Social Sciences, University of Tabriz, Iran

Khadije Khalili
Research Institute for Applied Physics and Astronomy (RIAPA), University of Tabriz, Iran

Rabin Dhakal, Yung Huh, David Galipeau and Xingzhong Yan
Department of Electrical Engineering and Computer Science, Department of Physics, South Dakota State University, Brookings SD 57007, USA

Cigdem Yumusak
Department of Physics, Faculty of Arts and Sciences, Yildiz Technical University, Davutpasa Campus, Esenler, Istanbul, Turkey Linz Institute for Organic Solar Cells (LIOS), Physical Chemistry, Johannes Kepler University of Linz, Linz, Austria

Daniel A. M. Egbe
Linz Institute for Organic Solar Cells (LIOS), Physical Chemistry, Johannes Kepler University of Linz, Linz, Austria

Enwei Zhu, Linyi Bian, Jiefeng Hai and Weihua Tang
Nanjing University of Science and Technology, People's Republic of China

Fujun Zhang
Beijing Jiaotong University, People's Republic of China

Taro Toyoda and Qing Shen
The University of Electro-Communications, Japan

Qun Ye and Chunyan Chi
Department of Chemistry, National University of Singapore, Singapore

Boussairi Bouzazi, Nobuaki Kijima, Yoshio Ohshita and Masafumi Yamaguchi
Toyota Technological Institute, Japan

Hidetoshi Suzuki
Myazaki University, Japan

Printed in the USA
CPSIA information can be obtained
at www.ICGtesting.com
JSHW011458221024
72173JS00005B/1126